PREBIOTIC FOODS FOR A HEALTHY GUT

Scientific Studies to Manage Metabolic Disorders

PREBIOTIC FOODS FOR A HEALTHY GUT

Scientific Studies to Manage Metabolic Disorders

Mini K. Sheth, PhD
Shruti Dwivedi, PhD
Editors

AAP | APPLE ACADEMIC PRESS

First edition published 2025

Apple Academic Press Inc.
1265 Goldenrod Circle, NE,
Palm Bay, FL 32905 USA

760 Laurentian Drive, Unit 19,
Burlington, ON L7N 0A4, CANADA

CRC Press
2385 NW Executive Center Drive,
Suite 320, Boca Raton FL 33431

4 Park Square, Milton Park,
Abingdon, Oxon, OX14 4RN UK

Library and Archives Canada Cataloguing in Publication

CIP data on file with Canada Library and Archives

Library of Congress Cataloging-in-Publication Data

CIP data on file with US Library of Congress

ISBN: 978-1-77491-852-4 (hbk)
ISBN: 978-1-77491-853-1 (pbk)
ISBN: 978-1-00356-880-3 (ebk)

About the Editors

Mini K. Sheth, PhD
Head, Department of Foods and Nutrition, Faculty of Family and Community Sciences, The Maharaja Sayajirao University of Baroda, Vadodara, Gujarat, India

Mini K. Sheth, PhD, is the Head of the Department of Foods and Nutrition at the Faculty of Family and Community Sciences, Maharaja Sayajirao University of Baroda, Vadodara, Gujarat, India. She has over 35 years of teaching and research experience and specializes in prebiotics and gut health, clinical dietetics, and food safety. She has published more than 80 papers in peer-reviewed journals. She has applied for patents for prebiotic cakes and synbiotic beverages and holds copyrights for booklets on fructooligosaccharides-added sweets, prebiotics, and diabetes management. She has over 10 professional memberships, including the Probiotic Association of India, Indian Dietetic Association, Nutrition Society of India, and Association of Food Scientist and Technologists, India. She has held important positions including RD board member of the Indian Dietetic Association (2015–2019); President of the Indian Dietetic Association (Gujarat Chapter) from 2010 to Dec 2014; Deputy Director of a WHO collaborating center, FN Department (2010–2013); and Deputy Coordinator of a UGC/DSA SAP III program. She has undertaken projects sponsored by DBT, Government of India, and TCL on FOS and Obesity. She has guided over 20 PhD students and has received more than 25 awards and recognitions. She has a PhD in Foods and Nutrition (1994) and has been a Registered Dietitian since 1986.

Shruti Dwivedi, PhD
Founder of Nutrideck, Researcher, and Health Coach

Shruti Dwivedi, PhD, is a researcher and health coach. She has been awarded a Young Scientist Award by the Universal Society of Foods and Nutrition and the Swedish South Asian Network for Fermented Foods for her research work on the gut-brain axis. Her other accomplishments include UGC NET, SRF and ICAR AIEEA. She has membership in the Indian Dietetic Association and Nutrition Society of India. She has published papers in various peer-reviewed international and national journals. Her work on the gut and mental health connection has been featured in *The Times of India, Gujarat Samachar, Sandesh Daily,* and various peer-reviewed journals. As a consultant at MAMTA Health Institute of Mother and Child and a health coach at NGOs such as Kalam and Mahima, she has supervised various health camps and activities in rural as well as urban settings. She is on a mission to make people aware of mindful fitness. As founder of Nutrideck, she counsels people, both individually and in collaboration with organizations such as Mahila Patanjali Yog Samiti and Kaka Ayurvedic Chiktsalya, on lifestyle management in order to combat noncommunicable diseases and raise awareness about healthy living. She received a doctorate degree in Public Health and Nutrition from the Maharaja Sayajirao University of Baroda, Vadodara, India.

Contents

Contributors

Kankona Dey
Department of Foods and Nutrition, The Maharaja Sayajirao University of Baroda, Gujarat, India

Geeta Dharmatti
Geeta Nutri Heal Consultancy, Pune, Maharashtra, India

Shruti Dwivedi
Department of Foods and Nutrition, The Maharaja Sayajirao University of Baroda, Gujarat, India

Archana Gattupalli
Department of Microbiology, The Maharaja Sayajirao University of Baroda, Gujarat, India

Nareshkumar Gattupalli
Department of Microbiology, The Maharaja Sayajirao University of Baroda, Gujarat, India

Harsha Hirdyani
Unique Palate Sdn Bhd, Kuala Lumpur, Malaysia

A. H. Jana
Dairy Processing and Operations, SMC College of Dairy Science, Anand, Gujarat, India

Urvashi Mall
P.G. Department of Home Science, Sardar Patel University, Anand, Gujarat, India

Meghana Patel
S.B.K.S. Medical Institute and Research Center, Gujarat, India

V. H. Patel
P.G. Department of Home Science, Sardar Patel University, Anand, Gujarat, India

Vijayata Sanegar
Department of Foods and Nutrition, The Maharaja Sayajirao University of Baroda, Gujarat, India

Arpi Shah
International Institute of Sports and Management, Mumbai, Maharashtra, India

Mini K. Sheth
Department of Foods and Nutrition, The Maharaja Sayajirao University of Baroda, Gujarat, India

Renu M. Sitlani
Sun Pharma Advanced Research Company Limited, Vadodara, Gujarat, India

Hati Subrota
Department of Dairy Microbiology, SMC College of Dairy Science, Anand, Gujarat, India

Sujoya Sukul
Nutrition Consultant, Jaipur, Rajasthan, India

Abnita Thakuria
Assistant Professor, Department of Food Science and Nutrition, College of Community Science, Assam Agricultural University, Jorhat, Assam, India

Ruchi Vaidya
Ramaiah International Center for Public Health Innovations, Bangalore, Karnataka, India

Preface

Researching in the area led to clinical experimentation with prebiotics, a magical component that has proved its worth in the management of various metabolic disorders and other diseases, driving us to compile the research-based data for this book.

The subject material is presented in 18 chapters, ranging from introducing prebiotics as a food component to the potential of various prebiotics and their safety, their proven health implications, mechanism of action, and use in day-to-day consumption.

Synopsis of Chapter 1 will provide readers with an overview of the highly dynamic gut molecular ecology and diversity. Chapter 2 is designed to emphasize the role of potential factors that define the healthy gut: prebiotics – their types, classification, structure, sources, properties, and benefits. Chapter 3 provides insight into recent developments in shaping the gut microbiota for improved human health. Chapters 4 to 12 deal with the prevalence, etiology, and pathophysiology of various metabolic disorders and diseases. Efforts have been made in this edition to include recent research-based results on the management of disorders and diseases which include diarrhea, constipation, IBS, oral health, depression, Alzheimer's disease, diabetes, obesity, cancer, and immunity. Chapter 13 covers different in vivo and in vitro techniques to evaluate prebiotic properties. Xylooligosaccharide is coming up as an excellent prebiotic, exerting various health benefits. Chapter 14 details the extraction and its mechanism of action in various metabolic disorders and diseases. In Chapter 15, the role played by various prebiotics in synbiotic fermented dairy foods is discussed. Chapter 16 introduces prebiotics as a functional food, along with the effect on sensory acceptance, concerned regulatory issues, consumer attitude, commercialization, scaling up, and the market potential of such therapeutic products. Prebiotics provide exciting avenues for functional foods. Chapter 17 covers all the standardized recipes formulated using prebiotics as part of dissertations of doctoral and master's research work, namely by Dr. Harsha Hirdiyani, Dr. Abnita Thakuria, Dr. Neha Gupta, Dr. Aparna Assudani, Dr. Nidhi Jain, Ms. Heena Shah, Ms. Viral Lodaya, and Late Akanksha Mahendra.

The formulation of the manuscript of this book required considerable support and guidance from eminent professionals. We are grateful and wish to acknowledge the subject experts for commencing the chapter drafts. Thanks are due to Prof. Archana Gattupalli and Prof. Nareshkumar Gattupalli, Department of Microbiology; Dr. Vijayata Sengar, Ms. Konkana Dey, Department of Foods and Nutrition, The Maharaja Sayajirao University of Baroda, Vadodara; Dr. Renu Gurudasani, Sun Pharma Industries, Vadodara; Ms. Arpi Shah, Parul University, Vadodara; Dr. Ruchi Vaidya, Ramaiah International Center for Public Health Innovations, Bengaluru; Dr. Geeta Dharmatti, Nutri Heal Consultancy, Pune; Dr. Sujoya Sukul, Nutrition Consultant, Jaipur; Dr. Kejal Joshi, Symbiosis International University, Pune; Prof. Vinayak Patel and Ms. Urvashi Mall, P.G. Department of Home Science, Sardar Patel University, Anand; Dr. Abnita Thakuria, Citrus Research Station, Assam Agriculture University, Assam; Dr. Harsha Hirdyani, Unique Palate Sdn Bhd, Kuala Lumpur.

Special thanks are due to Prof. Atanu Jana, Department of Dairy Processing & Operations, and Dr. Hati Subrota, Department of Dairy Microbiology, both from SMC College of Dairy Science, Anand, who have not only contributed a chapter to this book but have consistently encouraged us in the publication process.

Each chapter has been the primary responsibility of one or two author(s) in giving shape to this invaluable text. However, each chapter has been read, reviewed, and critiqued by professional associates. We take collective responsibility for the complete content of this text.

—Mini K. Sheth, PhD

Shruti Dwivedi, PhD

CHAPTER 1

Molecular Ecology of Human Gut

MINI K. SHETH

Department of Foods and Nutrition, The Maharaja Sayajirao University of Baroda, Gujarat, India

> ***The colon is by far the primary location of microbial colonization in mammalian gastrointestinal tract. The complexity of this microbiota has been thoroughly researched over the last 20–25 years.***

1.1 INTRODUCTION

The human gastrointestinal tract is a highly dynamic ecosystem that has evolved both physiologically and microbiologically over time. It is the largest surface area in the body of 300 m^2 that interacts with the external environment (Bjorksten, 2006). A series of complex organs ranging from the stomach to the distal colon harbors approximately 10^{14} microbial cells, which is ten times the number of cells in our body (Sekirov et al., 2010).

The human GI tract is made up of specially designed areas for mediating its various activities, which have a significant effect on host health and well-being. A few major important roles of the gut microflora include trophic effects on intestinal epithelia, immunological structure and function, and defense of the colonized host against invasion by foreign bacteria.

Dysbiosis of the gut microflora is often associated with bacterial translocation mainly due to intestinal barrier failure, which promotes overgrowth of pathogenic bacteria that secrete proinflammatory cytokines, amyloids, and

Prebiotic Foods for a Healthy Gut: Scientific Studies to Manage Metabolic Disorders.
Mini K. Sheth and Shruti Dwivedi (Eds.)

lipopolysaccharides, associated with the pathogenesis of various metabolic diseases and infections affecting the liver, respiratory tract, gastrointestinal cancers, and neurodegenerative diseases such as AD and depression (De Whelan et al., 2018). On the contrary, increased density of good bacteria, particularly *Bifidobacteria* and *Lactobacilli*, are markers of the stable gut microflora. They produce strong acids, i.e., butyric acid, acetic acid, and lactic acid, which reduce intestinal pH and neutralize the activity of dietary carcinogens, such as nitrosamines, inhibiting the growth of pathogens and preventing several diseases (Tojo et al., 2014).

1.2 COMPOSITION OF GUT MICROFLORA DURING LIFE CYCLES

Newborns' gastrointestinal tracts begin to colonize shortly after birth. The pattern may initially be affected by the mode of delivery (birth canal versus cesarean section) and food (breastfeeding versus formula feeding) (Harmsen et al., 2000; Hooper, 2001). The initial colonization has a significant impact on the final make-up of the adult permanent flora (Guarner and Malagelada, 2003). Breast-fed infants' microbiota is dominated by *Bifidobacteria,* whereas formula-fed infants' microbiota is more complex, resembling the adult gut in that *Bacteroides*, *Clostridia, Bifidobacteria, Lactobacilli*, gram-positive cocci, coliforms, and other groups are all represented in roughly equal proportions (Guaraldi and Salvatori, 2012).

There is a major change in the gut bacteria community as the infant switches to a more solid and varied diet. According to a recent study, the age-related microbiota population shifts from 3 days to 2 years after birth, with significant differences between 2 years and adulthood (Naseribafrouei et al., 2014). The most well-studied dietary bioactive compounds are polyunsaturated fatty acids (PUFAs), the -3 and -6 fatty acids, and phytochemicals which have an impact on various aspects of immunity and metabolism. Their metabolic products may inhibit pathogenic bacteria while enabling the proliferation of beneficial bacteria exerting prebiotic-like effects.

1.3 THE HUMAN GASTROINTESTINAL TRACT AND ITS MICROBIOTA: A STATE OF ART

The types and numbers of bacterial populations differ significantly between the colon, small intestine, and stomach. Transit time, secretions, and nutrient availability are the main causes of this. The GI tract begins with the oral cavity where complex microbiota exists. These include members of the genera *Prevotella, Porphyromonas, Peptostreptococcus, Bacteroides, Fusobacterium, Eubacterium,* and *Desulfovibrio*. As they encounter the stomach, these numbers fall dramatically to $<10^3$ CFU, due to the strong acidic and peristaltic nature of the stomach, including *Lactobacilli*, yeasts, and *Helicobacter pylori* (Zhu et al., 2006). Due to rapid transit and the secretion of intestinal fluids, which create a hostile environment, the duodenum also has low microbial populations (Fooks and Gibson, 2004). However, there is a progressive increase in both number and species along the jejunum and the ileum.

Enterococci, Enterobacteria, Lactobacilli, Bacteroides, and *Clostridium* are found in the small intestine. They multiply quickly from 10^4–10^6 CFU mL^{-1} in the small intestine to 10^{11}–10^{12} CFU mL-1 in the large intestine (Sanford, 2005).

The most numerically predominant bacteria are non-spore forming anaerobes: *Bacteroides* and *Bifidobacterium*, *Eubacterium*, *Clostridium*, *Lactobacillus*, *Fusobacterium*, and various gram-positive cocci. Bacteria present in lower numbers include Enterococcus, Enterobacteriaceae, methanogens, and dissimilatory sulfate-reducing bacteria (Sanford, 2005). Obligate anaerobes present in the gut include *Bifidobacterium, Clostridium, Eubacterium, Fusobacterium, Peptococcus, Peptostreptococcus, Bacteroides*, and facultative anaerobes *Lactobacillus, Bacillus, Streptococcus, Staphylococcus, E. coli, Klebsiella*, and *P. aeruginosa* (Papoff et al., 2012). *Bifidobacterium* accounts for approximately 80% of cultivable bacteria in infants and 25% in adults.

1.4 MICROBIAL DIVERSITY AND IMPORTANCE OF MAINTAINING A BALANCE IN THE GUT

Humans can be considered as "*super organisms*" with an internal ecosystem of diverse microorganisms. The host and its microbial components interact to maintain homeostatic balance (Rezzi, 2007). A balanced intestinal flora is a precondition for a fairly stable ecosystem in which both host-related factors and antagonistic interactions among intestinal bacteria play a role. The stability of the human intestinal microflora is indicated by the presence of the proper concentrations of *Bifidobacteria* and *Lactobacilli* (Turroni et al., 2008; Van et al., 2007; Harmsen et al., 2002). These helpful bacteria might serve as guards, controlling the activity of other disease-causing bacteria such as *Salmonella*, *Shigella*, *Clostridia*, *Staphylococcus aureus*, *Candida albicans*, *Campylobacter jejuni, Escherichia coli, Veillonella*, and Klebsiella (Ventura et al., 2009). Research has shown the potential of beneficial bacteria, particularly *Bifidobacteria* and *Lactobacilli*, in preventing several disease-related dysfunctions related to an imbalanced GI tract (Heeney et al., 2018).

Reducing or eliminating more of the healthy gut microflora, like *Bifidobacteria*, has its consequences. When the human diet influences the species composition and metabolic characteristics of the intestinal microflora, toxic metabolite production is affected, such as the conversion of pro-carcinogens to active carcinogens. In addition to producing toxic metabolites, several harmful bacteria, such as *Salmonella, Shigella, Listeria, Bacteroides, Proteus, E. coli, Clostridium perfringens,* and *Vibrio cholerae* also have associations with diarrhea, infections, liver damage, carcinogenesis, and intestinal putrefaction. Pathogenic effects associated with harmful intestinal microflora such as *E. coli* not only include colonic disorders but also have implications with possible vaginal infections and systemic disorders. Major factors in the biology of these disorders are the overgrowth of pathogenic bacteria such as clostridia, *E. coli.* As well as parasites, viral infections, extensive burn injury, post-operative stress, and antibiotic therapy. These disorders are often associated with bacterial translocation due to intestinal barrier failure. *Lactobacillus* and *Bifidobacteria* produce strong acids, i.e., acetic and lactic acid. The production of these acids reduces intestinal pH, which results in lower blood ammonia levels and a reduced hepatic load (Zhang et al., 2015).

Scientific investigations of the gut microbial environment have identified over 1000 species and possibly over 7000 strains, the vast majority (80%) of which are uncultured (Zoetendal et al., 2008; Blaut and Clavel, 2007; Bäckhed et al., 2005).

Functional aspects of microbial diversity in the human large intestine are now being discussed by many scientists (Dethlefsen et al., 2008; Flint, 2006; Ley et al., 2006). Recent studies using metagenomics have shown that the gut microbiome's capacity for coding far outpaces that of the human genome (Kurokawa et al., 2007; Turnbaugh et al., 2007; Gill et al., 2006; Ley et al., 2006; Bäckhed et al., 2005). Today, a great challenge is to expand and discover newer arenas of complex relationships and mechanisms related to responsiveness, diversity, and resilience of the gut ecology, and its interactions with diet and related aspects of the human host which affect each and every aspect of life.

1.5 MAIN FUNCTIONS OF COLONIC MICROBIOTA IN THE GUT

Evidence based on many studies suggests that gut microflora have specific functions that are very important and crucial for maintaining a balance and equilibrium in the gut ecology. The presence of gut microbiota has influenced the evolution of humans, as they cannot carry out some major essential intestinal functions without them. Germ-free animal models have provided useful insights into the extensive roles of the microflora and the extent of interaction with the host. The gut microbiota can be thought of as a microbial organ within a human organ, as the processes performed by this diverse population are extensive; it can communicate with itself and with the host. It is also a site of energy consumption, transformation, and distribution. Functions of gut microbiota are divided into three main categories: Metabolic, trophic, and protective functions.

1.5.1 METABOLIC FUNCTIONS

The fermentation of indigestible dietary waste and endogenous mucus produced by the epithelia is a key metabolic activity of colonic microflora (Guarner and Malagelada, 2003). Enzymes and biochemical pathways that are different from the host's own constitutive resources are provided by the microbial community's gene diversity. Overall, this intricate metabolic activity recovers the host's metabolic energy and absorbable substrates while also supplying the bacteria with the energy and nutrients needed to multiply and grow. In the colon, the fermentation of carbohydrates serves as a significant energy source. Non-digestible carbohydrates include large polysaccharides (such as resistant starches, cellulose, hemicellulose, pectins, and gums), some oligosaccharides that are not digested, and unabsorbed sugars and alcohols. Short-chain fatty acid production is the metabolic end point.

Colonic microorganisms are also involved in the synthesis of vitamins. The caecum's ability to absorb ions is enhanced by the fermentation of carbohydrates and short-chain fatty acids, particularly acetate, propionate, and butyrate. These fatty acids play significant roles in the physiology of the host. Butyrate is a major source of energy for colonocytes and is almost entirely consumed by the colonic epithelium. The liver eventually metabolizes acetate and propionate, which are found in portal blood (Younes et al., 2001).

1.5.2 TROPIC FUNCTIONS

Two types of tropic major functions performed by the colonic microbiota are epithelial cell growth and differentiation and interactions between host immunity and gut bacteria.

1.5.2.1 EPITHELIAL CELL GROWTH AND DIFFERENTIATION

Interaction with local microorganisms has a significant impact on how epithelial cells differentiate (Hooper et al., 2001). In the large and small bowel *in vivo*, all three of the major short-chain fatty acids promote epithelial cell proliferation and differentiation. Butyrate, on the other hand, inhibits cell division and promotes cell differentiation in epithelial cell lines with neoplastic origin. Additionally, butyrate encourages cells to switch from neoplastic to non-neoplastic phenotypes (Siavoshian, 2000). Short-chain fatty acids play a vital role in the prevention of human pathological conditions, including colonic carcinogenesis and chronic ulcerative colitis.

1.5.2.2 GUT BACTERIA AND HOST IMMUNITY INTERACTIONS

The gut microbiota prevents pathogens from entering the body through competitive exclusion mechanisms such as attachment site occupation, nutrient consumption, and antimicrobial substance production. It also induces the host to produce a variety of antibacterial substances (Sekirov et al., 2010). Several studies (Qin et al., 2010; Ashida et al., 2012; DuPont and DuPont, 2011) have discovered abnormal and impaired microbiota in a variety of illnesses, such as inflammatory bowel disease, colorectal cancer, irritable bowel syndrome, metabolic syndrome, and non-alcoholic fatty liver disease, supporting the notion that gut microflora is an essential organ for our body's perfect homeostasis. According to Schley and Field (2002), a mucosal immune system (consisting of innate and acquired immune system defenses) is strategically placed in areas where pathogens and antigens may enter the body, including GALT (Gut Associated Lymphoid Tissue), which protects the intestine, and mucosal-associated lymphoid tissues, which shield areas like the urinary, respiratory, and reproductive tracts. It is essential that protective immune responses are made to potential pathogens, but it is also crucial to minimize hypersensitivity reactions to dietary antigens because the intestine is the body's first line of defense against the environment.

1.5.3 PROTECTIVE FUNCTIONS: THE BARRIER EFFECT

Multifaceted intestinal barrier maintains homeostasis by monitoring nutrient and electrolyte absorption and averting the penetration of uncontrolled luminal content like pathogens and toxic substances into the circulation (Farhadi et al., 2003). This efficient barrier is preserved by physical, biochemical, and immunological components. The seven intestinal epithelial cells (IECs) – enterocytes, goblet cells, Paneth cells, microfold cells (M cells), enteroendocrine cells, cup cells, and tuft cells – are distributed across the surface area of 400 m^2 (Peterson and Artis, 2014).

At least 90% of crypt cells or villus cells are enterocytes. 40 TJ proteins, including occludin, claudins, junctional adhesion molecule A, and tricellulin, seal the contact between IECs (Braniste et al., 2014). The epithelium is lined by a highly glycosylated mucin that contains immunoglobulin (Ig) A and antimicrobial peptides. By the eighth week of gestation, this intestinal barrier has fully developed. From week ten on, epithelial cells with microvilli, enteroendocrine cells, and tight junctions are visible. In the mucus layers, antimicrobial proteins (AMPs) regulate the colonization of commensal bacteria. Reciprocally, the microbiota and their metabolites control the production of some AMPs. Immunoglobulin is thought to interact with commensal bacteria in the intestinal mucosa to protect against pathogens (Woof and Russell, 2011). Additionally, the production of SCFA by the gut microbiota serves as the epithelial cells' energy source (Krajmalnik et al., 2012). The property of AMP required for physically separating commensal bacteria from intestinal epithelium and controlling bacterial overgrowth is restored by pre- and probiotic administration (Yan et al., 2011).

1.6 MODULATING THE COMPOSITION OF MICROFLORA: PROBIOTICS

The intestinal habitat of an individual contains 300–500 different species of bacteria. Of which, some have potential pathogenic properties and some bacteria (probiotics) have positive health effects in terms of prevention of gut infection or a reinforcement of innate defenses (Andrew et al., 2008).

1.6.1 PROBIOTICS: TYPES AND BENEFITS

Since its inception 50 years ago, probiotics, a word derived from Latin and Greek meaning literally "for life," has been defined in a variety of ways. According to a joint FAO/WHO report (2001), probiotics are "living microorganisms that, when administered in adequate amounts, confer health benefits on the host." This definition corresponds to that of functional foods. Probiotics are typically bacterial components of the normal human intestinal flora, such as *Lactobacilli* and *Bifidobacteria*, that produce short-chain fatty acids lactate, acetate, and butyrate as end products of metabolism (Hill et al., 2014). Organizations and agencies such as Codex (which falls under the FAO/WHO umbrella), Health Canada, the World Gastroenterology Organization, the European Food Safety Authority (EFSA), and the Institute of Food Technologists use this definition when referring to probiotics. The definition distinguishes between live microbes that are administered primarily for their health benefits from those that are used as processing aids or sources of useful compounds. This definition also implies the distinction between commensal microorganisms and probiotics.

The genus, species, subspecies (if applicable), and an alphanumeric designation, serve as a means of identification for a specific strain of probiotics (World Gastroenterology Organization, 2007). *Lactobacillus, Bifidobacterium, Saccharomyces, Streptococcus, Enterococcus, Escherichia,* and *Bacillus* are the seven core genera of microorganisms that are most frequently used in probiotic products.

Recent research has shown evidence-based effects of prebiotics and probiotics in the management of disorders and diseases which include diarrhea, constipation, IBS, oral health, depression, Alzheimer's disease, diabetes, obesity, cancer, immunity management, and other non-communicable diseases.

1.7 MECHANISM OF ACTION OF PROBIOTICS FOR IMPROVED HEALTH

A probiotic's mode of action may involve modulating the host's microflora through interactions between orally administered viable microbes and the microflora in the lumen, as well as the host's metabolic activities, such as stabilizing the pattern of digestive enzymes, and immunomodulation through the induction and control of mucosa-associated and systemic immune system responses (Tripathi and Giri, 2015). Additionally, the modes of action depend on the strain. The intestinal microflora offers defense against a variety of pathogens, including yeasts like *Candida albicans*, certain types of Clostridia, *Escherichia coli*, Shigella, and Pseudomonas. Compounds made by *Lactobacillus* casei subsp. rhamnosus (*Lactobacillus* GG) prevent the growth of a number of gram-positive and gram-negative bacteria. Examples include pyroglutamate and hydrogen peroxide. Several other *Lactobacilli* can produce substances that are similar to these. The production of short-chain fatty acids lowers the colonic pH, which favors the growth of organisms with less pathogenicity (Tripathi and Giri, 2015). Probiotics may also compete with pathogenic organisms for nutrients that they would otherwise consume. For instance, *Clostridium difficile,* which depends on monosaccharides for growth, may be inhibited from growing if a probiotic consumes monosaccharides. The capacity of a probiotic to endure digestion and flourish in the gastrointestinal tract is a key determinant of its efficacy. Most studies looking for effective probiotics concentrate on the challenges of colonization (Argyri, et al., 2013; Kimoto et al., 2003). Due to their numerous advantages, such as the treatment of lactose intolerance, high cholesterol, cardiac diseases, and body weight management, probiotics are becoming more and more popular. The market potential for probiotics is enormous given the current emphasis on disease prevention and the pursuit of optimal health at all ages (Tripathi and Giri, 2015).

KEYWORDS

- **bacteria**
- **blood**
- **fatty acids**
- **health benefits**
- **intestinal diseases**
- **metabolism**
- **parasites**
- **pathogens**
- **probiotics**
- **respiratory tract diseases**

REFERENCES

Andrew, L. W., Delphind, M. A., Saulnier, D., & Gibson, G. R. (2008). Gastrointestinal microflora and interactions with gut mucosa. In G. R. Gibson & M. B. Roberfroid (Eds.), *Handbook of prebiotics* (pp. 11–38). CRC Press, Taylor and Francis Group USA.

Argyri, A. A., Zoumpopoulou, G., Karatzas, K. A. G., Tsakalidou, E., Nychas, G. J. E., Panagou, E. Z., & Tassou, C. C. (2013). Selection of potential probiotic lactic acid bacteria from fermented olives by in vitro tests. *Food Microbiology, 33*(2), 282–291.

Arrieta, M. C., Stiemsma, L. T., Amenyogbe, N., Brown, E., & Finlay, B. (2014). The intestinal microbiome in early life: Health and disease. *Frontiers in Immunology, 5*, Article 427. https://doi.org/10.3389/fimmun.2014.00427.

Ashida, H., Ogawa, M., Kim, M., Mimuro, H., & Sasakawa, C. (2012). Bacteria and host interactions in the gut epithelial barrier. *Nature Chemical Biology, 8*, 36–45.

Bäckhed, F., Ley, R. E., Sonnenburg, J. L., Peterson, D. A., & Gordon, J. I. (2005). Host-bacterial mutualism in the human intestine. *Science, 307*, 1915–1920.

Balcazar, J. (2007). Probiotics in health maintenance: Do they really work? *British Journal of Infection Control, 8*(3), 87–93.

Björkstén, B. (2006). The gut microbiota: A complex ecosystem. *Clinical and Experimental Allergy, 36*(10), 1215–1217.

Braniste, V., Al-Asmakh, M., Kowal, C., Anuar, F., Abbaspour, A., Tóth, M., Korecka, A., Bakocevic, N., Guan, N. L., Kundu, P., Gulyás, B., Halldin, C., Hultenby, K., Nilsson, H., Hebert, H., Volpe, B. T., Diamond, B., & Pettersson, S. (2014). The gut microbiota influences blood-brain barrier permeability in mice. *Science Translational Medicine, 6*(263). https://doi.org/10.1126/scitranslmed.3009759.

De Whelan, K., Rossi, M., Morrison, M., Holtmann, G., Kelly, J. T., Shanahan, E. R., Staudacher, H. M., & Campbell, K. L. (2018). Dietary fiber intervention on gut microbiota composition in healthy adults: A systematic review and meta-analysis. *The American Journal of Clinical Nutrition, 107*(6), 965–983. https://doi.org/10.1093/ajcn/nqy041.

Dethlefsen, L., Huse, S., Sogin, M. L., & Relman, D. A. (2008). The pervasive effects of an antibiotic on the human gut microbiota, as revealed by deep 16S rRNA sequencing. *PLoS Biology, 6*(11), 2383–2400.

Duncan, S. H., Louis, P., & Flint, H. J. (2007). Cultivable bacterial diversity from the human colon. *Letters in Applied Microbiology, 44*, 343–350.

DuPont, A. W., & DuPont, H. L. (2011). The intestinal microbiota and chronic disorders of the gut. *Nature Reviews Gastroenterology & Hepatology, 8*, 523–531.

Dykstra, N. S., Hyde, L., Mackenzie, A., & Mack, D. R. (2011). *Lactobacillus plantarum* 299v prevents caspase-dependent apoptosis in vitro. *Probiotics and Antimicrobial Proteins, 3*, 21–26.

Farhadi, A., Banan, A., Fields, J., & Keshavarzian, A. (2003). Intestinal barrier: An interface between health and disease. *Journal of Gastroenterology and Hepatology, 18*(5), 479–497. https://doi.org/10.1046/j.1440-1746.2003.03032.x.

Flint, H. J. (2006). The significance of prokaryote diversity in the human gastrointestinal tract. In N. A. Logan, H. M. Lappin-Scott, & P. C. F. Oyston (Eds.), *Prokaryotic Diversity: Mechanisms and Significance* (pp. 65–90). Cambridge University Press.

Fooks, L. J., & Gibson, G. R. (2002). Probiotics as modulators of the gut flora. *British Journal of Nutrition, 88*(S1), S39–S49.

Gill, S. R., Pop, M., Deboy, R. T., Eckburg, P. B., Turnbaugh, P. J., & Samuel, B. S. (2006). Metagenomic analysis of the human distal gut microbiome. *Science, 312*, 1355–1359.

Guaraldi, F., & Salvatori, G. (2012). Effect of breast and formula feeding on gut microbiota shaping in newborns. *Frontiers in Cellular and Infection Microbiology, 2*, 94. https://doi.org/10.3389/fcimb.2012.00094.

Guarner, F., & Malagelada, J. R. (2003). Gut flora in health and disease. *The Lancet, 361*(9356), 512–519.

Harmsen, H. J., Raangs, G. C., He, T., Degener, J. E., & Welling, G. W. (2002). Extensive set of 16S rRNA based probes for detection of bacteria in human feces. *Applied and Environmental Microbiology, 68*, 2982–2990.

Heeney, D. D., Gareau, M. G., & Marco, M. L. (2018). Intestinal Lactobacillus in health and disease: A driver or just along for the ride? *Current Opinion in Biotechnology, 49*, 140. https://doi.org/10.1016/j.copbio.2017.08.004.

Hill, C., Guarner, F., Reid, G., Gibson, G. R., Merenstein, D. J., Pot, B., & Calder, P. C. (2014). The international scientific association for probiotics and prebiotics consensus statement on the scope and appropriate use of the term probiotic. *Nature Reviews Gastroenterology & Hepatology, 11*, 506–514.

Hooper, L. V., Wong, M. H., Thelin, A., Hansson, L., Falk, P. G., & Gordon, J. I. (2001). Molecular analysis of commensal host-microbial relationships in the intestine. *Science, 291*, 881–884.

Kimoto, H., Nomura, M., Kobayashi, M., Mizumachi, K., & Okamoto, T. (2003). Survival of lactococcal during passage through mouse digestive tract. *Canadian Journal of Microbiology, 49*(11), 707–711.

Krajmalnik-Brown, R., Ilhan, Z. E., Kang, D. W., & DiBaise, J. K. (2012). Effects of gut microbes on nutrient absorption and energy regulation. *Nutrition in Clinical Practice, 27*(2), 201–214. https://doi.org/10.1177/0884533611436116.

Kurokawa, K., Itoh, T., Kuwahara, T., Oshima, K., Toh, H., & Toyoda, A. (2007). Comparative metagenomics revealed commonly enriched gene sets in human gut microbiomes. *DNA Research, 14*, 169–181.

Ley, R. E., Turnbaugh, P. J., Klein, S., & Gordon, J. I. (2006). Microbial ecology: Human gut microbes associated with obesity. *Nature, 444*, 1022–1023.

Naseribafrouei, A., Hestad, K., Avershina, E., Sekelja, M., Linløkken, A., Wilson, R., & Rudi, K. (2014). Correlation between the human fecal microbiota and depression. *Neurogastroenterology and Motility: The Official Journal of the European Gastrointestinal Motility Society, 26*(8), 1155–1162. https://doi.org/10.1111/nmo.12378.

Papoff, P., Ceccarelli, G., d'Ettorre, G., Cerasaro, C., Caresta, E., Midulla, F., & Moretti, C. (2012). Gut microbial translocation in critically ill children and effects of supplementation with pre- and probiotics. *International Journal of Microbiology, 2012(2), 151393.*

Peterson, L. W., & Artis, D. (2014). Intestinal epithelial cells: Regulators of barrier function and immune homeostasis. *Nature Reviews Immunology, 14*(3), 141–153. https://doi.org/10.1038/nri3608.

Qin, J., Li, R., Raes, J., Arumugam, M., Burgdorf, K., Manichanh, C., & Bork, P. (2010). A human gut microbial gene catalogue established by metagenomic sequencing. *Nature, 464*, 59–65.

Rezzi, S., Ramadan, Z., Fay, L. B., & Kochhar, S. (2007). Nutritional metabonomic: Applications and perspectives. *Journal of Proteome Research, 6*(2), 513–525. https://doi.org/10.1021/pr060522z.

Sanford, P. A. (2005). *Digestive System Physiology* (2nd ed.). Edward Arnold.

Schley, P. D., & Field, C. J. (2002). The immune-enhancing effects of dietary fibers and prebiotics. *British Journal of Nutrition, 87*(Suppl 2), S221–S230.

Sekirov, I., Russell, S. L., Antunes, L. C. M., & Finlay, B. B. (2010). Gut microbiota in health and disease. *Physiological Reviews, 90*(3), 859–904.

Siavoshian, S., Segain, J. P., Kornprobst, M., Bonnet, C., Cherbut, C., Galmiche, J. P., & Blottière, H. M. (2000). Butyrate and trichostatin A effects on the proliferation/differentiation of human intestinal epithelial cells: Induction of cyclin D3 and p21 expression. *Gut, 46*, 507–514.

Tojo, R., Suárez, A., Clemente, M. G., De Los Reyes-Gavilán, C. G., Margolles, A., Gueimonde, M., & Ruas-Madiedo, P. (2014). Intestinal microbiota in health and disease: Role of bifidobacteria in gut homeostasis. *World Journal of Gastroenterology, 20*(41), 15163–15176. https://doi.org/10.3748/wjg.v20.i41.15163.

Tripathi, M. K., & Giri, S. K. (2015). *Probiotic Food Technology, Safety and Human Health.* Jaya Publishing House.

Turnbaugh, P. J., Ley, R. E., Hamady, M., Fraser-Liggett, C. M., Knight, R., & Gordon, J. I. (2007). The human microbiome project. *Nature, 449*, 804–810.

Turroni, F., Ribbera, A., Foroni, E., Van Sinderen, D., & Ventura, M. (2008). Human gut microbiota and bifidobacteria: From composition to functionality. *Antonie van Leeuwenhoek, 94*, 35–50.

Van der Waaij, L. A. (1999). The ecology of the human intestine and its consequences for overgrowth by pathogens such as *Clostridium difficile. Annual Review of Microbiology, 43*, 69–87.

Ventura, M., Turroni, F., Canchaya, C., & Vaughan, E. E. (2009). Microbial diversity in the human intestine and novel insights from metagenomics. *Frontiers in Bioscience, 14*, 3214–3221.

Woof, J. M., & Russell, M. W. (2011). Structure and function relationships in IgA. *Mucosal Immunology, 4*(6), 590–597. https://doi.org/10.1038/mi.2011.39.

Yan, F., Cao, H., Cover, T. L., Whitehead, R., Washington, M. K., & Polk, D. B. (2007). Soluble proteins produced by probiotic bacteria regulate intestinal epithelial cell survival and growth. *Gastroenterology, 132*, 562–575.

Younes, H., Coudray, C., Bellanger, J., Demigné, C., Rayssiguier, Y., & Rémésy, C. (2001). Effects of two fermentable carbohydrates (inulin and resistant starch) and their combination on calcium and magnesium balance in rats. *British Journal of Nutrition, 86*, 479–485.

Zhang, Y. J., Li, S., Gan, R. Y., Zhou, T., Xu, D. P., & Li, H. B. (2015). Impacts of gut bacteria on human health and diseases. *International Journal of Molecular Sciences, 16*(4), 7493–7519. https://doi.org/10.3390/ijms16047493.

Zhu, H., Hart, C. A., Sales, D., & Roberts, N. B. (2006). Bacterial killing in gastric juice: Effect of pH and pepsin on *Escherichia coli* and *Helicobacter pylori. Journal of Medical Microbiology, 55*(9), 1265–1270. https://doi.org/10.1099/jmm.0.46611-0.

CHAPTER 2

Prebiotics

MINI K. SHETH

Department of Foods and Nutrition, The Maharaja Sayajirao University of Baroda, Gujarat, India

Evidences provide a treasure of information on prebiotics through experimental, animal, and human studies, as well as their organoleptic properties in food industries have been studied. However, data is still scarce. New research must be conducted with the goal of understanding the mechanism of action and elucidating their beneficial health effects on the human host.

2.1 INTRODUCTION

Gibson and Roberfroid in 1995 defined Prebiotics as a non-digestible food component that enhances the health of the host by deliberately promoting the growth or activity of one or a small number of bacteria in the colon. Later, a redefined definition of a prebiotic which focused on selective targeting of specific microorganisms was proposed as "A selectively fermented ingredient that allows specific changes, both in composition and/or activity in the gastrointestinal microflora that confers benefits" (Gibson et al., 2004). Current applications of this knowledge include the formation of fresh nutraceutical products and their potential as natural remedies for various diseases. These definitions have generated a lot of interest in the field of nutrition, both in terms of scientific and culinary applications.

Prebiotic Foods for a Healthy Gut: Scientific Studies to Manage Metabolic Disorders.
Mini K. Sheth and Shruti Dwivedi (Eds.)

2.2 CRITERIA FOR FOOD TO BE TERMED AS PREBIOTICS

A food ingredient can be termed as 'prebiotic' when it can be classified under following criteria

- Resist gastric acidity;
- Is not hydrolyze by mammalian enzymes;
- Is not absorbed by the upper gastrointestinal tract;
- The intestinal microflora ferments it;
- Encourages the development and activity of intestinal bacteria that may be beneficial to health and well-being (Gibson and Roberfroid, 1995).

Although there are certain similarities in the mode of action of probiotics and prebiotics in the human body, they differ in composition and metabolism. Prebiotics are found naturally in some plants and are also produced enzymatically from disaccharides.

The prebiotic property has now been adequately demonstrated for many food ingredients such as non-digestible carbohydrates, which are often described as soluble fibers, non-starch polysaccharides, resistant starches, and soluble oligosaccharides. The classification of natural and some synthetic prebiotic types is listed in Table 2.1.

TABLE 2.1 Classification of Prebiotics

Classification	Origin/ Manufacturing process
Disaccharides	from lactose, synthetic
Lactulose	from lactose, synthetic
Lacticol	
Oligosaccharides	Legumes, vegetables, cereals
Fructose Oligosaccharides (FOS)	Extraction/hydrolysis
Soyabean Oligosaccharides	Soyabean
(Trans) Galactooligosaccharides	Extraction/hydrolysis
Inulin	From lactose, Synthetic
	Legumes, vegetables, cereals
	Extraction
Polysaccharides	Legumes, vegetables, cereals
Resistant Starch	Extraction

2.3 EVOLUTION OF FRUCTOOLIGOSACCHARIDE (FOS) AS A PREBIOTIC

Inulin and oligofructose have the potential to elevate indigenous *Bifidobacteria* and *Lactobacilli* levels in the colon and thus influence the whole body's physiology and consequently health and well-being. Several fruits and vegetables naturally contain inulin-type fructans. The most popular dietary sources are leek, wheat, onion, banana, and garlic. Inulin-type fructans are a linear polydisperse carbohydrate material made up primarily of -(2)1) fructosyl-fructose linkages. Chicory plant of the compositae family is most commonly used in the industry for the extraction of inulin-type fructans. Native chicory inulin is non-fractionated and derived from the fresh roots (De Leenheer, 1996). They are classified as "nondigestible" oligosaccharides because of the configuration of the anomeric C2 in their fructose monomers. Inulin-type fructans are resistant to hydrolysis by human small intestinal digestive enzymes that are specific for a-glycosidic bonds (Delzenne and Roberfroid, 1994; Roberfroid et al., 2000).

2.3.1 FRUCTOOLIGOSACCHARIDE FORMATION FROM INULIN

2.3.1.1 PARTIAL ENZYMATIC HYDROLYSIS OF INULIN USING AN ENDOINULINASE

Degree of polymerization (DP) of Chicory inulin ranges from 2 to 60 units. Oligofructose is produced by the partial enzymatic hydrolysis of inulin with an endoinulinase, which is a mixture of GpyFn and FpyFn molecules with DPs ranging from 2 to 7 (Roberfroid, 2005). It primarily consists of lower-DP oligosaccharides like 1,1,1-kestopentaose, 1-kestotriose, and 1-kestotetraose as well as inulobiose, inuloriose, and 1-kestotetraose.

2.3.1.2 ENZYMATIC SYNTHESIS (TRANSFRUCTOSYLATION)

Oligofructose can be synthesized enzymatically (transfructosylation) using the fungal enzyme fructosidase, which can be obtained from a variety of sources. Sucrose is used as a substrate in this reaction, and additional fructose units are added by forming new linkages. DP in this synthetic compound ranges from 2 to 4. It is also possible to use physical separation techniques to remove all oligomers with DP 10 to produce a high molecular-weight inulin-type fructan or inulin. Oligofructose-enriched inulin is a mixture of two distinct populations of low-molecular-weight oligofructose and high-molecular-weight inulin (Roberfroid, 2005). Fructooligosaccharide (FOS) is undeniably a component of the dietary fiber complex, and it must be labeled as such on consumer food products. However, it is important to note that we do not need to limit FOS as dietary fiber, which is commonly labeled on various food products, because it possesses specific fermentative properties that improve the host's health.

2.3.2 SAFETY AND TOLERANCE OF FRUCTOOLIGOSACCHARIDE (FOS)

Inulin and Fructooligosaccharide are staples in the diets of the vast majority of people. Daily intakes in the United States and Europe have been estimated to be up to 10 g, with the 97th percentile in the United States receiving 1–4 g. Because inulin and oligofructose are both macro ingredients, traditional toxicology tests are difficult to apply. Some high-dose animal tests have been conducted; no toxic effects have been discovered. Many legal authorities around the world have evaluated the suitability of oligofructose and inulin for use in food. As a result, most countries accept inulin and oligofructose as food ingredients that can be used in food formulations without restriction. High dose consumption in some cases has reported intestinal discomfort and increased flatulence. However, the acceptance rate varies greatly from person to person. Both inulin and oligofructose are gradually becoming recognized as "dietary fibers" in the majority of nations around the world in terms of labeling. Some nations have made it compulsory to state the "bifidogenic effect" on food labels. FDA estimates that dietary intake of inulin at the 90th percentile level would be approximately 6 grams per day for infants under one year of age, approximately 15 grams per day for infants one year of age, and approximately 20 grams per day for the general population, according to the U.S. FDA notice on GRAS of FOS with notice number GRAS Notice No. GRN 000118 (Kruger, 2003).

2.3.3 CALORIC VALUE OF FRUCTOOLIGOSACCHARIDE (FOS)

The caloric value of an oligofructose fructosyl unit is calculated to be 30 to 40% of a digested fructose molecule, or between 1–1.5 kcal/g (Roberfroid et al., 1993). Oligofructose has a caloric value of 1.48 kcal/g. Using six healthy human subjects, Molis et al. in 1996 defined the energy value of fructooligosaccharide (44% GF2, 46% GF3, and 10% GF4). The fructooligosaccharide calculated mean energy value was 9.5 ± 0.6 kJ/g (range: 8.3–11.7 kJ/g), i.e., about 2 kcal/gram. Roberfoid in 1999 further stated inulin, oligofructose, and all nondigestible oligosaccharides that are primarily fermented in the colon should be given a caloric value of 1.5 kcal/g (6.3 kJ/g) for nutrition labeling purposes.

2.3.4 FRUCTOOLIGOSACCHARIDE (FOS) AND ITS TECHNOLOGICAL FUNCTIONS

The growing consumer concern about nutrition and health is an important determinant in food choice. As a result, the food industry can help by reducing the number of products that are high in energy, improving the nutrient profile, and developing health-focused functional foods (Nehir and Simsek, 2011). Consumers are becoming more aware of maintaining 'internal health' by modifying gastrointestinal flora. Over the last 20 years, there has been significant interest in the production and consumption of prebiotics in the daily diet by both consumers and food manufacturers.

The main objective of prebiotics is to improve gut microflora through dietary means. The prebiotic approach involves the interaction of a non-digestible food ingredient and beneficial microorganisms such as *bifidobacteria* and *lactobacillus* in the human colon. Using prebiotics in the food formulation process has some advantages over the probiotic strategy as they have been shown to reduce the problem of keeping the organisms alive during transit through the upper gastrointestinal tract as well as during storage (Crittenden, 1999).

Among the prebiotics researched so far, fructooligosaccharide and inulin hold a key position in the food industry because of their interesting technological characteristics. Refined native inulin powder from chicory is white, amorphous, and slightly hygroscopic, with a specific gravity of about 1.35 and an average molecular weight of about 1,600. It is neutral in odor and taste. Due to a small amount of naturally occurring mono- and disaccharides, commercial inulin has a marginally sweet taste. Oligofructooligosaccharide is soluble in water, with solubility dependent on the temperature of the water, degree of polymerization, distribution of the molecular chains, degree of molecular branching, and how the molecule is processed. FOS exhibit minimal influence on the organoleptic characteristics of a product and possess nutritional benefits and health claims.

2.3.5.1 FOS AS A FAT SUBSTITUTE

Specific prebiotics, such as oligosaccharides and inulin, have been developed as fat substitutes and texture modifiers because they can (1) reduce total fat or partial fat content, (2) alter smoothness and creaminess, (3) increase perception of body and richness, and (4) improve overall eating quality and an acceptable appearance. Oligofructose is a well-known prebiotic due to its ability to replace fat in the production of low-calorie foods. When oligofructose is mixed with water, it forms gels composed of a three-dimensional gel network of insoluble sub-micron crystals. This gel has the texture and mouthfeel of fat (Silva, 1996; Franck, 2000).

Most water-based foods, such as dairy products, frozen desserts, dressings, table spreads, sauces, soups, and even meat products, use oligofructose and inulin to replace fat (Murphy, 2001). In most foods, 1 g of fat can be replaced by 0.35 g of FOS or inulin (Coussement, 1999). Combining OFS and inulin in foods helps to maximize freeze-thaw stability and minimize emulsion separation (Bishay, 1998). Additionally, OFS gives liquid products and spreads a richer texture and gives extruded snacks and cereals crispness and expansion.

The food industry has developed traditional food products to contain less fat in response to the rising demand for low-fat food. Numerous studies have been done to examine the impact of replacing fat or oil in foods with oligofructose and inulin in this regard. In 2011, Meyer et al. developed ice cream using inulin corn syrup as a fat replacer at 50% and 100% levels. The results showed that adding inulin corn syrup instead of fat improved the texture and chewiness of the ice cream and extended its shelf life by 6 weeks. An additional study sought to create low-fat biscuits with useful characteristics. FOS-incorporated biscuits had a 25% lower fat content (Gallagher et al., 2003).

Numerous researchers have also looked into the function of FOS as a fat substitute in meat products. A study on the characteristics of low-fat cooked meat sausages showed a 40% reduction in fat and a low-calorie content of the product. Oligofructose at various concentrations replaced the fat in the German-style mortadella sausage formula, resulting in a 6–22% reduction in fat content (Nowak et al., 2007). A couple of studies revealed that FOS-incorporated fermented and cooked sausages exhibited around a 15%-50% reduction in fat with increased sensory properties (Salazar et al., 2009; Santos, 2012). Farzanmehr and Abbasi (2009) carried out a study on inulin as a bulking agent and fat replacer in milk chocolate. The results of the study showed that chocolate made from inulin was accepted with better sensory and textural attributes and replaced fat by 5% compared to the control chocolate. A more recent study on handmade chocolate and nut bar replaced FOS with fat at different levels and revealed good acceptance and textural properties (Folly et al., 2013). All these findings together indicate that FOS and inulin are good fat replacers and can be used in a variety of food products.

2.3.5.2 FOS AS A SUGAR REPLACER

A variety of non-digestible oligosaccharides and polyols can be used as sugar substitutes because of their physiological traits, such as their ability to perform bulking functions and lower energy intake (Franck, 2000).

Oligofructose is being used as a sugar substitute in dairy and bakery goods like chocolate-filled biscuits, chewing gum, confections, dairy desserts, and ice cream. It also provides some functional properties similar to those of glucose syrup (Franck, 2000). Additionally, oligofructose functions similarly to sugar as a binder in nutrition bars and lowers the freezing point of frozen desserts. Oligofructose is more soluble than sucrose, but it has a sweetness that is only about 30% that of sucrose. Oligofructose can provide the desired sweetness and a better flavor profile when combined with other sweeteners like aspartame (Kaur and Gupta, 2002). More importantly, because they have a low glycemic index, these low-calorie ingredients have advantages over traditional digestible carbohydrates like sucrose, glucose, and fructose (Schumann, 2002).

FOS as a sugar substitute has been tried out in a variety of food products like peanut brittle, chocolate chips, chocolate coating, nougat, ginger snap cookies, cheesecake, and all the products adjudged to be satisfactory (Barndt et al., 2003). An Indian study on the preparation of low-calorie gulab jamun revealed that FOS is a potent low-calorie healthier alternative to sucrose without altering any of the sensory characteristics (Renuka et al., 2004). Another study carried out by Renuka et al. (2009) prepared fruit juice beverage with fortification of FOS as a low-calorie prebiotic sweetener and showed that FOS did not affect the sensory attributes of the beverages, and even the shelf life increased with an addition of FOS. Choonhahirun and Akesowan (2012) prepared *Thai p*andanus custard replaced with coconut milk and sugar with soy milk and inulin, and were compared with the quality characteristics of the regular formulation (control). Custard replaced with 5% inulin as sugar replacement presented desirable characteristics in relation to the regular formulation. The product has about a 20.7% reduction in energy, and storage time was also increased by 3 weeks.

2.3.5.3 FOS AS A TEXTURE MODIFIER

As FOS is more water-soluble than 'classical fibers,' it can be used to fortify dairy products such as milk drinks, yogurt, cheeses, and desserts, which have traditionally been difficult to fortify (Niness, 1999). The effect of oligofructose or inulin on water solutions at various solid levels determines its functionality. It increases viscosity significantly at low concentrations and can be used as a rheology modifier. At a concentration of 40–45%, a firm gel or fatty creamy is formed. This cream has numerous advantages. Water is tightly bound in the gel, enhancing freeze-thaw stability and inhibiting syneresis. In this form, oligofructose and inulin are stable in acidic or high-temperature environments (Silva, 1996).

Inulin and *L. acidophilus*-containing plain yogurt had a significantly lower pH than other yogurts, have good flavor, have less synergism, and had better body and texture, according to a 2007 study by Aryana et al. Preparation of gluten-free bread with a combination of inulin, oligosaccharide syrup, and chicory flour with 3%, 5%, and 8% additions exhibits the best effects on sensory features with a medium level of 5% incorporation (Korus et al., 2008). A most recent study on the development of prebiotic yogurt revealed that the addition of 2%, 4%, 6%, and 8% FOS increased physiochemical, rheological, and microbiological characteristics of low-calorie yogurt (Cruz, 2013).

Inulin and *L. paracasei* were added to chocolate mousse, which was advantageous and potentially effective as it improved the product's textural and sensory qualities. The storage time also extended by 28 days with the addition of inulin (Cardarelli et al., 2008).

Oligofructose is used in food formulations to enhance the nutritional value of the finished product by improving the product's textural and organoleptic qualities, dietary fiber content, calorie and fat content, and bifidus-promoting abilities. The sensory characteristics of orange cakes that had inulin and fructooligosaccharide (3 g per 60 g serving) had a better crust compared to sugar-added cakes (Larissa et al., 2012). These ingredients provide a special synergy of technological and nutritional benefits. As a result, they are frequently interpreted as concrete examples of active food ingredients for "functional foods."

2.4 ROLE OF FRUCTOOLIGOSACCHARIDE FERMENTATION IN THE GUT

Inulin and oligosaccharides specifically promote the growth of *Lactobacillus* and *Bifidobacteria*, which improves the nutritive value of the food and holds therapeutic value.

2.4.1 ROLE OF FRUCTOOLIGOSACCHARIDE IN ENHANCING THE GUT MICROBIOTA

The fermentability of FOS and inulin by fecal bacteria has been extensively investigated in several *in vitro* models (Langlands et al., 2004; Vuyst and Leroy, 2011). Wang and Gibson (1993) determined *in vitro* the prebiotic efficacy of FOS and inulin compared to a range of reference carbohydrates (starch, polydextrose, fructose, and pectin) in 12-hour batch cultures with mixed populations of gut bacteria. Data revealed preferential fermentation by *Bifidobacteria* while *Escherichia coli* and *Clostridium perfringens* populations remained at relatively low levels, demonstrating the prebiotics' bifidogenic properties.

Non-fat dry milk containing 0.5, 1.0, 3.0, and 5.0% (w/v) fructooligosaccharide (FOS), galactooligosaccharide (GOS), and inulin was used to culture two commercial strains of *Bifidobacterium* spp. (Bf-1 and Bf-6). Inoculated samples underwent a 48-hour anaerobic incubation period at 37°C. The cultures' development and activities were identified. Each strain's viability was evaluated four weeks after being stored in a refrigerator at 4°C. When *Bifidobacteria* Bf-1 and Bf-6 were grown in the presence of FOS, growth promotion, activity enhancement, and viability retention were most enhanced, followed by GOS and inulin. With increasing carbohydrate concentration, the effects of oligofructose and inulin became more pronounced, reaching their peak at 5% (w/v) (Shin et al., 2000).

The level of oligofructose polymerization has a significant impact on the growth and viability of bacteria. For instance, the gene for β-fructofuranosidase from *Bifidobacterium* lactis was discovered and described in a study. High identity was found between this gene and a related gene in *Bifidobacterium* longum. The enzyme's maximum activity was towards oligofructose, followed by a lesser degree of activity towards inulin and a minimum amount of activity towards long-chain inulin. The characterized enzyme is highly selective for oligofructose and has a high affinity for fructosyl-linkages, and as the degree of fructans polymerization (DP) increases, its specificity decreases (Janer et al., 2004).

In pre-hypertensive subjects with type 2 diabetes mellitus, synbiotic supplementation reduced hs-CRP levels by 48.7%, fasting blood sugar (FBS) by 4.67%, HbA1c by 14.8%, systolic blood pressure (SBP) by 6.2%, and diastolic blood pressure (DBP) by 8.4% (Sheth et al., 2016).

Anthropometric, biochemical, and microbial profiles of obese subjects based on dietary fiber intake show that obese subjects who consumed more than 15 gm/day of total dietary fiber had a lower BMI ($p < 0.05$), a higher GLP-1 value, and a lower LPS level ($p < 0.05$). High fiber intake showed a positive correlation with the colonization of *Bifidobacteria* ($p < 0.05$) and a negative correlation with enteric pathogen count. FOS supplementation resulted in a significant reduction in weight, BMI ($p < 0.001$), WC, WHR ($p < 0.05$), and percent body fat ($p < 0.01$). The colonization of *Lactobacillus*, *Bifidobacteria*, and gut hormones (GLP-1) increased, affecting satiety levels in the hypothalamus region and resulting in reduced food intake (Sheth et al., 2014).

FOS supplementation for 30 days over a 45-day period benefited undernourished children by increasing *Bifidobacteria* and Lactic acid bacteria colonization by 28.6% and 2.56%, respectively. A decrease in the colonization of *E. coli* by 2.38%. There was a 79.6% reduction in diarrheal episodes and an 82% reduction in common cold episodes (Tanu Shree and Sheth, 2014).

2.4.2 ROLE OF FRUCTOOLIGOSACCHARIDE IN PRODUCTION OF SHORT CHAIN FATTY ACID

When FOS reaches the large intestine, it is preferentially utilized by a group of healthy bacteria called *Bifidobacteria* and *Lactobacilli*. Each gram of intestinal content contains about 10^11 CFU (colony forming units), making the gastrointestinal tract a highly complex ecosystem. This large bacterial population is critical to the host's nutrition and health (Roberfroid, 2000). The colonic microbiota ferments indigestible organic material in the upper gut (Topping and Clifton, 2001). Acetate (C2:0), propionate (C3:0), and butyrate (C4:0) are primarily the linear SCFAs that are produced when fructans are metabolized in the colon (Topping and Clifton, 2001). Other fermentation products, such as lactate, succinate, and ethanol, may also occur (Roberfroid et al., 1998), though these are typically only intermediates in the microbiota's overall process of fermenting carbohydrates into SCFAs. Gases produced by fructans fermentation include CO_2, CH_4, and H_2 (Flamm et al., 2001; Topping and Clifton, 2001). Both non-digestible carbohydrates and SCFAs can significantly change the colon's physiology for the better (Wong et al., 2006). The production of SCFAs is in the order C2:0 > C3:0 > C4:0 in a molar ratio of about 60:20:20, primarily in the proximal and distal colon. Additionally, an increase in SCFA synthesis causes the gut to become more acidic, which is essential *in vivo* for resistance to pathogen colonization (Roberfroid, 2001). There are numerous variables that affect the production of SCFA, such as the source of the substrate, specifically its chemical makeup as a fermentable substrate, the quantity of substrate that is available, and its physical form (e.g., particle size, solubility, association with indigestible complexes such as lignin) (Macfarlane and Macfarlane, 2007), the bacterial species composition of the microbiota (Roberfroid, 2005), ecological factors (competitive and cooperative interactions between different groups of bacteria), and intestinal transit time (Cummings et al., 1997). The caecum and proximal colon are the primary sites where fermentation occurs.

There are three major sites in the body where SCFAs are metabolized: 1) cells of the caecum-colonic epithelium, which use C4:0 as a major substrate for maintenance; 2) liver cells; and 3) muscle cells, which generate energy from the oxidation of residual C2:0 (Van et al., 1995). C2:0 is the most abundant SCFA produced in the colon; it is easily absorbed and transported to the liver (Cook and Sellin, 1998). As acetyl-CoA synthetase is present in the cytosol of adipose and mammary glands, C2:0 can be used for lipid metabolism and lipogenesis once it enters the systemic circulation (Wong et al., 2006). The primary substrate for cholesterol synthesis is C2:0. It may also be absorbed and used by peripheral tissues in the host. C3:0 is a substrate for hepatic gluconeogenesis and has been shown to inhibit cholesterol synthesis in hepatic tissue. The C3:0 to C2:0 ratio in the colon is important because it reduces cholesterol synthesis. C4:0 is the preferred fuel of colonic epithelial cells, and it plays an important role in cell proliferation and differentiation regulation (Topping and Clifton, 2001). Although approximately 95% of the C4:0 produced by colonic bacteria is transported across the epithelium, concentrations in portal blood are typically undetectable due to rapid utilization (Pryde et al., 2002). C4:0 production may also occur as a result of the use of other fermentation products such as C2:0 or lactate, which can act as C4:0 precursors (García et al., 2013) (Figure 2.1).

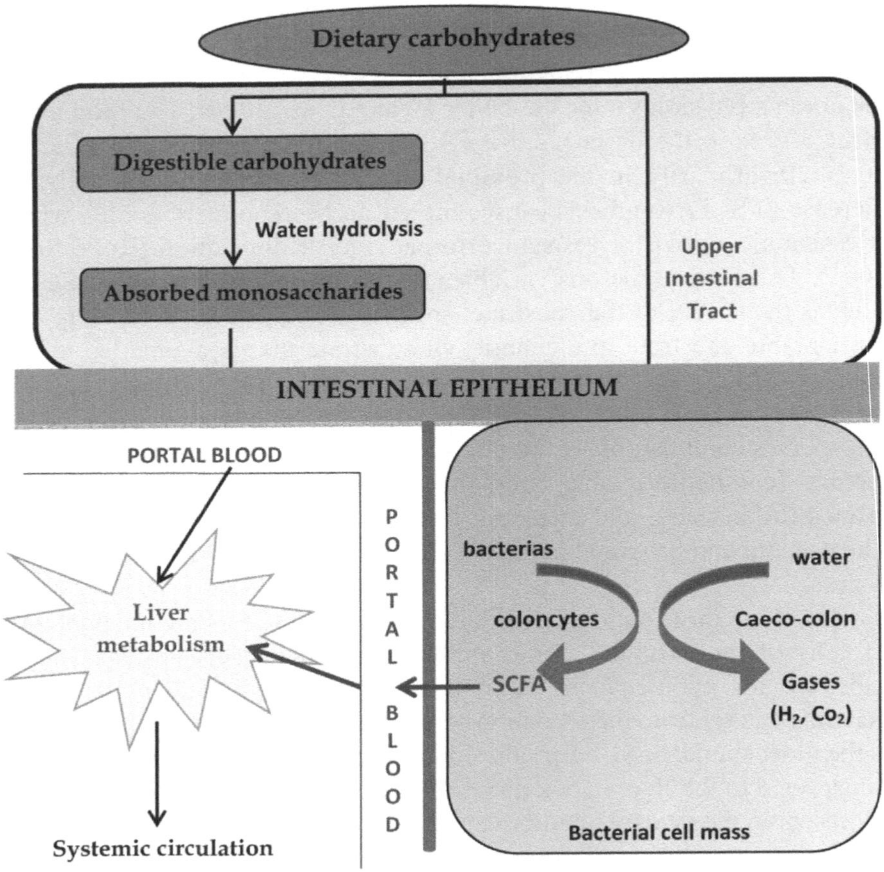

FIGURE 2.1 Schematic of production and bioavailability of SCFA from FOS/inulin.

According to most *in vitro* and *in vivo* studies, oligofructose is an excellent substrate for the production of short-chain fatty acids. Few studies found the highest levels of C3:0 acid in the caecum, proximal and distal colons of rats fed inulins, while the highest levels of C4:0 acid were found in the caecum, proximal and distal colons of rats fed FOS (Nilsson and Nyman, 2005). Licht in 2006 obtained comparable results when they fed rats different dietary carbohydrates. These researchers concluded that C3:0 acid concentrations reached statistical significance in animals fed inulin, whereas C4:0 acid concentrations were significantly higher in animals fed FOS. *In vitro* fermentation of FOS and inulin with fecal inoculum was compared by Van et al. in 2007, who found that FOS and inulin increased the production of SCFAs by about 30%. A rise in C3:0 and C4:0 acids was primarily responsible for this increase in SCFA production. Whereas inulin demonstrated a greater (almost 2-fold) increase in C3:0 acid production than FOS. According to the authors, these variations were well correlated with the structural variations between FOS and inulin.

2.5 HEALTH IMPLICATIONS OF FOS

FOS has recently been used as a food ingredient, which has sparked a lot of research to understand how it works and what impact it has on human health. Its potential for assisting in the prevention and control of some diseases has received a lot of attention (Conterno et al., 2011).

2.5.1 EFFECT OF FRUCTOOLIGOSACCHARIDE IN IMMUNE FUNCTION

As the gut matures during infancy, so does the immune system, and this process is markedly affected by dietary patterns. Human breast milk is a rich source of prebiotics, including '*bifidus* factor'. As a result, the gut flora of breast-fed infants is dominated by *Lactobacillus* and *Bifidobacteria*, whereas formula-fed infants have a more complex composition of gut flora (Gibson and Delzenne, 2008). This clearly indicates that immunity builds up from the very start of life. The immune system consists of two types: innate immune system and acquired immune system. Innate immunity is present from birth, whereas acquired immunity is developed in an individual after being exposed to a pathogen or through vaccination.

In the past decade, FOS has been studied extensively in humans as well as in animal models for its modulating immune properties. In a small randomized, double-blind controlled human trial with ulcerative colitis, supplementation of *B. longum* and OFS resulted in an improvement of chronic inflammation. Furthermore, intestinal mRNA levels of the proinflammatory cytokines (IL-1β) and tumor necrosis factor (TNF-α) were significantly reduced (Furrie et al., 2006).

Consumption of formula containing 0.8 g prebiotics/100 ml (FOS+GOS) in a randomized controlled trial with 259 infants at high risk of atopy resulted in a significantly lower incidence of atopic dermatitis, indicating that these prebiotics altered postnatal immune development (Moro et al., 2006). Another randomized controlled trial revealed increased fecal SIgA secretion is associated with significantly faster intestinal clearance of pathogenic bacteria (Bakker et al., 2006). Possible mechanisms by which FOS might cause immune changes are:

1. Selective up- or down-regulation of particular bacteria that affect cytokine and antibody production.
2. An increase in the production of intestinal SCFA and improved SCFA binding to leukocytes' G-coupled protein receptors.
3. Prebiotics that are partially absorbed come into contact with the immune system both locally and systemically.
4. FOS and leukocyte carbohydrate receptor interaction.

2.5.2 EFFECT OF FRUCTOOLIGOSACCHARIDE IN COLON CANCER

Colon cancer is thought to develop as a result of interactions between diet, gut microflora, immune system, and other environmental and disease factors. Oligosaccharides are being studied for decreasing cancer risk by mediating the colonization of specific microbial populations, thus triggering a biological chain of events (Rafter et al., 2007). Although data from animal studies supports the role of oligofructose and inulin in prevention of the disease, trials on human beings are still in the early stages. In a large multi-centric European study, biomarkers of colonic cancer risk, such as cell proliferation and toxicity of feces (fecal water), improved when participants were given a synbiotic (FOS+*L. rhamnosus* and *B. lactis*) for 12 weeks (Rafter et al., 2007). Oligosaccharides may play an important role in reducing colon cancer risk through a decrease in colonic pH and by shifting the resident microflora and production of SCFA. In chemo-preventive studies in colon cancer-induced rat models, an increase in cecum contents weight and a reduction of cecal pH were reported for inulin-like fructans (Jacobsen et al., 2006).

Evidence suggests that FOS can prevent the development of aberrant crypt foci, a putative colonic preneoplastic lesion that mimics early carcinogenic events and is characterized by enlarged cells and thickening of the colonic epithelial lining in animals. As fermentation of FOS increases, intestinal permeability rises via the dose-dependent production of SCFA. In turn, these SCFA stimulate more mucin production and protect the GI lumen against cellular damage. Studies also report that bacteria of the gut microflora possess a variety of enzymes such as azoreductase, nitroreductase, or β-glucuronidase that can metabolize exogenous and endogenous compounds associated with colon cancer.

2.5.3 *EFFECT OF FRUCTOOLIGOSACCHARIDE IN MINERAL ABSORPTION*

Several food ingredients have drawn interest as potential boosters of mineral absorption. Prebiotics, such as inulin-type fructans, have a strong track record of increasing the absorption of a wide range of minerals in experimental animals, including calcium, magnesium, iron, and zinc (Scholz-Ahrens et al., 2001, 2002). A study on a rat model has shown that ingestion of fructooligosaccharide resulted in increased calcium and magnesium absorption, reduced the occurrence of post-gastrectomy osteoporosis, and improved bone mineralization. These effects are associated with an increase in the weight of cecal contents, an increase in the amount of cecal short-chain fatty acids, and a decrease in cecal pH (Ohta et al., 1998; Delzenne et al., 1999; Lopez, 2000; Coudray, 2000, 2003, 2005).

Results of several studies have revealed improvement in calcium absorption, bone mineral content, and bone mineral density and reduction in ovariectomy-induced bone loss after supplementation with inulin-type fructans (Coxam, 2005; Holloway et al., 2007; Tahiri, 2003). The findings imply that inulin-type fructans target not only the absorption of minerals but also other facets of bone health, particularly bone mineralization, bone density, and bone accretion and resorption (Yamada, 1999; Griffin et al., 2003; Griffin et al., 2005).

Prebiotics' impact on mineral absorption is explained by a number of different mechanisms. According to the first theory, prebiotics that are not absorbed pass into the large intestine undigested and are then fermented into short-chain fatty acids. These acids increase the solubility of minerals in the luminal contents, which increases passive concentration-dependent calcium absorption in the colon. These acids also lower the pH of the large intestinal contents (Scholz-Ahrens, 2002). The second theory hypothesizes that the effect of prebiotics on calcium absorption is through a tropic effect on the gut, increasing absorptive surface area and thus passive calcium absorption (Scholz-Ahrens, 2002). According to the third theory, increased calcium absorption can be attributed to increased availability by calcium transfer from the small intestine to the large intestine and the osmotic effect of inulin-type fructans, which transfers water into the large intestine, allowing it to become more soluble (Kaur and Gupta, 2002).

2.5.4 EFFECT OF FRUCTOOLIGOSACCHARIDE IN GASTROINTESTINAL TRACT

Inflammatory bowel disease (IBD), pouchitis, Crohn's disease, and ulcerative colitis are chronic conditions that are characterized by persistent mucosal inflammation at various levels of the gastrointestinal tract (O'Hara and Shanahan, 2007). There is evidence that the microbiota of IBD patients differs from that of healthy subjects (Guarner, 2005; Manichanch et al., 2006; Ott et al., 2004). Prebiotics such as inulin and oligofructose have been shown in numerous studies to increase saccharolytic activity in the gut and promote *Bifidobacteria* growth. Inulin and oligofructose may improve barrier function in IBD and prevent mucosal colonization by increasing the number of 'friendly' bacteria on the mucosal surface. Both human and animal data have shown a powerful impact on improved gastrointestinal tract diseases by inulin-type fructans.

Daily oral administration of inulin increased the counts of indigenous *Lactobacilli* in the cecal lumen and decreased intra-colonic pH; it also resulted in the extension of saccharolytic areas of acidic environment. A subsequent study looked at the effects of oligofructose and inulin on the DSS model alone and in combination with probiotic *Bifidobacteria*. Both the prebiotic alone and the synbiotic significantly improved disease activity indexes and decreased colonic myeloperoxidase activity, as well as inflammatory mediator expression (Osman et al., 2006). It was concluded that these beneficial bacteria might have resulted in the production of short-chain fatty acids and it's likely to be the principal mechanism mediating the anti-inflammatory effect. Almost similar results have been drawn from clinical human trials. In a double-blind crossover design, twenty patients with mild gastrointestinal disease were supplemented 24 g/d of inulin (dissolved in 200 ml of commercially available milk) for 3 weeks, which resulted in increased fecal butyrate and a decrease in the counts of bacteriods in feces (Welters, 2002). FOS administered to patients with ulcerative colitis and Crohn's disease with different doses ranging from 12 g/d to 15 g/d for the duration of 3–4 weeks reported overall improvement and decreased mucosal inflammation (Furrie et al., 2005; Lindsay et al., 2006). From these clinical studies, it can be concluded that inulin and OFS can offer an opportunity to prevent or mitigate gastrointestinal disease and their symptoms.

2.5.5 EFFECT OF FRUCTOOLIGOSACCHARIDE ON OBESITY

Recent research suggests that the gut microbiota may play a role in obesity-related weight gain and altered energy metabolism (Harris et al., 2012). Many studies have been conducted to investigate the compositional changes of the human gut microbiota in response to weight loss. Ley et al. in 2006 investigated the fecal gut microbiota of obese subjects who took part in a year-long weight-loss program with restricted diets. After losing weight, the proportion of *Bacteroidetes* increased while the number of *firmicutes* decreased. Studies have also demonstrated that a high-fat diet is a triggering factor for the development of obesity and inflammation. Lipopolysaccharide (LPS) triggers the secretion of pro-inflammatory cytokines and binds to the complex of cluster of differentiation 14 (CD14) and the toll-like receptor 4 (TLR 4) at the surface of innate immune cells (Wright et al., 1990). LPS is continuously produced within the gut by the death of gram-negative bacteria and is physiologically carried into intestinal capillaries towards target tissues (Neal et al., 2006; Tomita et al., 2004). High fat and LPS promote low-grade inflammation-induced metabolic disorders.

Inulin-type fructans are well studied and clearly effective in humans and animal models to stimulate the growth of health-promoting species belonging to *Bifidobacterium* and *Lactobacillus* (Macfarlane, 2006; Flamm, 2001). A study reported that oligofructose (20 g/day) significantly increased plasma GLP-1 after a meal (Piche et al., 2003). Furthermore, a study demonstrated that in healthy humans, consumption of FOS (16 g/day) promoted satiety accompanied by a significant 10% lower total energy intake (Cani et al., 2006).

Reduction in LPS activity was noted by intake of higher levels of fructans, which reduced endotoxemia (Cani et al., 2007). Excessive production of fasting-induced adipocyte factor (FIAF) suppresses hepatic *de novo* lipogenesis and inhibits triglyceride storage in white adipose tissues (Backhed et al., 2007). It also inhibits lipoprotein lipase (LPL), thereby blocking the disassociation of fatty acids from triglycerides for uptake into tissues (Mandard et al., 2004). Production of SCFA (Xu and Gordon, 2003) activates gut peptides GLP-1, GIP, and Ghrelin, which have been proposed as important modulators of food intake and energy expenditure (Knauf et al., 2008).

2.5.6 EFFECT OF FRUCTOOLIGOSACCHARIDE ON LIPID METABOLISM

Another physiological effect of inulin-type fructans is lipid metabolism modification, which affects triglyceridemia and cholesterolemia as well as the distribution of lipids among different lipoproteins in favor of a healthier pattern (Moshfegh et al., 1999). Studies using animal models have indicated that a moderate amount of inulin and oligofructose may reduce blood cholesterol levels. Rat studies have repeatedly shown that inulin and OFS consumption of 9–20 g/d induces a significant reduction (40%) in serum and liver triglyceride,

LDL, and total cholesterol concentrations (Delzenne et al., 2002; Fava et al., 2006; Rault et al., 2006). Bruserolls et al. (2003) revealed that OFS is protective against pro-oxidative stress, lowers heart lipid oxidation, and thus contributes to being cardioprotective. A study reported by Moroti et al. (2012) investigated the effect of synbiotic shake (*L. acidophilus*, B. bifidum, and 2% FOS in 200 ml shake) supplementation in 20 hypercholesterolemic patients and found a reduction in total cholesterol and triglycerides and an increment in HDL cholesterol. Studies have also shown a significant serum TG and TC reduction in hypercholesterolemic and hyperlipidemic subjects consuming 20 g inulin and OFS (Causey et al., 2000). On the contrary, some studies show that dietary supplementation with 15–20 g/d FOS for 4 weeks had no effect on serum TC or TG levels in type 2 diabetic subjects (Alles et al., 1999).

Various mechanisms have been suggested by scientists regarding the reduction in blood lipid levels via fructans. Ingestion of dietary fructans increases the beneficial microflora in the gut. Kiebling et al., (2002) reported a significant increase in HDL levels with the administration of FOS and *LAB* in humans. Intestinal breakdown of FOS leads to the production of substantial amounts of SCFA, which are almost completely absorbed along the digestive tract. Butyrate is widely metabolized by enterocytes, while propionate and acetate can reach the liver through the portal vein (Demigne et al., 1999). When acetate enters the hepatocyte, it is mainly activated by the cytosolic acetyl CoA synthetase 2, and then enters the cholesterogenesis and lipogenesis pathways. This effect has been proposed as a rationale behind the hypercholesterolemic effect of fructans. Conversely, propionate is a competitive inhibitor of the protein devoted to the entrance of acetate into the liver cells (Delzenne et al., 2002), a phenomenon which contributes to a decrease in lipogenesis and cholesterogenesis. Thus, the production of a high concentration of propionate has been proposed as an explanation for the reduction in serum and hepatic cholesterol through fructans (Lopez et al., 2001; Delzenne and Kok, 2001). Another mechanism related to SCFA is an increase in acetate, which activates AMP kinase in the liver. This, in turn, inhibits *de novo* lipogenesis through a corresponding reduction in the activity of all lipogenic enzymes, leading to reduced VLDL-triglyceride levels (Sakakibara et al., 2006). The effect of inulin (10 g/day for 3 weeks) supplementation on hepatic lipogenesis and cholesterogenesis has been analyzed in normal subjects in a double-blind, placebo-controlled crossover study, which reported a decrease in hepatic *de novo* lipogenesis (Letexier et al., 2003).

Underlying mechanisms related to lipid reduction involve the ability of probiotic bacteria to reduce the resorption of bile acids through enterohepatic circulation. This is because conjugated bile acids are directly absorbed back into the liver. *Lactobacillus* and *Bifidobacteria* cells can hydrolyze the conjugated bile acids, excreting them faster and reducing the extent to which they are absorbed (Ooi and Liong, 2010). Nutrient-related selection of specific gut microbiota species may lower intestinal endotoxin levels and promote mucosal barrier function. The supply of monosaccharides and SCFA may enhance hepatic *de novo* lipogenesis and triglyceride (TG) accumulation. Altered supply of gut-derived nutrients may also affect transcription factors such as ChREBP and SREBP-1 and modify AMP-activated protein kinase (AMPK) activity in both the liver and skeletal muscle. Gut microbiota may suppress gut-derived fasting-induced adipose factor (Fiaf), a circulating lipoprotein lipase (LPL) inhibitor, leading to increased LPL activity and lipolysis.

2.5.7 ROLE OF FRUCTOOLIGOSACCHARIDE ON GLUCOSE HOMEOSTASIS

Prebiotics have historically been the subject of research centered on digestive health. But in recent years, researchers have been looking into the potential function of fructooligosaccharide in the management of various non-communicable diseases, including the maintenance of blood glucose homeostasis.

In the latter part of the 19th century, Thorne and Howard (1925) recommended including Jerusalem artichokes with a high oligofructose content in the diets of diabetic patients. Glycosuria decreased and blood levels dropped when these were used in place of other carbohydrate-rich foods (Kim and Shin, 1996). Yamashita and colleagues in 1984 studied diabetic subjects who took 8 grams of fructooligosaccharides every day for 14 days. By the end of the study, their fasting blood glucose levels dropped by 15 mg/dL compared to the control subjects. The findings of this study suggest that foods containing inulin may be useful carbohydrate replacements in diabetic diets.

According to Kaur and Gupta (2002), inulin and oligofructose modulate insulin hormone levels as well as carbohydrate metabolism due to their ability to lower blood sugar levels. A study on the effect of FOS on postprandial insulin found an increase in insulin levels after 3 weeks of 10 g/d FOS supplementation (Kaufhold et al., 2001).

Voss et al. (2008) evaluated the postprandial glycemic, insulinemic, and GLP-1 responses of 48 individuals with type 2 diabetes, after the consumption of 10.7 g tube feed formula consisting of FOS, a significant reduction in postprandial blood glucose and an increment in plasma insulin and GLP-1 levels was witnessed. Consumption of 20 g/d of oligofructose by type

2 diabetic subjects demonstrated a substantial reduction in blood glucose levels after 6 weeks (Cicek et al., 2009). A more recent study revealed that consumption of 200 ml synbiotic shake (2% FOS, *LAB*, and *Bifidobacteria*) resulted in a significant reduction in fasting glycemia in diabetic and hyperlipidemic subjects (Moroti et al., 2012).

Both animal and human studies have unveiled potential mechanisms through which reduction in blood glucose levels could be explained. The first mechanism suggests that the reduction in glucose levels may be attributed to transit time. Rats fed 10% and 20% FOS in their diet for 6 weeks had a 25% and 50% reduction in mouth-to-anus time, respectively. This decrease in transit time confirms a dose-dependent effect that may be similar to that of other dietary fibers: OFS and inulin affect macronutrient absorption, particularly carbohydrate absorption, by delaying gastric emptying and small intestine transit time (Oku et al., 1984; Bornet, 1994).

Numerous studies provide proof that the microbial colonies of diabetics and healthy humans differ from one another. Intestinal microbiota in type 2 diabetic patients was characterized by Larsen et al. (2011), who also noted that the proportions of the phylum *Firmicutes* and the class Clostridia were lower in the diabetic group compared to controls. This suggested that compositional changes in the intestinal microbiota are related to type 2 diabetes in humans. Similar studies found that diabetic subjects had lower *Bifidobacteria* colony establishment, which decreases intestinal permeability and increases endotoxemia (Cani et al., 2007; Cani, 2008).

Lowering the lipopolysaccharide (LPS) reduces the level of endotoxemia and hence reduces the severity of diabetes. Animal studies showed that bacterial LPS derived from gram-negative bacteria triggers pro-inflammatory cytokine production when it binds to the CD14/TLR4 complex at the surface of immune and gut epithelium cells, resulting in body weight gain, fat mass gain, liver steatosis, and diabetes (Cani et al., 2007; Creely et al., 2007; Wellen and Hotamisligil, 2006).

A significant increase in *Bifidobacteria* and *Lactobacilli*, a slight increase in plasma insulin and GLP-1 levels, and a reduction in severe glycosuria have also been demonstrated in a study in diabetic rats when supplemented with fructooligosaccharides. A study reported that the intake of 10 g/d of short-chain fructans increased *Bifidobacteria* counts in 30 diabetic subjects (Giacco et al., 2004). An increase in fermentable bacterial counts gives rise to SCFA such as propionate, acetate, and butyrate (Alles et al., 1999). Le Blay et al. (1999) observed that when 9% fructan was present in the diet of rats, intestinal butyrate concentration was doubled with an increase in both acetate and propionate levels. Moreover, the profile of SCFA in terms of proportions of butyrate, acetate, and propionate differs with varying degrees of polymerization of fructans in rat models (Nyman, 2002).

Furthermore, a mechanism has been postulated which states that an increase in SCFA concentration (especially butyrate) increases the gut incretins which have a direct effect on increasing insulin levels, thereby affecting overall blood glucose homeostasis. Kok et al. (1998) observed that OFS feeding leads to an increase in total cecal GLP-1 and jejunum GIP concentrations in rats. Therefore, it was postulated that the modulation of gut peptides could be a key process in mediating the effect of FOS on glucose metabolism. However, the putative mechanism by which FOS exerts its effects on glucose metabolism was shown to be mainly dependent on GLP-1 (Cani et al., 2005).

By understanding the new ecological strategies based on prebiotics, it is evident that Fructooligosaccharide is unequivocally associated with glycemic, insulinemic, and lipemic responses. In this context, FOS is a substantial nutraceutical approach in the management of metabolic disorders like obesity, cardiovascular diseases, and type 2 diabetes. Multidisciplinary research based on animal studies firmly supports the above evidence of consumption of FOS in combating these diseased conditions. Nevertheless, a tremendous lack of data limits the current knowledge of the complex mechanisms linking diet, gut microbiota, and metabolic disorders in humans. Therefore, long-term human studies will constitute new knowledge in determining the effects of FOS consumption on the metabolic syndrome and the meta dynamics behind it.

KEYWORDS

- **bacteria**
- **diabetes**
- **enzymes**
- **lactulose**
- **metabolic disorders**
- **metabolism**
- **microbiome**
- **obesity**
- **prebiotics**
- **probiotics**

REFERENCES

Aryana, K. J., Plauche, S., Rao, R. M., McGrew, P., & Shah, N. P. (2007). Fat-free plain yogurt manufactured with inulin's of various chain lengths and *Lactobacillus acidophilus*. *Journal of Food Science, 72*(3), M79–M84. https://doi.org/10.1111/j.1750-3841.2007.00334.x.

Bäckhed, F., Manchester, J. K., Semenkovich, C. F., & Gordon, J. I. (2007). Mechanisms underlying the resistance to diet-induced obesity in germ-free mice. *Proceedings of the National Academy of Sciences USA, 104*, 979–984. https://doi.org/10.1073/pnas.0605374104.

Bakker-Zijlstra, A. M. (2006). Fecal SIgA secretion in infants fed on pre- or probiotic infant formula. *Pediatric Allergy and Immunology, 17*, 134–140. https://doi.org/10.1111/j.1399-3038.2006.00376.x.

Barndt, R. L., Liao, S., Merkel, C. M., Chapello, W. J., & Navia, J. L. (2003). Low calorie palatable fiber containing sugar substitute. *United States Patent Application Publication*, Pub No. US 2003/0054080A1.

Bishay, I. E. (1998). Rheological characterization of inulin. In P. A. Williams & G. O. Phillips (Eds.), *Gums and Stabilizers for The Food Industry* (pp. 201–210). Cambridge, UK: Royal Society of Chemistry.

Bornet, F. R. J. (1994). Indigestible sugars in food products. *American Journal of Clinical Nutrition, 59*(1), 763S–769S. https://doi.org/10.1093/ajcn/59.1.763S.

Busserolls, J., et al. (2003). Oligofructose protects against the hyperlipidemic and pro-oxidative effects of a high fructose diet in rats. *Journal of Nutrition, 133*, 1903.

Cani, P. D., Bibiloni, R., Knauf, C., et al. (2008). Changes in gut microbiota control metabolic endotoxemia-induced inflammation in high-fat diet-induced obesity and diabetes in mice. *Diabetes, 57*, 1470–1481. https://doi.org/10.2337/db07-1403.

Cani, P. D., Daubioul, C. A., Reusens, B., Remacle, C., Catillon, G., & Delzenne, N. M. (2005). Involvement of endogenous glucagon-like peptide-1(7–36) amide in the glycaemia-lowering effect of oligofructose in streptozotocin-treated rats. *Journal of Endocrinology, 185*, 457–465. https://doi.org/10.1677/joe.1.06092.

Cani, P. D., Joly, E., Horsmans, Y., & Delzenne, N. M. (2006). Oligofructose promotes satiety in healthy humans: A pilot study. *European Journal of Clinical Nutrition, 60*, 567–572. https://doi.org/10.1038/sj.ejcn.1602370.

Cardarelli, H. R., Aragon-Alegro, L. C., Alegro, J. H., Castro de, I. A., & Saad, M. I. (2008). Effect of inulin and *Lactobacillus paracasei* on sensory and instrumental texture properties of functional chocolate mousse. *Journal of the Science of Food and Agriculture, 88*(8), 1318–1324. https://doi.org/10.1002/jsfa.3170.

Causey, J. L., et al. (2000). Effects of dietary inulin on serum lipids, blood glucose, and the gastrointestinal environment in hypercholesterolemic men. *Nutrition Research, 20*, 191–198. https://doi.org/10.1016/S0271-5317(00)00005-1.

Choonhahirun, A., & Akesowan, A. (2013). Quality assessment of reduced-calorie Thai mung bean marzipan made with erythritol-sucralose blend and soy milk. *Life Science Journal, 10*(2), 2129–2134.

Conterno, L., Fava, F., Viola, R., & Tuohy, K. M. (2011). Obesity and the gut microbiota: Does up-regulating colonic fermentation protect against obesity and metabolic disease? *Genes & Nutrition, 6*, 241–260. https://doi.org/10.1007/s12263-010-0214-0.

Coussement, P. (1999). Inulin and oligofructose as dietary fiber: Analytical, nutritional and legal aspects. In *Complex Carbohydrates in Foods* (pp. 203–204). Eastern Hemisphere Distribution Publishing Group.

Coxam, V. (2005). Inulin-type fructans and bone health: State of the art and perspectives in the management of osteoporosis. *British Journal of Nutrition, 93*(1), S111–S123. https://doi.org/10.1079/BJN20051433.

Creely, S. J., McTernan, P. G., Kusminski, C. M., Fisher, M., Da Silva, N. F., Khanolkar, M., Evans, M., Harte, A. L., & Kumar, S. (2007). Lipopolysaccharide activates an innate immune system response in human adipose tissue in obesity and type 2 diabetes. *American Journal of Physiology-Endocrinology and Metabolism, 292*(3), E740–E747. https://doi.org/10.1152/ajpendo.00455.2006.

Crittenden, R. G. (1999). Prebiotics. In G. W. Tannock (Ed.), *Probiotics: A critical review* (pp. 141–156). Wymondham: Horizon Scientific Press.

Cruza, A. G., Cavalcantia, R. N., Guerreiroa, L. M. R., Sant'Anac, A. S., Nogueirab, L. C., & Oliveira, C. A. F., et al. (2013). Developing a prebiotic yogurt: Rheological, physico-chemical and microbiological aspects and adequacy of survival analysis methodology. *Journal of Food Engineering, 114*(3), 323–330. https://doi.org/10.1016/j.jfoodeng.2012.08.007.

De Leenheer, L. (1996). Production and use of inulin: Industrial reality with a promising future. In H. Van Bekkum, H. Roper, & F. Voragen (Eds.), *Carbohydrates as Organic Raw Materials III* (pp. 67–92). New York: V.C.H. Publishers.

Delzenne, N. M., Daubioul, A., Neyrinck, A. M., & Taper, H. S. (2002). Inulin and oligofructose modulate lipid metabolism in animals: Review of biochemical events and future prospects. *British Journal of Nutrition, 87*(2), S255–S259. https://doi.org/10.1079/BJN2002563.

Delzenne, N., & Kok, N. N. (1999). Dietary fructooligosaccharides modify lipid metabolism in the rat. *Journal of Nutrition, 129*(7), 1467S–1469S. https://doi.org/10.1093/jn/129.7.1467S.

Delzenne, N., & Roberfroid, M. B. (1994). Physiological effects of nondigestible oligosaccharides. *Lebensmittel-Wissenschaft & Technologie, 27*, 1–6. https://doi.org/10.1006/fstl.1994.1001.

Demigne, C., Remesy, C., & Morand, C. (1999). Short chain fatty acids. In G. R. Gibson, et al. (Eds.), *Colonic Microbiota, Nutrition and Health* (pp. 55–63). Kluwer Academic Publishers.

Farzanmehr, H., & Abbasi, S. (2009). Effects of inulin and bulking agents on some physicochemical, textural and sensory properties of milk chocolate. *Journal of Texture Studies, 40*(5), 536–553. https://doi.org/10.1111/j.2042-4568.2009.00166.x.

Fava, F., Lovegrove, R., Gitau, K., Jackson, K. G., & Tuohy, K. M. (2006). The gut microbiota and lipid metabolism: Implications for human health and coronary heart disease. *Current Medical Chemistry, 13*(25), 3005–3021. https://doi.org/10.2174/092986706778606401.

Flamm, G., Glinsmann, W., Kritchevsky, D., Prosky, L., & Roberfroid, M. (2001). Inulin and oligofructose as dietary fiber: A review of evidence. *Critical Reviews in Food Science and Nutrition, 41*(5), 353–362. https://doi.org/10.1080/20014091094241.

Flamm, G., Glinsmann, W., Kritchevsky, D., Prosky, L., & Roberfroid, M. (2001). Inulin and oligofructose as dietary fiber: A review of evidence. *Critical Reviews in Food Science and Nutrition, 41*(5), 353–362. https://doi.org/10.1080/20014091094241.

Folly de, G., Silva da, E. N., Verner, F. V., Silva dos, F., & Volp, C. P. (2013). Acceptance of handmade products containing nuts and fructooligosaccharides. *Nutrition Hospitalaria, 28*(1), 86–92. https://doi.org/10.3305/nh.2013.28.1.6062.

Franck, A. M. (2000). Inulin and oligofructose. In G. R. Gibson & F. Angus (Eds.), *LFRA Ingredient Handbook: Prebiotics and Probiotics* (pp. 1–18). Surrey: Leatherhead Publishing.

Furrie, E., et al. (2006). Synbiotic therapy (*Bifidobacterium longum*/Synergy 1) initiates resolution of inflammation in patients with active ulcerative colitis: A randomized controlled pilot trial. *Gut, 54*, 242–249. https://doi.org/10.1136/gut.2004.060451.

Gallagher, E., Brein, C. M. O., Scannell, A. G. M., & Arendt, E. K. (2003). Use of response surface methodology to produce functional short dough biscuits. *Journal of Food Engineering, 56*(2–3), 269–271. https://doi.org/10.1016/S0260-8774(02)00288-5.

García, A. H., & Lopez, G. (2013). Metabolism of short chain fatty acids in the colon and faeces of mice after a supplementation of diets with agave fructans. In *Lipid metabolism* (pp. 163–182). https://doi.org/10.5772/51248

Giacco, R., Clemente, G., Luongo, D., Lasorella, G., Fiume, I., Brouns, F., et al. (2004). Effects of short-chain fructo-oligosaccharides on glucose and lipid metabolism in mild hypercholesterolemia individuals. *Clinical Nutrition, 23*(3), 331–340. https://doi.org/10.1016/j.clnu.2003.07.004.

Gibson, G. R., & Delzenne, N. M. (2008). Inulin and oligofructose: New scientific developments. *Nutrition Today, 43*(2), 54–59. https://doi.org/10.1097/01.NT.0000315146.36879.42.

Gibson, G. R., & Roberfroid, M. B. (1995). Dietary modulation of the human colonic microbiota: Introducing the concept of prebiotics. *The Journal of Nutrition, 125*(6), 1401–1412. https://doi.org/10.1093/jn/125.6.1401.

Gibson, G. R., Probert, H. M., Loo, J. V., Rastall, R. A., & Roberfroid, M. B. (2004). Dietary modulation of the human colonic microbiota: Updating the concept of prebiotics. *Nutrition Research Reviews, 17*(2), 259–275. https://doi.org/10.1079/NRR200479.

Griffin, I. J., & Abrams, S. A. (2005). Methodological considerations in measuring human calcium absorption: Relevance to studies of the effects of inulin-type fructans. *British Journal of Nutrition, 93*(Suppl 1), S105–S110. https://doi.org/10.1079/BJN20051491.

Griffin, I. J., Hicks, P. M. D., Heaney, R. P., & Abrams, S. A. (2003). Enriched chicory inulin increases calcium absorption in girls with lower calcium absorption. *Nutrition Research, 23*, 901–909. https://doi.org/10.1016/S0271-5317(03)00057-6.

Guarner, F. (2005). The intestinal flora in inflammatory bowel disease: Normal or abnormal? *Current Opinion in Gastroenterology, 21*, 414–418. https://doi.org/10.1097/01.mog.0000177525.78907.9e.

Hara, A. M., & Shanahan, F. (2007). Gut microbiota: Mining for therapeutic potential. *Clinical Gastroenterology and Hepatology, 5*(3), 274–284. https://doi.org/10.1016/j.cgh.2006.10.003.

Harris, K., Kassis, A., Major, G., & Chou, C. J. (2012). Is the gut microbiota a new factor contributing to obesity and its metabolic disorders? *Journal of Obesity*, *2012*, 1–14. https://doi.org/10.1155/2012/879151.

Hidaka, H., Tashiro, Y., & Eida, T. (1991). Proliferation of bifidobacteria by oligosaccharides and their useful effect on human health. *Microflora, 10*, 65–79.

Holloway, L., Moynihan, S., Abrams, S. A., Kent, K., Hsu, A. R., & Friedlander, A. L. (2007). Effects of oligofructose-enriched inulin on intestinal absorption of calcium and magnesium and bone turnover markers in postmenopausal women. *British Journal of Nutrition, 97*(2), 365–372. https://doi.org/10.1017/S0007114507339573.

Hotamisligil, G. S. (2006). Inflammation and metabolic disorders. *Nature, 444*(7121), 860–867. https://doi.org/10.1038/nature05485.

Jacobsen, B. A., Fallingborg, J., Rasmussen, H. H., Nielsen, K. R., Drewes, A. M., Puho, E., Nielsen, G. L., & Sørensen, H. T. (2006). Increase in incidence and prevalence of inflammatory bowel disease in northern Denmark: A population-based study, 1978–2002. *European Journal of Gastroenterology & Hepatology, 18*(6), 601–606. https://doi.org/10.1097/00042737-200606000-00005.

Janer, C., Rohr, L. M., Pelaez, C., Laloi, M., Cleusix, V., Requena, T., et al. (2004). Hydrolysis of oligofructose by recombinant β-fructosidase from *Bifidobacterium lactis*. *Systematic and Applied Microbiology, 27*(4), 279–285. https://doi.org/10.1078/0723-2020-00219.

Kaufhold, J., & Hammon, H. M. (2001). Fructo-oligosaccharide supplementation: Effects on metabolic, endocrine, and hematological traits in veal calves. *Journal of Veterinary Medicine, 47*(1), 17–29. https://doi.org/10.1046/j.1439-0442.2001.00316.x.

Kaur, N., & Gupta, K. (2002). Applications of inulin and oligofructose in health and nutrition. *Journal of Biosciences, 27*(7), 703–714. https://doi.org/10.1007/BF02703494.

Kiebling, G., Schneider, J., & Jahreis, G. (2002). Long-term consumption of fermented dairy products over 6 months increases HDL cholesterol. *European Journal of Clinical Nutrition, 56*(9), 843–849. https://doi.org/10.1038/sj.ejcn.1601391.

Kim, M., & Shin, H. K. (1996). The water-soluble extract of chicory reduces glucose uptake from the perfused jejunum in rats. *Journal of Food Science, 126*(6), 2236–2242. https://doi.org/10.1111/j.1365-2621.1996.tb15305.x.

Knauf, C., Cani, P. D., Kim, D. H., Iglesias, M. A., Chabo, C., Waget, A., et al. (2008). Role of central nervous system glucagon-like peptide-1 receptors in enteric glucose sensing. *Diabetes, 57*(10), 2603–2612. https://doi.org/10.2337/db08-0267.

Kok, N. N., Morgan, L. M., Williams, C. M., Roberfroid, M. B., Thissen, J. P., & Delzenne, N. M. (1998). Insulin, GLP-1, GIP and IGF-I as putative mediators of the hypolipidemic effect of oligofructose in rats. *Journal of Nutrition, 128*(7), 1099–1103. https://doi.org/10.1093/jn/128.7.1099.

Korus, J., Achremowicz, K. G., & Sabat, R. (2012). Influence of prebiotic additions on the quality of gluten-free bread and on the content of inulin and fructooligosaccharides. *European Food Research and Technology, 235*(3), 545–554. https://doi.org/10.1007/s00217-012-1821-6.

Kruger, C. L. (2003). Agency Response Letter GRAS Notice No. GRN 000118. *CFSAN/Office of Food Additive Safety*. Retrieved from: FDA

Langlands, S. J., Hopkins, M. J., Coleman, N., & Cummings, J. H. (2004). Prebiotics carbohydrates modify the mucosa associated microflora of the human large bowel. *Gut, 53*(11), 1601–1616. https://doi.org/10.1136/gut.2003.037660.

Larissa, F. V. P., Fabiana, R. S., & Silva, A. C. C. (2012). Sensory profile and preference mapping of orange cakes with addition of prebiotics inulin and oligofructose. *Food Science and Technology, 41*(1), 37–42. https://doi.org/10.1590/S0101-20612012000100006.

Larsen, P. J., Fledelius, C., Knudsen, L. B., & Tang-Christensen, M. (2001). Systemic administration of the long-acting GLP-1 derivative NN2211 induces lasting and reversible weight loss in both normal and obese rats. *Diabetes, 50*(11), 2530–2539. https://doi.org/10.2337/diabetes.50.11.2530.

Le Blay, G., et al. (1999). Prolonged intake of fructooligosaccharide induces a short-term elevation of lactic-acid producing bacteria and a persistent increase in cecal butyrate in rats. *Journal of Nutrition, 129*(12), 2231–2236. https://doi.org/10.1093/jn/129.12.2231.

Letexier, D., Diraison, F., & Beylot, M. (2003). Addition of inulin to a moderately high carbohydrate diet reduces hepatic lipogenesis and plasma triglycerol concentrations in humans. *American Journal of Clinical Nutrition, 77*(3), 559–564. https://doi.org/10.1093/ajcn/77.3.559.

Ley, R. E., Turnbaugh, P. J., Klein, S., & Gordon, J. I. (2006). Microbial ecology: Human gut microbes associated with obesity. *Nature, 444*(7122), 1022–1023. https://doi.org/10.1038/4441022a.

Licht, T., Hansen, M., Poulsen, M., & Dragsted, L. O. (2006). Dietary carbohydrate source influences molecular fingerprints of the rat fecal microbiota. *BMC Microbiology, 6*, 98. https://doi.org/10.1186/1471-2180-6-98.

Lindsay, J. O., Whelan, K., Stagg, A. J., Gobin, P., Hassi, H. O., Rayment, N., et al. (2006). Clinical, microbiological, and immunological effects of fructooligosaccharide in patients with Crohn's disease. *Gut, 55*(3), 348–355. https://doi.org/10.1136/gut.2005.077952.

Lopez, H. W., Coudray, C., Levrat-Verny, M. A., Feillet-Coudray, C., Demigne, C., & Remesy, C. (2000). Fructooligosaccharide enhances mineral apparent absorption and counteracts the deleterious effect of phytic acid on mineral homeostasis in rats. *Journal of Nutrition Biochemistry, 11*(9), 500–508. https://doi.org/10.1016/S0955-2863(00)00090-7.

Lopez, H. W., et al. (2001). Class II resistant starches lower plasma and liver lipids and improve mineral retention in rats. *Journal of Nutrition, 131*(5), 1283–1288. https://doi.org/10.1093/jn/131.5.1283.

Macfarlane, S., Macfarlane, G. T., & Cummings, J. H. (2006). Review article: Prebiotics in the gastrointestinal tract. *Alimentary Pharmacology & Therapeutics, 24*(6), 701–714. https://doi.org/10.1111/j.1365-2036.2006.03182.x.

Mandard, S., Zandbergen, F., Nguan, S. T., et al. (2004). The direct peroxisome proliferator-activated receptor target fasting induced adipose factor (FIAF/PGAR/ANGPTL4) is present in blood plasma as a truncated protein that is increased by fenofibrate treatment. *The Journal of Biological Chemistry, 279*(33), 34411–34420. https://doi.org/10.1074/jbc.M403445200.

Manichanh, C., Rigottier, G. L., Bonnaud, E., Gloux, K., Pelletier, E., Frangeul, L., Nalin, R., et al. (2006). Reduced diversity of fecal microbiota in Crohn's disease revealed by a metagenomic approach. *Gut, 55*(2), 205–211. https://doi.org/10.1136/gut.2005.073817.

Meyer, D., Bayarri, S., Tárrega, A., & Costell, E. (2011). Inulin as texture modifier in dairy products. *Food Hydrocolloids, 25*(8), 1881–1890. https://doi.org/10.1016/j.foodhyd.2011.04.012.

Molis, C., Flourie, B., Ouarne, F., Failing, M. F., Lartigue, S., Guibert, A., Bornet, F., & Galmiche, J. P. (1996). Digestion, excretion, and energy value of fructooligosaccharides in healthy humans. *American Journal of Clinical Nutrition, 64*(3), 324–328. https://doi.org/10.1093/ajcn/64.3.324.

Moro, G., et al. (2006). A mixture of prebiotic oligosaccharides reduces the incidence of atopic dermatitis during the first six months of age. *Archives of Disease in Childhood, 91*(9), 814–819. https://doi.org/10.1136/adc.2005.087261.

Moroti, C., Souza Magri, L. F., Costa, M. R., Cavallini, D. C. U., & Sivieri, K. (2012). Effect of the consumption of a new symbiotic shake on glycemia and cholesterol levels in elderly people with type 2 diabetes mellitus. *Lipids in Health and Disease, 11*, 29. https://doi.org/10.1186/1476-511X-11-29.

Moshfegh, A. J., Friday, J. E., Goldman, J. P., & Chug-Ahuja, J. K. (1999). Presence of inulin and oligofructose in the diets of Americans. *Journal of Nutrition, 129*(7 Suppl), 1407S–1411S. https://doi.org/10.1093/jn/129.7.1407S.

Murphy, O. (2001). Non-polyol low-digestible carbohydrates: Food applications and functional benefits. *British Journal of Nutrition, 85*(1), 47–53. https://doi.org/10.1079/BJN2000286.

Neal, M. D., Leaphart, C., Levy, R., Prince, J., Billiar, T. R., Watkins, S., Li, J., Cetin, S., Ford, H., Schreiber, A., & Hackam, D. J. (2006). Enterocyte TLR4 mediates phagocytosis and translocation of bacteria across the intestinal barrier. *Journal of Immunology, 176*(6), 3070–3079. https://doi.org/10.4049/jimmunol.176.6.3070.

Nehir, S. E., & Simsek, S. (2011). Food technological applications for optimal nutrition: An overview of opportunities for the food industry. *Comprehensive Reviews in Food Science and Food Safety, 11*(1), 2–13. https://doi.org/10.1111/j.1541-4337.2011.00153.x.

Nilsson, U., & Nyman, M. (2005). Short-chain fatty acid formation in the hindgut of rats fed oligosaccharides varying in monomeric composition, degree of polymerization, and solubility. *British Journal of Nutrition, 94*(5), 705–713. https://doi.org/10.1079/BJN20051542.

Niness, K. R. (1999). Inulin and oligofructose: What are they? *Journal of Nutrition, 129*(7 Suppl), 1402S–1406S. https://doi.org/10.1093/jn/129.7.1402S.

Nowak, B., Von Mueffling, T., Grotheer, J., Klein, G., & Watkinson, B. M. (2007). Energy content, sensory properties, and microbiological shelf life of German Bologna-type sausages produced with citrate or phosphate and with inulin as fat replacer. *Journal of Food Science, 72*(9), S629–S638. https://doi.org/10.1111/j.1750-3841.2007.00662.x.

Nyman, M. (2002). Fermentation and bulking capacity of indigestible carbohydrates: The case of inulin and oligofructose. *British Journal of Nutrition, 87*(Suppl 2), S163–S168. https://doi.org/10.1079/BJN2002532.

O'Hara, A. M., & Shanahan, F. (2006). The gut flora as a forgotten organ. *EMBO Reports, 7*(7), 688–693. https://doi.org/10.1038/sj.embor.7400731.

Ohta, A., Ohtsuki, M., Hosono, A., Adachi, T., Hara, H., & Sakata, T. (1998). Dietary fructooligosaccharides prevent osteopenia after gastrectomy in rats. *Journal of Nutrition, 128*(1), 106–110. https://doi.org/10.1093/jn/128.1.106.

Oku, T., Tokunaga, T., & Hosoya, N. (1984). Non-digestibility of a new sweetener, "Neosugar," in the rat. *Journal of Nutrition, 114*(9), 1574–1581. https://doi.org/10.1093/jn/114.9.1574.

Ooi, L. G., & Liong, M. T. (2010). Cholesterol-lowering effects of probiotics and prebiotics: A review of in vivo and in vitro findings. *International Journal of Molecular Sciences, 11*(6), 2499–2522. https://doi.org/10.3390/ijms11062499.

Osman, N., Adawi, D., Molin, G., Ahrne, S., Berggren, A., & Jeppsson, B. (2006). Bifidobacterium infants' strains with and without a combination of oligofructose and inulin attenuate inflammation in DSS-induced colitis in rats. *BMC Gastroenterology, 6*, 31. https://doi.org/10.1186/1471-230X-6-31.

Ott, S. J., Musfeldt, M., Wenderoth, D. F., Hampe, J., Brant, O., Folsch, U. R., et al. (2004). Reduction in diversity of the colonic mucosa-associated bacterial microflora in patients with active inflammatory bowel disease. *Gut, 53*(5), 685–693. https://doi.org/10.1136/gut.2003.025403.

Piche, T., des Varannes, S. B., Sacher-Huvelin, S., Holst, J. J., Cuber, J. C., & Galmiche, J. P. (2003). Colonic fermentation influences lower esophageal sphincter function in gastroesophageal reflux disease. *Gastroenterology, 124*(4), 894–902. https://doi.org/10.1053/gast.2003.50060.

Pool-Zobel, B. L. (2005). Inulin-type fructans and reduction in colon cancer risk: Review of experimental and human data. *British Journal of Nutrition, 93*(Suppl 1), S73–S90. https://doi.org/10.1079/BJN20051491.

Pryde, S. E., Duncan, S. H., Hold, G. L., Stewart, C. S., & Flint, H. J. (2002). The microbiology of butyrate formation in the human colon. *FEMS Microbiology Letters, 217*(2), 133–139. https://doi.org/10.1016/S0378-1097(02)01080-8.

Rafter, J., Bennen, M., Caderni, G., Clune, Y., Hughes, R., Karlsson, P. C., Klinder, A., O'Riordan, M., O'Sullivan, G. C., et al. (2007). Dietary symbiotic reduce cancer risk factors in polypectomies and colon cancer patients. *American Journal of Clinical Nutrition, 85*(2), 488–496. https://doi.org/10.1093/ajcn/85.2.488.

Rault, N., Gueux, C., Demougeot, C., Demigne, C., & Mazur, A. (2006). Inulin attenuates atherosclerosis in apolipoprotein E deficient mice. *British Journal of Nutrition, 96*(5), 840–844. https://doi.org/10.1079/BJN20061816.

Renuka, B., Kulkarni, S. G., Vijyanand, P., & Prapulla, S. G. (2009). Fructooligosaccharide fortification of selected fruit juice beverages: Effect on the quality characteristics. *Food Science and Technology, 42*(5), 1031–1033. https://doi.org/10.1016/j.lwt.2008.09.008.

Roberfroid, M. (2000). Fructo-oligosaccharide malabsorption: Benefit for gastrointestinal functions. *Current Opinion in Gastroenterology, 16*(2), 173–177. https://doi.org/10.1097/00001574-200004000-00010.

Roberfroid, M. (2005). Introducing inulin-type fructans. *British Journal of Nutrition, 93*(1), 13S–25. https://doi.org/10.1079/BJN20051416.

Sakakibara, S., et al. (2006). Acetic acid activates hepatic AMPK and reduces hyperglycemia in diabetic KK-A(y) mice. *Biochemistry and Biophysics Research Communications, 344*(2), 597–602. https://doi.org/10.1016/j.bbrc.2006.03.066.

Salazar, P., García, M. L., & Selgas, M. D. (2009). Short-chain fructooligosaccharides as potential functional ingredients in dry fermented sausages with different fat levels. *International Journal of Food Science and Technology, 44*(6), 1100–1107. https://doi.org/10.1111/j.1365-2621.2008.01845.x.

Santos, B. A. D., Paulo, C. B. C., Pacheco, M. T., & Pollonio, M. A. R. (2012). Fructooligosaccharides as a fat replacer in fermented cooked sausages. *International Journal of Food Science and Technology, 47*(6), 1183–1192. https://doi.org/10.1111/j.1365-2621.2011.02993.x.

Scholz-Ahrens, K. E., & Schrezenmeir, J. (2002). Inulin, oligofructose and mineral metabolism: Experimental data and mechanisms. *British Journal of Nutrition, 87*(2), S179–S186. https://doi.org/10.1079/BJN2002537.

Scholz-Ahrens, K. E., Schaafsma, G., Van Den E, G., & Schrezenmeir, J. (2001). Effects of prebiotics on mineral absorption. *American Journal of Clinical Nutrition, 73*(2), 459S–464S. https://doi.org/10.1093/ajcn/73.2.459S.

Schumann, C. (2002). Medical, nutritional and technological properties of lactulose: An update. *European Journal of Clinical Nutrition, 41*(1), 17–11. https://doi.org/10.1038/sj.ejcn.1601552.

Sheth, M. (2014). *International Journal of Applied Biology and Pharmaceutical Technology, 5*(3), 256–264.

Sheth, M. (2016). *International Journal of Advanced Research, 4*(5), 1888–1899.

Shin, H. S., Lee, J. H., Pestka, J. J., & Ustunol, Z. (2000). Growth and viability of commercial Bifidobacterium strains in skim milk containing oligosaccharides and inulin. *Food Microbiology and Safety, 65*(5), 884–887.

Silva, R. F. (1996). Use of inulin as a natural texture modifier. *Cereal Foods World, 41*(10), 792–795.

Singh, T., & Sheth, M. (2014). *International Journal of Applied Biology and Pharmaceutical Technology, 5*(3), 99–102.

Tahiri, M., Tressol, J. C., Arnaud, J., Bornet, F. R. J., Bouteloup-Demange, C., Feillet-Coudray, C., Brandolini, M., Ducros, V., Pépin, D., et al. (2003). Effect of short-chain fructooligosaccharides on intestinal calcium absorption and calcium status in postmenopausal women: A stable-isotope study. *American Journal of Clinical Nutrition, 77*(2), 449–457. https://doi.org/10.1093/ajcn/77.2.449.s

Thorne, M. C., & Carpenter, H. F. (1928). The utilization of Jerusalem artichokes by a patient with diabetes. *Archives of Internal Medicine, 42*(1), 64–73. https://doi.org/10.1001/archinte.1928.00130190067007.

Tomita, M., Ohkubo, R., & Hayashi, M. (2004). Lipopolysaccharide transport system across colonic epithelial cells in normal and infective rats. *Drug Metabolism and Pharmacokinetics, 19*(1), 33–40. https://doi.org/10.2133/dmpk.19.33.

Topping, D. L., & Clifton, P. M. (2001). Short-chain fatty acids and human colonic function: Roles of resistant starch and non-starch polysaccharides. *Physiological Reviews, 81*(3), 1031–1063. https://doi.org/10.1152/physrev.2001.81.3.1031.

Voss, A. C., Maki, K. C., Garvey, W. T., Hustead, D. S., Alish, C., Fix, B., & Mustad, V. A. (2008). Effect of two carbohydrate-modified tube-feeding formulas on metabolic responses in patients with type 2 diabetes. *Nutrition, 24*(10), 990–997. https://doi.org/10.1016/j.nut.2008.07.005.

Vuyst, L., & Leroy, F. (2011). Cross-feeding between bifidobacteria and butyrate-producing colon bacteria explains bifidobacterial competitiveness, butyrate production, and gas production. *International Journal of Food Microbiology, 149*(1), 73–80. https://doi.org/10.1016/j.ijfoodmicro.2011.05.015.

Wellen, K. E., & Hotamisligil, G. S. (2005). Inflammation, stress, and diabetes. *Journal of Clinical Investigation, 115*(5), 1111–1119. https://doi.org/10.1172/JCI24505.

Welters, C. F. M., Heineman, E., Thunnissen, B. J. M., Van den Bogaard, A. E. J. M., Soeters, P. B., & Baeten, C. G. M. I. (2002). Effect of dietary inulin supplementation on inflammation of pouch mucosa in patients with an ileal pouch-anal anastomosis. *Diseases of the Colon & Rectum, 45*(5), 621–627. https://doi.org/10.1007/s10350-001-0353-5.

Wong, J., De Souza, R., Kendall, C., Emam, A., & Jenkins, D. J. A. (2006). Colonic health: Fermentation and short-chain fatty acids. *Journal of Clinical Gastroenterology, 40*(3), 235–243. https://doi.org/10.1097/01.mcg.0000197034.17820.8c.

Wright, S. D., Ramos, R. A., Tobias, P. S., Ulevitch, R. J., & Mathison, J. C. (1990). CD14, a receptor for complexes of LPS and LPS-binding protein. *Science, 249*(4975), 1431–1433. https://doi.org/10.1126/science.2200121.

Xu, J., & Gordon, J. I. (2003). Honor thy symbionts. *Proceedings of the National Academy of Sciences of the United States of America, 100*(18), 10452–10459. https://doi.org/10.1073/pnas.1833769100.

Yamada, S. (1994). 45Ca kinetics and balance study: A useful method for analyzing the effects of drugs on calcium metabolism. In H. Ogura (Ed.), *Pharmacological Approach to Study the Formation and The Resorption Mechanism of Hard Tissues* (pp. 75–92). Tokyo: EuroAmerica Publishers.

CHAPTER 3

Factors Associated in Shaping the Gut Microbiota for Improved Health

ARCHANA GATTUPALLI AND NARESHKUMAR GATTUPALLI

Department of Microbiology, The Maharaja Sayajirao University of Baroda, Gujarat, India

Human gut microbiome has about 10 times a greater number of microbes than human cells, belonging to 1000 different genera, cumulatively harboring 150 times the genes of humans with varying diversity and is emerging as an essential organ associated with normal and disease physiological status.

3.1 INTRODUCTION

Microbiome is dominantly represented by bacteria and lesser numbers of archaea, fungi. Bacteriophages constitute a major component of the gut virome. The human gut microbiome is highly dynamic in nature, representing two different categories: autochthonous (resident) and allochthonous (transient) microbes (Nava and Stappenbeck, 2011). Alternatively, the human microbiome could be considered as an ecosystem with interactions with each other in a network (Loftus et al., 2021). Interestingly, two unrelated individuals share 43% of microbial species. Based on the nature of host interactions, microorganisms have been classified into commensal, symbiotic, and pathogenic. The microbiome also changes with growth stages, starting from low diversity in infants, reaching high levels in adults, and then losing diversity in aged individuals.

Prebiotic Foods for a Healthy Gut: Scientific Studies to Manage Metabolic Disorders.
Mini K. Sheth and Shruti Dwivedi (Eds.)

Specific stages like adolescence and pregnancy are associated with major transitions in the gut microbiome. Interestingly, psychological status also governs the gut microbiome, resulting in the functioning gut-brain-liver axis.

3.2 DIVERSITY AND CORE GUT MICROBIOME

Human microbiome is variable at the population as well as the individual level (Priya and Blekhman, 2019). Hence, defining the normal microbiome associated with healthy individuals is very challenging. Defining microbiome at a higher taxonomic level may be better than the species level. Additionally, the abundance of certain taxa and their activity may also be significant in microbiome interactions with the host. A novel method of bioorthogonal noncanonical amino acid tagging (BONCAT) coupled with fluorescently activated cell sorting and sequencing (FACS-Seq) was developed for monitoring translational activity (Taguer et al., 2021). Half of the microbiota was translationally active and similar to the microbial community. High nucleic acid-containing bacteria were also half of the total with more damaged DNA but were distinct. The human microbiome is influenced not only by genetics and age but also by lifestyle factors such as sleep patterns, physical activity, nature, and amount of diet (Figure 3.1). Gut microbiome variations have been investigated with respect to eubiosis and dysbiosis. This chapter discusses the recent findings on how the gut microbiome is influenced by certain factors and the remedial actions to restore the healthy interaction of the microbiome and host upon dysbiosis.

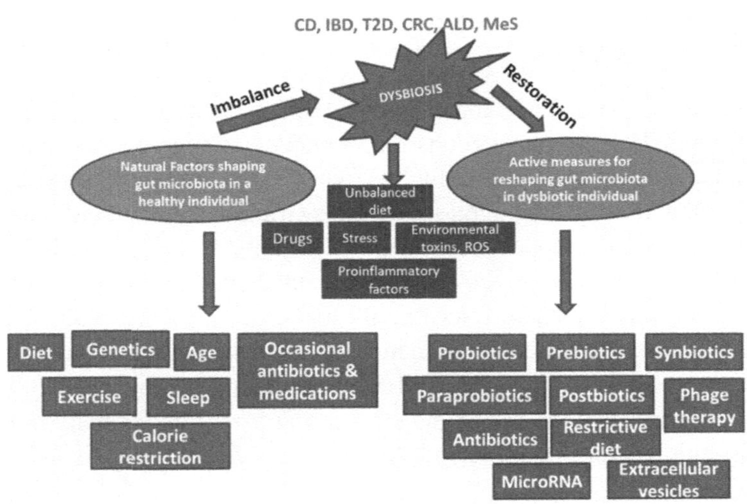

FIGURE 3.1 Factors that shape the healthy gut microbiome and therapeutic treatments for its restoration in dysbiosis. Abbreviations as mentioned in the text.

3.3 SLEEP AND GUT MICROBIOME

Irregular and deprived sleep has been associated with metabolic diseases (Tatti et al., 2019; Chasens et al., 2021). Short sleepers had less abundance of *Lachnoclostridium, Sutterella, Alistipes, Bilophila, Phascolarctobacterium, UBA1819* and *Paraprevotella* and increased *Pseudomonas* and *Eubacterium siraeum* (Agrawal et al., 2021). Lower abundance of *Sutterella* and *Phascolarctobacterium* is associated with depression and postpartum depressive disorder, respectively. Sleep quality is negatively associated with *Lachnospiraceae*, *Corynebacterium*, *Blautia*, *Ruminococcus*, and *Prevotella* (Smith et al., 2019; Grosicki et al., 2020). Moreover, fragmented sleep increased *Lachnospiraceae* and *Ruminococcaceae* families and decreased *Bacteroidetes, Actinobacteria, Bifidobacteriaceae* and *Lactobacillaceae* family.

About 10% of the human gut microbiome has diurnal oscillations (Matenchuk et al., 2020; Pearson et al., 2021). The gut microbiota is also affected by jet lag, involving a decrease in the abundance of *Christensenellaceae, Dorea, Anaeroplasmatales, Anaeroplasmataceae, Anaeroplasma, Lactobacillus, Lactococcus, RF32, Alphaproteobacteria, Proteobacteria, Ruminococcus,* and *Lactobacillus ruminis.* Reduction in *Fusobacteria* and *Christensenellaceae* in jet lag individuals is associated with opportunistic pathogen infection and weight gain, respectively. It will be interesting to determine the combined effect of melatonin and probiotics in restoring a healthy gut microflora in sleep and circadian rhythm disorders.

3.4 EXERCISE AND GUT MICROBIOME

Exercise in individuals with high respiratory exchange ratio increased the abundance of gut microbiome containing genera of *Prevotella, Romboutsia*, and *Dialister* while individuals with low respiratory exchange ratio had increased genera of *Bacteroides*, *Bacteroides* B, and *Parabacteroides* (Bycura et al., 2021). Similarly, individuals experience three repetition maximum (3RM) squat change had gut flora dominated by *Ruminococcus* and *Lachnospiraceae* and increased median abundance of *Turicibacter* and *Clostridium* while individuals with low 3RM had dominated by *Siccibacter*, *Bacteroides*, and *Bacteroides* B and median abundance of *Alistipes*, *Oscillibacter*, and ER4. Elite athletes had increased genera of *Coprococcus, Parasutterella*, and the family *Ruminococcaceae,* while inactive individuals had genera of *Dialister, Odoribacter,* and *Phascolarctobacterium* (Moitinho-Silva et al., 2021). Similarly, sports persons with vigorous activity had increased abundance of *Faecalibacterium, Lachnospira*, and *Akkermansia*

(Aya et al., 2021). Higher abundance of *Ruminococcaceae, Faecalibacteria, Prevotella*, and lower abundance of C. difficile and genus *Megasphaera* were associated with exercise, while the *Firmicutes*/Bacteroidetes ratio is influenced by many factors (Dorelli et al., 2021). Additionally, *Lachnospiraceae, Paraprevotellaceae, Ruminococcaceae*, and *Veillonellaceae* are associated with beneficial metabolic functions. Exercise-induced gut microbiome alterations are implicated in the improvement of behavior and mental health (Nay et al., 2021). However, very rigorous exercise induces gastrointestinal disorders mediated by leaky gut (Ribeiro et al., 2021). Although exercise enriches beneficial microbiome, the nature of exercise and the extent of physiological significance need to be established.

3.5 CALORIE RESTRICTION AND GUT MICROBIOME

Calorie restriction is employed in humans as a nutritional intervention strategy with beneficial effects in aging and metabolic diseases (Most et al., 2017, 2021). Calorie restriction led to decrease in Order *Clostridiales* containing families *Eubacteriaceae, Lachnospiraceae, Peptostreptococcaceae* and *Erysipelotrichaceae* while *Rikenellaceae, Porphyromonadaceae* and *Bacteroidaceae* members were increased (Zheng et al., 2018). Calorie restriction diet of humans reduced the body weight by 12.5% in 12 weeks (von Schwartzenberg et al., 2021). Growth of host glycan utilizing *Akkermansia* and pathogen *Clostridiodes difficile* were enriched while plant saccharide utilizing *Roseburia, Ruminococcus* and *Eubacterium* were depleted.

Very low-calorie diets decreased short-chain fatty acid secretion and decreased resistance against pathogen colonization. On the other hand, calorie-restricted high- and low-protein diets had beneficial colonization of *Akkermansia* spp. and *Bifidobacterium* spp. and decreased pathogenic *Prevotella* spp (Dong et al., 2020). Mild calorie restriction decreased body weight, decreased visceral fat, and increased gut microbial diversity, which is related to the activation of amino acid metabolism (Ben Lassen et al., 2021). However, the population of *Bacteroides* 2 enterotype associated with a severe metabolic phenotype remained unaltered. Interestingly, calorie restriction effects were dependent on the nature of the gut microbiome, whether enriched *Bacteroides* or *Prevotella* spp., in individuals of similar age and body mass index (Zou et al., 2020). Weight loss was more in individuals with enriched *Prevotella* spp. Recently, some concerns were observed suggesting that calorie restriction may prime pathogens (Greathouse and Johnson, 2021).

3.6 GUT MICROBIOTA DYSBIOSIS

Gut microbiome dysbiosis is associated with a decrease in beneficial bacteria, a decrease in overall diversity, and an increase in pathogens (Levy et al., 2017; Meedt et al., 2021). Microbial dysbiosis is chiefly dependent on statistical associations without any supportive causal relations (Brussow, 2020). Hence, emphasis needs to be made on therapy rather than associations (Reid, 2019). Intestinal epithelial damage or inflammation generates high oxygen levels facilitating blooms of *Enterobacteriaceae* members, and their lipopolysaccharides further aggravate the intestinal damage and increase permeability (Zeng et al., 2017). Circulating lipopolysaccharides contribute to tissue dysfunctions mediated by Toll-like receptor 4 signaling. Additionally, dead intestinal cells release phosphatidylcholine and phosphatidylethanolamine which support the growth of *Firmicutes, Actinobacteria, Proteobacteria* members along with *Salmonella* and *Pseudomonas* pathogens. Oral microbial dysbiosis has been implicated in triggering gut microbial dysbiosis and thereby causing gastrointestinal diseases (Khor et al., 2021). Broad-spectrum antibiotics induce microbiota dysbiosis and increase intestinal permeability (Feng et al., 2019). In humans, altered microbiota gets partial recovery after fluoroquinolone withdrawal (Levast et al., 2021).

3.7 DIET AND DIETARY SUPPLEMENTS AND GUT MICROBIOTA EUBIOSIS

Diet is one of the major factors influencing the diversity and dynamics of the gut microbiome (Wilson et al., 2020; Beam et al., 2021; Leeming et al., 2021). Alcohol consumption and bowel movement quality are associated with gut microbiota in the American population (Vujkovic-Cvijin et al., 2020). However, dairy products appear to be more significant compared to other diet components (Shui et al., 2021). The Mediterranean diet is known to be associated with a protective effect on cardiometabolic disease correlated with the abundance of *Prevotella copri* (Wang et al., 2021). The metabolites in the serum are related to the gut microflora, which in turn is governed by diet (Asnicar et al., 2021; Bastiaanssen et al., 2021; Shui et al., 2021). The association of diet and gut microbiome is gaining prominence, and personalized nutrition is being considered as a strategy for the treatment of gastrointestinal tract diseases (Kviatcovsky et al., 2021). Several supplements and the changes influenced by them on the gut flora are listed in Table 3.1.

Tryptophan rich food, e.g., fish and dairy products, are catabolized by *Peptostreptococcus russellii, Clostridium sporogenes*, and *Lactobacillus* spp.

producing indole derivatives improving the intestinal epithelial integrity by acting on aryl hydrocarbons receptors (Molzer et al., 2021). Branched-chain amino acids such as leucine, isoleucine, and valine are essential amino acids, produced by *Prevotella copri* and *Bacteroides vulgatus* (Man et al., 2020).

Folate is synthesized by the phylum *Bacteroidetes (Prevotella, Bacteroides, Porphyromonas*) and is produced and absorbed in the colon. Folate-secreting probiotic lactic acid bacilli, *Streptococcus* (Strep.) thermophilus CRL 808 and *Strep. thermophilus* CRL 415, and *Bifidobacteria* producing folates are effective in providing folate requirements for humans (Rossi et al., 2011; de LeBlanc et al., 2018).

Vitamin D regulation of the gut microbiome of older men is becoming clear (Akimbekov et al., 2020; Thomas et al., 2020). Vitamin D inhibits Th1 and Th17 cells by altering the gut microbiome (Yamamoto and Jørgensen, 2020). However, fecal microbiome analysis in humans showed no consistent variations in response to Vitamin D supplementation, which raised the limitations of fecal microbiome analysis. Alternatively, *Streptomyces griseolus, Ruminococcus torques, Mycobacterium tuberculosis, Nocardia, Sphingomonas,* and *Amycolata* possess Vitamin D hydroxylation activity. Vitamin D also improves intestinal epithelial barrier integrity by increasing the expression of tight junction proteins.

Polyphenols possess antioxidant and antibacterial activity leading to alteration of gut microbiome towards beneficial commensals (Makarewicz et al., 2020). *Bifidobacterium* and *Lactobacillus* maintain gut barrier protection, *Akkermansia muciniphila* and *Faecalibacterium prausnitzii* inhibit the activation of NF-kB, and *Roseburia* sp. produce butyrate. Phenolic acids *Eubacterium limosum* and *E. ramulus* catalyze demethylation of isoxanthohumol into 8-prenylnaringenin which prevents bone loss, possesses antiandrogenic activity, inhibits angiogenesis and metastasis. Quercetin is present in food as a glycoside and its bioavailability is very low (Tan et al., 2021). However, quercetin is known to prevent high fat-induced obesity and type 2 diabetes by increasing the population of *Akkermansia* and decreasing Firmicute/Bacteroidetes ratio.

Curcumin decreased the population of 36 harmful bacteria belonging to *Ruminococcus, Coriobacterales, Prevotellaceae, Enterobacteria,* and *Enterococci* while *Bacteroidaceae, Lactobacilli, Rikenellaceae,* and *Alistipes* increased with beneficial health effects (Scazzocchio et al., 2020; Jabczyk et al., 2021). Chitosan and its derivatives increase the population of beneficial *Bifidobacterium, Lactobacilli, Akkermansia,* and *Parabacteroides* (Yan et al., 2020). *Lactobacillus plantarum* SeZi strain with efficient uptake of Se and Zn has high antioxidant activity and reduced equol-producing *Adlercreutzia equolifaciens* which was negatively associated with the expression of tight junction proteins (Kang et al., 2020). Additionally, equol has very potent estrogenic activity.

TABLE 3.1 Changes in the Gut Microflora in Response to Dietary Supplements

Supplement	Changes in the gut microflora		Reference
	Increase	Decrease	
Tryptophan	*Peptostreptococcus russellii, Clostridium sporogenes*, and *Lactobacillus* spp		Molzer et al., 2021
Branched chain amino acids	*Prevotella copri* and *Bacteroides vulgatus*		Man et al., 2020
Folate	*Prevotella, Bacteroides, Porphyromona,* lactic acid bacilli, *Streptococcus* (Strep.) *thermophilus* CRL 808 and *Strep. Thermophilus* CRL 415, *Bifiodobacteria*		Rossi et al., 2011; de LeBlanc et al., 2018
Vitamin D	*Streptomyces griseolus, Ruminococcus torques, Mycobacterium tuberculosis, Nocardia, Sphinogmonas,* and *Amycolata*		Yamamoto and Jørgensen, 2020
Quercetin	*Akkermansia*	Firmicute/ Bacteroidetes ratio	Tan et al., 2021
Polyphenols	*Bifidobacterium, Lactobacillus, Akkermansia muciniphila* and *Faecalibacterium prausnitzii*		Makarewicz et al., 2020
Phenolic acids	*Eubacterium limosum* and *E. ramulus*		Makarewicz et al., 2020
Curcumin	*Bacteroidaceae, Lactobacilli, Rikenellaceae* and *Alistipes*	*Ruminococcus, Coriobacterales, Prevotellaceae, Enterobacteria* and *Enterococci*	Scazzocchio et al., 2020; Jabczyk et al., 2021
Chitosan	*Bifidobacterium, Lactobacilli, Akkermansia* and *Parabacteroides*		Yan et al., 2020
Se and Zn	*Lactobacillus plantarum* SeZi	*Adlercreutzia equolifaciens*	Kang et al., 2020

3.8 PROBIOTICS, PREBIOTICS, AND RELATED THERAPIES

Intake of specially screened, live beneficial bacteria or probiotics has been practiced for several decades (Tuohy et al., 2003). Mostly strains of

Lactobacillus spp., *Bifidobacterium longum, Streptococcus faecalis*, specific strains of *Escherichia coli* (e.g., *E. coli,* Nissle, 1917) and yeasts like *Saccharomyces boulardii* are available commercially as individual strains or a multispecies combination and are used for wellness as well as to treat several clinical conditions not restricted to gut disorders (Stavropoulou and Bezirtzoglou, 2020). Recent developments in probiotics research are focused on increasing their efficiency by improving stability in the stomach and in the commercial product, strain improvement for enhanced functional properties and for targeted action (Mathipa and Thantsha, 2017). Prebiotics, substances which stimulate the growth of beneficial microbes when consumed, are mostly oligosaccharides which reach the colon without being acted upon by stomach acids and hydrolyzing enzymes in the intestine. Products with combinations of probiotics and prebiotics, known as synbiotics, are considered superior to probiotics alone (Kvakova et al., 2021). Other related terms para-biotics and post-biotics refer to products of probiotic microorganisms or non-viable probiotics which have also been shown to be effective (Table 3.2). Recent approaches include the use of microbial products such as extracellular vesicles or miRNAs (Diez-Sainz et al., 2021). Since probiotics provide transient benefits and are not able to colonize the gut efficiently, fecal microbial transplants (FMT) have been envisaged as a better option wherein the entire population of microbes from healthy donors is used for therapeutic purposes albeit with risks of its own (Green et al., 2020).

Certain new bacterial species have been recently gaining attention as next-generation probiotics. *Akkermansia muciniphila* belongs to the Verrucomicrobia phylum and colonizes the mucus layer in the gastrointestinal tract. It represents 1 to 4% of the fecal microbiota and is envisaged as a future probiotic candidate. Supplementation of this bacterium live or in pasteurized form to obese individuals resulted in improved insulin sensitivity and promoted weight loss (Depommier et al., 2020). Another candidate bacterium, *Roseburia intestinalis*, a butyrate-producing firmicute, is a prominent xylan degrader and is involved in the metabolism of β-mannans (La Rosa et al., 2019). The abundance of this bacterium is reduced in individuals affected by inflammatory disease and colorectal cancer. Like *A. muciniphila, R. intestinalis* also preferentially colonizes the mucin layer and contributes to the local level of butyrate available for the colonic epithelial cells. A syntrophic association of this bacterium with another putative beneficial organism, *Faecalibacterium prausnitzii* (Lindstad et al., 2021), is beneficial, possibly since the latter has been shown to secrete anti-inflammatory peptides and relieve chemically induced colitis in mice (Breyner et al., 2017).

3.9 PHAGE THERAPY

Bacteriophages, the viruses that infect bacteria, comprise a major component of the intestinal virome (Li et al., 2021) and have been the subject of recent renewed interest as therapeutic agents to restore dysbiotic gut flora (Dahlman et al., 2021; Ganeshan and Hosseinidoust, 2019). Bacteriophages are naturally present free in the human gut as well as integrated in the genome of gut microbiota as prophages (Lugli et al., 2016). Predation of gut bacteria by phages can shape the gut microbiome and its metabolic activity by directly affecting the target bacterium and indirectly by triggering a cascade of other effects mediated through inter-bacterial interactions (Hsu et al., 2019). Dysbiosis resulted in alterations of enteric phage diversity as shown in inflammatory bowel disease (Norman et al., 2015), colorectal cancer (Nakatsu et al., 2018). The demonstration that sterile bacteria-free filtered fecal matter transfer from a healthy individual reduced recurrent *Clostridium difficile* infection in patients indicated a possible therapeutic role of phages (Ott et al., 2017). Galtier et al. (2017) showed phages against adherent invasive *Escherichia coli* diminished its population and reduced symptoms of colitis in the mouse model.

Fecal viral transplant was shown to restore the microbiome of antibiotic-induced dysbiosis in mice (Draper et al., 2020). Ramnussen et al. (2020) demonstrated the efficacy of fecal virome transplantation (FVT) from lean donors in alleviating obesity phenotype in mice. Phage therapy has shown success in the treatment of alcohol-related liver disease in which lytic phages against cytolysin-producing *Enterococcus faecalis* could reduce liver damage. Prophages can be induced from their temperate lifestyle into free viral particles by common drugs and OTC medications (Sutcliffe et al., 2021), dietary sugar and bacterial metabolites (Oh et al., 2019), bile salts (Hernandez et al., 2012), and also by changes in quorum sensing signals (Rossman et al., 2015). Prophage activation mechanisms could be used in the future as a means for beneficial alteration of gut microbial communities. Sublethal concentrations of antibiotics can be targeted against the eradication of specific lysogens (Al-Anany et al., 2021). On the other hand, a prophage-containing strain of probiotic *Streptococcus thermophilus* was more effective in inhibiting adhesion of pathogenic bacteria to human intestinal cell lines compared to the isogenic prophage-cured strain (Guigas et al., 2016), suggesting that temperate strains of probiotics may have superior probiotic potential.

TABLE 3.2 Some Recent Examples of Therapeutic Potential of Fecal Virome Transfer (FVT), Bacteriophages, Postbiotics, and Para Probiotics

Type of Treatment	Dysbiosis Syndrome	Bacteria	Model System	References
FVT	rCDI	Bacteria free filtrate	Human	Ott et al., 2017, Kaoet al., 2019
	T2D	Bacteria free filtrate	Murine	Rasmussen et al., 2020
	Antibiotic induced perturbation	Bacteria free filtrate	Murine	Draper et al., 2019
Bacteriophage	CRC	*Fusobacterium nucleatum*	Mouse	Dong et al., 2020
	CRC	*Enterococcus faecalis*	Human Cell line	Kabwe et al., 2021
	Alcoholic hepatitis	Cytolytic *Enterococcus faecalis*	Humanized mouse model	Duan et al., 2019
	Asymptomatic intestinal carriage	Enteroaggregative *Escherichia coli*	Mouse	Maura et al., 2012
	CD	Adherant invasive *E. coli*	Mouse	Galtier et al., 2017
Postbiotics	Obesity and Diabetes	*Akkermansia muciniphila* (Purified thermostable outer membrane protein)	Mouse	Plovier et al. 2017
	Colitis	Faecalibacterium prausnitzii Microbial anti-inflammatory molecule (MAM)	Mouse	Breyner et al., 2017
	serotonin signaling	*Akkermansia muciniphila* extracellular vesicles	Mouse and Human cell lines	Yaghoubfar et al., 2020
	serotonin signaling	*Akkermansia muciniphila* and Faecalibacterium prausnitzii extracellular vesicles	Caco-2 cell line	Yaghoubfar et al., 2021
Para probiotics	Obesity and Diabetes	*Akkermansia muciniphila* (Pasteurized cells)	Mouse	Plovier et al., 2017
			Human	Depommier et al., 2019

3.10 FUTURE STRATEGIES

Future strategies need to focus on regulating lipopolysaccharide-secreting gut microbiota as their population is proportional to the intestinal epithelial damage by diet, alcohol abuse, drugs, and other chemicals. Specific antibiotics against Gram-negative bacteria are being developed, which could be useful in controlling their population in gut microflora (Imai et al., 2019; Kaur et al., 2021; Munroz et al., 2021). Preventing chronic oxidative damage and inflammation of the gastrointestinal tract will also be effective in restoring eubiosis. Next-generation probiotics and their products will be of interest in the coming future, as will the application of bacteriophages. Genetically engineered probiotic strains with superior or novel traits are likely to be used as designer probiotics addressing specific disorders (Zhou et al., 2020; Nareshkumar and Archana, 2021). Proof of concept studies and meta-analysis will help ascertain the benefits across different large population groups. Disease-specific probiotics and prebiotics will lead to customized treatments, and gut microbiome profiling might become routine for both diagnostic and prognostic purposes.

KEYWORDS

- **bacteria**
- **bacteriophages**
- **diabetes**
- **diets**
- **efficiency (properties)**
- **microbiome**
- **pathogens**
- **probiotics**
- **viruses**
- **vitamin D**

REFERENCES

Agrawal, R., Ajami, N. J., Malhotra, S., Chen, L., White, D. L., Sharafkhaneh, A., Hoffman, K. L., Graham, D. Y., El-Serag, H. B., Petrosino, J. F., et al. (2021). Habitual sleep duration and the colonic mucosa-associated gut microbiota in humans—A pilot study. *Clocks & Sleep, 3*, 387–397. https://doi.org/10.3390/clockssleep3030025.

Akimbekov, N. S., Digel, I., Sherelkhan, D. K., Lutfor, A. B., & Razzaque, M. S. (2020). Vitamin D and the host-gut microbiome: A brief overview. *Acta Histochem Cytochem, 53*, 33–42. https://doi.org/10.1267/ahc.20011.

Al-Anany, A. M., Fatima, R., & Hynes, A. P. (2021). Temperate phage-antibiotic synergy eradicates bacteria through depletion of lysogens. *Cell Reports, 35*, 109172. https://doi.org/10.1016/j.celrep.2021.109172.

Asnicar, F., Berry, S. E., Valdes, A. M., et al. (2021). Microbiome connections with host metabolism and habitual diet from 1,098 deeply phenotyped individuals. *Nature Medicine, 27*, 321–332. https://doi.org/10.1038/s41591-020-01183-8.

Aya, V., Flórez, A., Pérez, L., & Ramírez, J. D. (2021). Association between physical activity and changes in intestinal microbiota composition: A systematic review. *PLoS ONE, 16*, e0247039. https://doi.org/10.1371/journal.pone.0247039.

Bastiaanssen, T. F. S., & Cryan, J. F. (2021). Dairy alters the microbiome, are we but skimming the surface? *EBioMedicine, 68*, 103417. https://doi.org/10.1016/j.ebiom.2021.103417.

Beam, A., Clinger, E., & Hao, L. (2021). Effect of diet and dietary components on the composition of the gut microbiota. *Nutrients, 13*, 2795. https://doi.org/10.3390/nu13082795.

Bel Lassen, P., Belda, E., Prifti, E., Dao, M. C., Specque, F., Henegar, C., Rinaldi, L., Wang, X., Kennedy, S. P., Zucker, J. D., Calame, W., Lamarche, B., & Clément, K. (2021). Protein supplementation during an energy-restricted diet induces visceral fat loss and gut microbiota amino acid metabolism activation: A randomized trial. *Scientific Reports, 11*, 15620. https://doi.org/10.1038/s41598-021-94916-9.

Breyner, N. M., Michon, C., de Sousa, C. S., Vilas Boas, P. B., Chain, F., Azevedo, V. A., & Chatel, J. M. (2017). Microbial anti-inflammatory molecule (MAM) from *Faecalibacterium prausnitzii* shows a protective effect on DNBS and DSS-induced colitis model in mice through inhibition of NF-κB pathway. *Frontiers in Microbiology, 8*, 114. https://doi.org/10.3389/fmicb.2017.00114.

Brüssow, H. (2020). Problems with the concept of gut microbiota dysbiosis. *Microbial Biotechnology, 13*, 423–434. https://doi.org/10.1111/1751-7915.13479.

Bycura, D., Santos, A. C., Shiffer, A., Kyman, S., Winfree, K., Sutliffe, J., Pearson, T., Sonderegger, D., Cope, E., & Caporaso, J. G. (2019). Impact of different exercise modalities on the human gut microbiome. *Sports, 9*(2), 14. https://doi.org/10.3390/sports9020014.

Chasens, E. R., Imes, C. C., Kariuki, J. K., Luyster, F. S., Morris, J. L., DiNardo, M. M., Godzik, C. M., Jeon, B., & Yang, K. (2021). Sleep and metabolic syndrome. *Nursing Clinics of North America, 56*, 203–217. https://doi.org/10.1016/j.cnur.2020.10.012.

Dahlman, S., Avellaneda-Franco, L., & Barr, J. J. (2021). Phages to shape the gut microbiota? *Current Opinion in Biotechnology, 68*, 89–95. https://doi.org/10.1016/j.copbio.2021.05.003.

Depommier, C., et al. (2019). Supplementation with *Akkermansia muciniphila* in overweight and obese human volunteers: A proof-of-concept exploratory study. *Nature Medicine, 25*, 1096–1103. https://doi.org/10.1038/s41591-019-0495-2.

Díez-Sainz, E., Milagro, F. I., Riezu-Boj, J. I., & Lorente-Cebrián, S. (2021). Effects of gut microbiota-derived extracellular vesicles on obesity and diabetes and their potential modulation through diet. *Journal of Physiology and Biochemistry*, *78*, 485–499. https://doi.org/10.1007/s13105-021-00837-6.

Dong, T. S., Luu, K., Lagishetty, V., Sedighian, F., Woo, S. L., Dreskin, B. W., Katzka, W., Chang, C., Zhou, Y., Arias-Jayo, N., Yang, J., Ahdoot, A., Li, Z., Pisegna, J. R., & Jacobs,

J. P. (2020). A high protein calorie restriction diet alters the gut microbiome in obesity. *Nutrients, 12*(10), 3221. https://doi.org/10.3390/nu12103221.

Dong, X., Pan, P., Zheng, D. W., Bao, P., Zeng, X., & Zhang, X. Z. (2020). Bioinorganic hybrid bacteriophage for modulation of intestinal microbiota to remodel tumor-immune microenvironment against colorectal cancer. *Science Advances, 6*, eaba1590. https://doi.org/10.1126/sciadv.aba1590.

Dorelli, B., Gallè, F., De Vito, C., Duranti, G., Iachini, M., Zaccarin, M., Preziosi Standoli, J., Ceci, R., Romano, F., Liguori, G., et al. (2021). Can physical activity influence human gut microbiota composition independently of diet? A systematic review. *Nutrients, 13*, 1890. https://doi.org/10.3390/nu13061890.

Draper, L. A., Ryan, F. J., Dalmasso, M., Casey, P. G., McCann, A., Velayudhan, V., Ross, R. P., & Hill, C. (2020). Autochthonous faecal virome transplantation (FVT) reshapes the murine microbiome after antibiotic perturbation. *BMC Biology, 18*, 173. https://doi.org/10.1186/s12915-020-00906-0.

Duan, Y., Llorente, C., Lang, S., Brandl, K., Chu, H., Jiang, L., & Schnabl, B. (2019). Bacteriophage targeting of gut bacterium attenuates alcoholic liver disease. *Nature, 575*, 505–511. https://doi.org/10.1038/s41586-019-1742-x.

Duar, R. M., Henrick, B. M., Casaburi, G., & Frese, S. A. (2020). Integrating the ecosystem services framework to define dysbiosis of the breastfed infant gut: The role of *Bifidobacterium infantis* and human milk oligosaccharides. *Frontiers in Nutrition, 7*, 33. https://doi.org/10.3389/fnut.2020.00033.

Falony, G., Joossens, M., Vieira-Silva, S., Wang, J., Darzi, Y., Faust, K., Kurilshikov, A., Bonder, M. J., Valles-Colomer, M., Vandeputte, D., Tito, R. Y., Chaffron, S., Rymenans, L., Verspecht, C., De Sutter, L., Lima-Mendez, G., D'hoe, K., Jonckheere, K., Homola, D., Garcia, R., Tigchelaar, E. F., Eeckhaudt, L., Fu, J., Henckaerts, L., Zhernakova, A., Wijmenga, C., & Raes, J. (2016). Population-level analysis of gut microbiome variation. *Science, 352*(6285), 560–564. https://doi.org/10.1126/science.aad3503.

Feng, Y., Huang, Y., Wang, Y., Wang, P., Song, H., & Wang, F. (2019). Antibiotics induced intestinal tight junction barrier dysfunction is associated with microbiota dysbiosis, activated NLRP3 inflammasome and autophagy. *PLoS ONE, 14*(5), e0218384. https://doi.org/10.1371/journal.pone.0218384.

Galtier, M., Sordi, L. D., Sivignon, A., De Vallée, A., Maura, D., Neut, C., & Debarbieux, L. (2017). Bacteriophages targeting adherent invasive *Escherichia coli* strains as a promising new treatment for Crohn's disease. *Journal of Crohn's and Colitis, 11*, 840–847. https://doi.org/10.1093/ecco-jcc/jjw224.

Ganeshan, D., & Hosseinidoust, Z. (2019). Phage therapy with a focus on the human microbiota. *Antibiotics, 8*(3), 131. https://doi.org/10.3390/antibiotics8030131.

Gong, L., Wen, T., & Wang, J. (2020). Role of the microbiome in mediating health effects of dietary components. *Journal of Agricultural and Food Chemistry, 68*, 12820–12835. https://doi.org/10.1021/acs.jafc.9b08231.

Greathouse, K. L., & Johnson, A. J. (2021). Does caloric restriction prime the microbiome for pathogenic bacteria? *Cell Host & Microbe, 29*, 1209–1211. https://doi.org/10.1016/j.chom.2021.07.009.

Green, J. E., Davis, J. A., Berk, M., Hair, C., Loughman, A., Castle, D., Athan, E., Nierenberg, A. A., Cryan, J. F., Jacka, F., & Marx, W. (2020). Efficacy and safety of fecal microbiota transplantation for the treatment of diseases other than *Clostridium difficile* infection: A systematic review and meta-analysis. *Gut Microbes, 12*(1), 1–25. https://doi.org/10.1080/19490976.2020.1854640.

Grosicki, G. J., Riemann, B. L., Flatt, A. A., Valentino, T., & Lustgarten, M. S. (2020). Self-reported sleep quality is associated with gut microbiome composition in young, healthy individuals: A pilot study. *Sleep Medicine, 73*, 76–81. https://doi.org/10.1016/j.sleep.2020.04.013.

Guigas, C., Faulhaber, K., Duerbeck, D., Neve, H., & Heller, K. J. (2016). Prophage-mediated modulation of interaction of *Streptococcus thermophilus* J34 with human intestinal epithelial cells and its competition against human pathogens. *Beneficial Microbes, 7*(2), 289–297. https://doi.org/10.3920/BM2015.0108.

Gupta, V. K., Paul, S., & Dutta, C. (2017). Geography, ethnicity or subsistence-specific variations in human microbiome composition and diversity. *Frontiers in Microbiology, 8*, 1162. https://doi.org/10.3389/fmicb.2017.01162.

Hernandez, S. B., Cota, I., Ducret, A., Aussel, L., & Casadesus, J. (2012). Adaptation and preadaptation of *Salmonella enterica* to bile. *PLoS Genetics, 8*(10), e1002459. https://doi.org/10.1371/journal.pgen.1002459.

Hsu, B. B., Gibson, T. E., Yeliseyev, V., Liu, Q., Lyon, L., Bry, L., & Gerber, G. K. (2019). Dynamic modulation of the gut microbiota and metabolome by bacteriophages in a mouse model. *Cell Host & Microbe, 25*(6), 803–814. https://doi.org/10.1016/j.chom.2019.05.004.

Iizuka, K., & Yabe, D. (2020). The role of metagenomics in precision nutrition. *Nutrients, 12*(6), 1668. https://doi.org/10.3390/nu12061668.

Imai, Y., Meyer, K. J., Iinishi, A., Favre-Godal, Q., Green, R., Manuse, S., & Lewis, K. (2019). A new antibiotic selectively kills Gram-negative pathogens. *Nature, 576*, 459–464. https://doi.org/10.1038/s41586-019-1791-1.

Kabwe, M., Meehan-Andrews, T., Ku, H., et al. (2021). Lytic bacteriophage EFA1 modulates *HCT116* colon cancer cell growth and upregulates ROS production in an *Enterococcus faecalis* co-culture system. *Frontiers in Microbiology, 12*, 650849. https://doi.org/10.3389/fmicb.2021.650849.

Kao, D. H., Roach, B., Walter, J., Lobenberg, R., & Wong, K. (2019). Effect of lyophilized sterile fecal filtrate vs lyophilized donor stool on recurrent *Clostridium difficile* infection (rCDI): Preliminary results from a randomized, double-blind pilot study. *Journal of the Canadian Association of Gastroenterology, 2*, 101–102. https://doi.org/10.1093/jcag/gwz006.050.

Kaur, H., Jakob, R. P., Marzinek, J. K., Green, R., Imai, Y., Bolla, J. R., Agustoni, E., Robinson, C. V., Bond, P. J., Lewis, K., Maier, T., & Hiller, S. (2021). The antibiotic darobactin mimics a β-strand to inhibit outer membrane insertase. *Nature, 593*(7857), 125–129. https://doi.org/10.1038/s41586-021-03455-w.

Khor, B., Snow, M., Herrman, E., Ray, N., Mansukhani, K., Patel, K. A., Said-Al-Naief, N., Maier, T., & Machida, C. A. (2021). Interconnections between the oral and gut microbiomes: Reversal of microbial dysbiosis and the balance between systemic health and disease. *Microorganisms, 9*(3), 496. https://doi.org/10.3390/microorganisms9030496.

Kurilshikov, A., Medina-Gomez, C., Bacigalupe, R., Radjabzadeh, D., Wang, J., Demirkan, A., et al. (2021). Large-scale association analyses identify host factors influencing human gut microbiome composition. *Nature Genetics, 53*, 156–165. https://doi.org/10.1038/s41588-020-00763-1.

Kvakova, M., Bertkova, I., Stofilova, J., & Savidge, T. C. (2021). Co-encapsulated synbiotics and immobilized probiotics in human health and gut microbiota modulation. *Foods, 10*(6), 1297. https://doi.org/10.3390/foods10061297.

Kviatcovsky, D., Zheng, D., & Elinav, E. (2021). Gut microbiome and its potential link to personalized nutrition. *Current Opinion in Physiology, 22*, 100439. https://doi.org/10.1016/j.cophys.2021.05.002.

La Rosa, S. L., Leth, M. L., Michalak, L., Hansen, M. E., Pudlo, N. A., Glowacki, R., & Westereng, B. (2019). The human gut Firmicute *Roseburia intestinalis* is a primary degrader of dietary β-mannans. *Nature Communications, 10*(1), 1–14. https://doi.org/10.1038/s41467-019-11252-3.

Lee, S. H., Yoon, S. H., Jung, Y., et al. (2020). Emotional well-being and gut microbiome profiles by enterotype. *Scientific Reports, 10*, 20736. https://doi.org/10.1038/s41598-020-77673-z.

Leeming, E. R., Louca, P., Gibson, R., Menni, C., Spector, T. D., & Le Roy, C. I. (2021). The complexities of the diet-microbiome relationship: Advances and perspectives. *Genome Medicine, 13*(1), 10. https://doi.org/10.1186/s13073-020-00813-7.

Levast, B., Benech, N., Gasc, C., Batailler, C., Senneville, E., Lustig, S., Pouderoux, C., Boutoille, D., Boucinha, L., Dauchy, F. A., Zeller, V., Maynard, M., Cazanave, C., Le Thi, T. T., Josse, J., Doré, J., Laurent, F., & Ferry, T. (2021). Impact on the gut microbiota of intensive and prolonged antimicrobial therapy in patients with bone and joint infection. *Frontiers in Medicine (Lausanne), 8*, 586875. https://doi.org/10.3389/fmed.2021.586875.

Levy, M., Kolodziejczyk, A. A., Thaiss, C. A., et al. (2017). Dysbiosis and the immune system. *Nature Reviews Immunology, 17*, 219–232. https://doi.org/10.1038/nri.2017.7.

Li, Y., Handley, S. A., & Baldridge, M. T. (2021). The dark side of the gut: Virome–host interactions in intestinal homeostasis and disease. *Journal of Experimental Medicine, 218*(5), e20201044. https://doi.org/10.1084/jem.20201044.

Lindstad, L. J., Lo, G., Leivers, S., Lu, Z., Michalak, L., Pereira, G. V., & La Rosa, S. L. (2021). Human gut *Faecalibacterium prausnitzii* deploys a highly efficient conserved system to cross-feed on β-mannan-derived oligosaccharides. *mBio, 12*(3), e03628–20. https://doi.org/10.1128/mBio.03628-20.

Loftus, M., Hassouneh, S. A., & Yooseph, S. (2021). Bacterial associations in the healthy human gut microbiome across populations. *Scientific Reports, 11*(1), 2828. https://doi.org/10.1038/s41598-021-81663-2.

Lugli, G. A., Milani, C., Turroni, F., Tremblay, D., Ferrario, C., Mancabelli, L., Duranti, S., Ward, D. V., Ossiprandi, M. C., Moineau, S., van Sinderen, D., & Ventura, M. (2016). Prophages of the genus *Bifidobacterium* as modulating agents of the infant gut microbiota. *Environmental Microbiology, 18*(7), 2196–2213. https://doi.org/10.1111/1462-2920.13154.

Mailing, L. J., Allen, J. M., Buford, T. W., Fields, C. J., & Woods, J. A. (2019). Exercise and the gut microbiome: A review of the evidence, potential mechanisms, and implications for human health. *Exercise and Sport Sciences Reviews, 47*(2), 75–85. https://doi.org/10.1249/JES.0000000000000183.

Matenchuk, B. A., Mandhane, P. J., & Kozyrskyj, A. L. (2020). Sleep, circadian rhythm, and gut microbiota. *Sleep Medicine Reviews, 53*, 101340. https://doi.org/10.1016/j.smrv.2020.101340.

Mathipa, M. G., & Thantsha, M. S. (2017). Probiotic engineering: Towards development of robust probiotic strains with enhanced functional properties and for targeted control of enteric pathogens. *Gut Pathogens, 9*(1), 1–17. https://doi.org/10.1186/s13099-017-0217-1.

Maura, D., Galtier, M., Le Bouguénec, C., & Debarbieux, L. (2012). Virulent bacteriophages can target *O104* enteroaggregative *Escherichia coli* in the mouse intestine. *Antimicrobial Agents and Chemotherapy, 56*(12), 6235–6242. https://doi.org/10.1128/AAC.00602-12.

Meedt, E., Hiergeist, A., Gessner, A., Dettmer, K., Liebisch, G., Ghimire, S., Poeck, H., Edinger, M., Wolff, D., Herr, W., Holler, E., & Weber, D. (2021). Prolonged suppression of butyrate-producing bacteria is associated with acute gastrointestinal graft-versus-host disease and transplant-related mortality after allogeneic stem cell transplantation. *Clinical Infectious Diseases, 27*, ciab500. https://doi.org/10.1093/cid/ciab500.

Moitinho-Silva, L., Wegener, M., May, S., Schrinner, F., Akhtar, A., Boysen, T. J., Schaeffer, E., Hansen, C., Schmidt, T., Rühlemann, M. C., Hübenthal, M., Rausch, P., Kondakci, M. T., Maetzler, W., Weidinger, S., Laudes, M., Süß, P., Schulte, D., Junker, R., Sommer, F., Weisser, B., Bang, C., & Franke, A. (2021). Short-term physical exercise impacts on the human holobiont obtained by a randomized intervention study. *BMC Microbiology, 21*, 162. https://doi.org/10.1186/s12866-021-02214-1.

Most, J., Tosti, V., Redman, L. M., & Fontana, L. (2017). Calorie restriction in humans: An update. *Ageing Research Reviews, 39*, 36–45. https://doi.org/10.1016/j.arr.2016.08.005.

Muñoz, K. A., & Hergenrother, P. J. (2021). Facilitating compound entry as a means to discover antibiotics for Gram-negative bacteria. *Accounts of Chemical Research, 54*(7), 1322–1333. https://doi.org/10.1021/acs.accounts.0c00895.

Nakatsu, G., Zhou, H., Wu, W. K. K., Wong, S. H., Coker, O. O., Dai, Z., & Yu, J. (2018). Alterations in enteric virome are associated with colorectal cancer and survival outcomes. *Gastroenterology, 155*(2), 529–541. https://doi.org/10.1053/j.gastro.2018.04.018.

Napoleão, A., Fernandes, L., Miranda, C., & Marum, A. P. (2021). Effects of calorie restriction on health span and insulin resistance: Classic calorie restriction diet vs. ketosis-inducing diet. *Nutrients, 13*(4), 1302. https://doi.org/10.3390/nu13041302.

Nareshkumar, G., & Archana, G. (2021). Potential of *Escherichia coli* probiotics for improved health and disease management. In *IntechOpen*. Available from: https://www.intechopen.com/online-first/78845 (accessed on 19 July 2024).

Nava, G. M., & Stappenbeck, T. S. (2011). Diversity of the autochthonous colonic microbiota. *Gut Microbes, 2*(2), 99–104. https://doi.org/10.4161/gmic.2.2.15416.

Nay, K., Smiles, W. J., Kaiser, J., McAloon, L. M., Loh, K., Galic, S., Oakhill, J. S., Gundlach, A. L., & Scott, J. W. (2021). Molecular mechanisms underlying the beneficial effects of exercise on brain function and neurological disorders. *International Journal of Molecular Sciences, 22*(8), 4052. https://doi.org/10.3390/ijms22084052.

Nguyen, T. T., Zhang, X., Wu, T. C., Liu, J., Le, C., Tu, X. M., Knight, R., & Jeste, D. V. (2021). Association of loneliness and wisdom with gut microbial diversity and composition: An exploratory study. *Frontiers in Psychiatry, 12*, 648475. https://doi.org/10.3389/fpsyt.2021.648475.

Norman, J. M., Handley, S. A., Baldridge, M. T., Droit, L., Liu, C. Y., Keller, B. C., Kambal, A., Monaco, C. L., Zhao, G., Fleshner, P., & Virgin, H. W. (2015). Disease-specific alterations in the enteric virome in inflammatory bowel disease. *Cell, 160*(3), 447–460. https://doi.org/10.1016/j.cell.2015.01.002.

Oh, J. H., Alexander, L. M., Pan, M., Schueler, K. L., Keller, M. P., Attie, A. D., Walter, J., & van Pijkeren, J. P. (2019). Dietary fructose and microbiota-derived short-chain fatty acids promote bacteriophage production in the gut symbiont *Lactobacillus reuteri*. *Cell Host & Microbe, 25*(2), 273–284.e276. https://doi.org/10.1016/j.chom.2018.11.016.

Ott, S. J., Waetzig, G. H., Rehman, A., Moltzau-Anderson, J., Bharti, R., Grasis, J. A., Cassidy, L., Tholey, A., Fickenscher, H., Seegert, D., et al. (2017). Efficacy of sterile fecal filtrate transfer for treating patients with *Clostridium difficile* infection. *Gastroenterology, 152*(4), 799–811.

Pearson, J. A., Voisey, A. C., Boest-Bjerg, K., Wong, F. S., & Wen, L. (2021). Circadian rhythm modulation of microbes during health and infection. *Frontiers in Microbiology, 12*, 721004. https://doi.org/10.3389/fmicb.2021.721004.

Plovier, H., Everard, A., Druart, C., Depommier, C., Van Hul, M., Geurts, L., et al. (2017). A purified membrane protein from *Akkermansia muciniphila* or the pasteurized bacterium improves metabolism in obese and diabetic mice. *Nature Medicine, 23*(1), 107–113. https://doi.org/10.1038/nm.4236.

Priya, S., & Blekhman, R. (2019). Population dynamics of the human gut microbiome: Change is the only constant. *Genome Biology, 20*(1), 150. https://doi.org/10.1186/s13059-019-1775-3.

Rasmussen, T. S., Mentzel, C. M. J., Kot, W., Castro-Mejía, J. L., Zuffa, S., Swann, J. R., & Nielsen, D. S. (2020). Faecal virome transplantation decreases symptoms of type 2 diabetes and obesity in a murine model. *Gut, 69*(11), 2122–2130. https://doi.org/10.1136/gutjnl-2019-320005.

Reid, G. (2019). The need to focus on therapy instead of associations. *Frontiers in Cellular and Infection Microbiology, 9*, 327. https://doi.org/10.3389/fcimb.2019.00327.

Ribeiro, F. M., Petriz, B., Marques, G., Kamilla, L. H., & Franco, O. L. (2021). Is there an exercise-intensity threshold capable of avoiding the leaky gut? *Frontiers in Nutrition, 8*, 627289. https://doi.org/10.3389/fnut.2021.627289.

Risely, A., Gillingham, M. A. F., Béchet, A., Brändel, S., Heni, A. C., Heurich, M., Menke, S., Manser, M. B., Tschapka, M., Wasimuddin, & Sommer, S. (2021). Phylogeny and abundance-based metrics allow for the consistent comparison of core gut microbiome diversity indices across host species. *Frontiers in Microbiology, 12*, 659918. https://doi.org/10.3389/fmicb.2021.659918.

Rossmann, F. S., et al. (2015). Phage-mediated dispersal of biofilm and distribution of bacterial virulence genes is induced by quorum sensing. *PLoS Pathogens, 11*(3), e1004653. https://doi.org/10.1371/journal.ppat.1004653.

Schneider, K. M., Elfers, C., Ghallab, A., et al. (2021). Intestinal dysbiosis amplifies acetaminophen-induced acute liver injury. *Cellular and Molecular Gastroenterology and Hepatology, 11*(4), 909–933. https://doi.org/10.1016/j.jcmgh.2020.11.002.

Sen, P., Molinero-Perez, A., O'Riordan, K. J., McCafferty, C. P., O'Halloran, K. D., & Cryan, J. F. (2021). Microbiota and sleep: Awakening the gut feeling. *Trends in Molecular Medicine, 27*(12), 1002–1012. https://doi.org/10.1016/j.molmed.2021.07.004.

Shuai, M., Zuo, L., Miao, Z., et al. (2021). Multi-omics analyses reveal relationships among dairy consumption, gut microbiota, and cardiometabolic health. *Ebiomedicine, 66*, 103284. https://doi.org/10.1016/j.ebiom.2021.103284.

Smith, R. P., Easson, C., Lyle, S. M., Kapoor, R., Donnelly, C. P., Davidson, E. J., Parikh, E., Lopez, J. V., & Tartar, J. L. (2019). Gut microbiome diversity is associated with sleep physiology in humans. *PLoS One, 14*(10), e0222394. https://doi.org/10.1371/journal.pone.0222394.

Stavropoulou, E., & Bezirtzoglou, E. (2020). Probiotics in medicine: A long debate. *Frontiers in Immunology, 11*, 2192. https://doi.org/10.3389/fimmu.2020.02192.

Sutcliffe, S. G., Shamash, M., Hynes, A. P., & Maurice, C. F. (2021). Common oral medications lead to prophage induction in bacterial isolates from the human gut. *Viruses, 13*(3), 455. https://doi.org/10.3390/v13030455.

Taguer, M., Shapiro, B. J., & Maurice, C. F. (2021). Translational activity is uncoupled from nucleic acid content in bacterial cells of the human gut microbiota. *Gut Microbes, 13*(1), 1–15. https://doi.org/10.1080/19490976.2021.1903289.

Tannock, G. W. (2021). Modulating the gut microbiota of humans by dietary intervention with plant glycans. *Applied and Environmental Microbiology, 87*(9), e02757–20. https://doi.org/10.1128/AEM.02757-20.

Tatti, P., Tahrani, A., Passali, D., Reutrakul, S., & Kanagasabai, T. (2019). The relationship between disturbed sleep, OSAS, and metabolic diseases. *Journal of Diabetes Research, 2019*, 1463045. https://doi.org/10.1155/2019/1463045.

Thomas, R. L., Jiang, L., Adams, J. S., Xu, Z. Z., Shen, J., Janssen, S., Ackermann, G., Vanderschueren, D., Pauwels, S., Knight, R., Orwoll, E. S., & Kado, D. M. (2020). Vitamin D metabolites and the gut microbiome in older men. *Nature Communications, 11*, 5997. https://doi.org/10.1038/s41467-020-19793-8.

Tuohy, K. M., Probert, H. M., Smejkal, C. W., & Gibson, G. R. (2003). Using probiotics and prebiotics to improve gut health. *Drug Discovery Today, 8*(15), 692–700. https://doi.org/10.1016/S1359-6446(03)02746-6.

Vemuri, R. C., Gundamaraju, R., Shinde, T., & Eri, R. (2017). Therapeutic interventions for gut dysbiosis and related disorders in the elderly: Antibiotics, probiotics, or faecal microbiota transplantation? *Beneficial Microbes, 8*(2), 179–192. https://doi.org/10.3920/BM2016.0115.

von Schwartzenberg, R. J., Bisanz, J. E., Lyalina, S., Spanogiannopoulos, P., Ang, Q. Y., Cai, J., Dickmann, S., Friedrich, M., Liu, S. Y., Collins, S. L., Ingebrigtsen, D., Miller, S., Turnbaugh, J. A., Patterson, A. D., Pollard, K. S., Mai, K., Spranger, J., & Turnbaugh, P. J. (2021). Caloric restriction disrupts the microbiota and colonization resistance. *Nature, 595*(7867), 272–277. https://doi.org/10.1038/s41586-021-03663-4.

Vujkovic-Cvijin, I., Sklar, J., Jiang, L., Natarajan, L., Knight, R., & Belkaid, Y. (2020). Host variables confound gut microbiota studies of human disease. *Nature, 587*(7834), 448–454. https://doi.org/10.1038/s41586-020-2881-9.

Wang, D. D., Nguyen, L. H., Li, Y., et al. (2021). The gut microbiome modulates the protective association between a Mediterranean diet and cardiometabolic disease risk. *Nature Medicine, 27*(2), 333–343. https://doi.org/10.1038/s41591-020-01223-3.

Wilson, A. S., Koller, K. R., Ramaboli, M. C., Nesengani, L. T., Ocvirk, S., Chen, C., Flanagan, C. A., Sapp, F. R., Merritt, Z. T., Bhatti, F., Thomas, T. K., & O'Keefe, S. J. D. (2020). Diet and the human gut microbiome: An international review. *Digestive Diseases and Sciences, 65*(3), 723–740. https://doi.org/10.1007/s10620-020-06112-w.

Yaghoubfar, R., Behrouzi, A., Ashrafian, F., Shahryari, A., Moradi, H. R., Choopani, S., & Siadat, S. D. (2020). Modulation of serotonin signaling/metabolism by *Akkermansia muciniphila* and its extracellular vesicles through the gut-brain axis in mice. *Scientific Reports, 10*, 1–12. https://doi.org/10.1038/s41598-020-79171-8.

Yaghoubfar, R., Behrouzi, A., Banadkoki, E. Z., Ashrafian, F., Lari, A., Vaziri, F., & Siadat, S. D. (2021). Effect of *Akkermansia muciniphila, Faecalibacterium prausnitzii*, and their extracellular vesicles on the serotonin system in intestinal epithelial cells. *Probiotics and Antimicrobial Proteins, 13*, 1546–1556. https://doi.org/10.1007/s12602-021-09786-4.

Zeng, M. Y., Inohara, N., & Nuñez, G. (2017). Mechanisms of inflammation-driven bacterial dysbiosis in the gut. *Mucosal Immunology, 10*(1), 18–26. https://doi.org/10.1038/mi.2016.75.

Zheng, X., Wang, S., & Jia, W. (2018). Calorie restriction and its impact on gut microbial composition and global metabolism. *Frontiers in Medicine, 12*(6), 634–644. https://doi.org/10.1007/s11684-018-0670-8.

Zhou, Z., Chen, X., Sheng, H., Shen, X., Sun, X., Yan, Y., Wang, J., & Yuan, Q. (2020). Engineering probiotics as living diagnostics and therapeutics for improving human health. *Microbial Cell Factories, 19*(1), 1–12. https://doi.org/10.1186/s12934-020-01363-3.

Zou, H., Wang, D., Ren, H., Cai, K., Chen, P., Fang, C., Shi, Z., Zhang, P., Wang, J., Yang, H., & Zhong, H. (2020). Effect of caloric restriction on BMI, gut microbiota, and blood amino acid levels in non-obese adults. *Nutrients, 12*(3), 631. https://doi.org/10.3390/nu12030631.

CHAPTER 4

Prebiotics and Oral Health

RENU M. SITLANI

Sun Pharma Advanced Research Company Limited, Vadodara, Gujarat, India

Nutrition and oral health are closely linked. Tooth decay is directly correlated with sugar consumption, and untreated tooth decay is strongly associated with low BMI in children. Extended periods of micronutrient deficiencies can lead to serious oral symptoms. Under- and malnutrition are co-factors for tooth decay.

4.1 INTRODUCTION

Globalization and urbanization have transformed the food system. Our agriculture, horticulture, storage, preparation, and eating practices have undergone a major revolution. In spite of this, undernutrition and malnutrition still prevail in one section of society, while the other section of society faces diseases like obesity, cardiovascular disease, and diabetes. Over the years, the alliance between nutrition and oral health has been proven. Overconsumption of sweets directly harms teeth; noma is an outcome of malnutrition; tooth decay has been linked to undernutrition. Therefore, maintaining good eating habits is crucial for maintaining oral health.

Prebiotic Foods for a Healthy Gut: Scientific Studies to Manage Metabolic Disorders.
Mini K. Sheth and Shruti Dwivedi (Eds.)

4.2 DENTAL DISEASES AND CONDITIONS

The mouth is a reflection of the body and frequently displays symptoms of underlying illnesses. Nutritional deficiencies and unhealthy habits, such as alcohol or tobacco use, can be identified through a mouth examination. HIV infection may first manifest as oral lesions, and severe eating disorders may be indicated by changes in tooth appearance. The majority of oral diseases have risk factors in common with non-communicable diseases (NCDs), including cancer, diabetes, cardiovascular disease, and respiratory disorders. Unhealthy diet patterns, particularly overuse of sugars, tobacco, and alcohol use, are among high-risk factors. They lead to disparities in the prevalence of oral and general diseases among various population groups. There are different oral health requirements for people of different ages. Therefore, knowledge of the connections between oral and general health as well as collaboration between oral and general health professionals are crucial for providing holistic care.

Tooth decay is triggered by acid production resulting from the breakdown of sugar. Despite the fact that tooth decay is largely preventable through simple and cost-effective interventions, it still remains one of the most common conditions in middle-income nations (Global Burden of Disease Study, 2010).

With the rise in sugar consumption, health systems are unable to provide adequate prevention or access to oral healthcare. Untreated tooth decay has serious effects on quality of life and social interactions, as well as a negative impact on nutrition and growth in children.

- Periodontal (gum) disease: Ignoring routine flossing increases plaque, causing bacteria and leading to inflammation of gums known as gingivitis. Symptoms include bleeding and painful gums, mouth sores, and an increase in gaps between teeth. Tooth loss is commonly seen when the tissues and bones are affected, leading to difficulties in chewing and speaking. Other significant risk factors include smoking, a poor diet, genetic predispositions, stress, and binge drinking. Additionally, systemic diseases like diabetes, cardiovascular disease, unfavorable pregnancy outcomes, and respiratory diseases can all be linked to periodontal disease.
- Cavities and tooth decay: The digestion process starts in the mouth. When sugar comes in contact with saliva, it breaks down and produces acid. Excessive consumption of sugar leads to tooth decay and cavities due to an excess of acid production. Symptoms

include localized pain, gray or brown spots, difficulty in chewing, and swollen gums. Incidences of tooth decay are higher in sections of society where sugar consumption is high and awareness of oral health is limited.

- Oral cancer: Throat, tongue, cheeks, and lips are affected in oral cancer, which is the only fatal oral disease. In simple terms, it is an irreversible oral ulcer. Symptoms include red or white ulcers, localized bleeding, difficulty in swallowing, and swollen lymph nodes. Risk factors include tobacco and alcohol use.
- Oral thrush: Lack of oral hygiene often leads to the growth of harmful Candida fungus in the mouth. This is typically seen during infancy and in immunocompromised individuals. It is characterized by milky white lesions on the tongue and cheeks. Symptoms include reduced appetite, burning sensation during swallowing, white sores, cracks, and irritation on the corners of the mouth.
- Halitosis: Poor dental hygiene along with smoking and respiratory tract infections leads to bad breath known as halitosis. It coexists with periodontitis, and symptoms include thick mucosa, bad breath, bitter taste, and cracked mouth.
- Noma: The soft and hard tissues of the mouth and face are destroyed, progressing quickly. Risk factors for noma include malnutrition, poor oral hygiene, and illnesses like HIV, malaria, and measles. Swollen gums, cheek lining, and ulcers are among the symptoms.

4.3 ORAL HEALTH AS A RISK FACTOR FOR CHRONIC DISEASES

Greater emphasis on promoting good dietary habits and a focus on reducing sugar consumption is essential for the integration of oral health. The risk factors for periodontal disease are similar to those for other serious NCDs, with a particular focus on the use of tobacco and alcohol, excessive sugar intake, obesity, and poor dietary habits. Diabetes and other systemic diseases may also be connected to it. Common gingivitis may develop into severe periodontal disease in 10 to 15% of patients; focus should be placed on identifying this high-risk population before their disease has advanced to the extent that leads to tooth loss.

With effective lifelong self-care and the appropriate kind of professional oral care, tooth loss linked to disease progression can be stopped, just like with all chronic diseases. HIV-related oral lesions can be used to identify

HIV infection, track the course of the disease, gauge immune function, and support prompt therapeutic intervention. Early in the course of the illness, more than half of HIV-positive people experience oral symptoms like fungal, bacterial, and viral infections; severe periodontitis; hairy leukoplakia; warts; dry mouth; Kaposi sarcoma; and lymphoma. These cause discomfort and pain, which makes it difficult to chew, swallow, and taste food. This greatly affects the quality of life. People with HIV/AIDS are still subjected to prejudice and stigma. The management and treatment of oral HIV lesions can significantly enhance well-being and quality of life. Dentists and other oral healthcare professionals can also make sure that patients with oral symptoms are referred for HIV/AIDS testing, receive the proper medical follow-up, and are kept track of for HAART compliance. By performing a quick, easy, and affordable oral examination as part of patient care, healthcare professionals can improve surveillance of oral lesions linked to HIV infection. This could be the first step in identifying, stopping, and treating this potentially fatal illness. Fair and equitable approaches to promoting better oral health are produced by a more comprehensive integrative strategy that takes into account the common risk factors and the primary determinants of health.

4.4 MICROORGANISMS AND ORAL HEALTH

A deeper understanding of the mode of action of oral microbiota helps in developing and prescribing prebiotic solutions to oral health issues:

- *Oral Microbiota in health:* The human mouth has been associated with 700 species; 30 to 100 species can be found in an individual's mouth (Aas et al., 2005, 2008). Certain sites in the mouth are home to a number of microorganisms. A complex climax community is formed when supragingival and subgingival plaques are formed by sequential and particular adhesive interactions (Socransky and Haffajee, 2000; Kolenbrander et al., 2000) which commonly reside on the dorsum of the tongue (Gibbons, 1989; Beighton, 1987) whereas oral bacteria, especially streptococci, are found within buccal epithelial cells (Rudney et al., 2005; Rudney et al., 2005). The resident microbiota does not play a passive role in health maintenance but rather actively contributes to it. The large, diverse resident microbial communities that colonize mucosal sites coexist with a host, with harmful effects occurring only when the host becomes immunocompromised. Gastrointestinal microorganisms prevent pathogen attack to the host (Mead and Barrow, 1990; Roos et al., 2001), by the immune system (Cebra, 1999) and cell structure and function (Hooper et al., 2000; Freitas et al., 2002). Similarly, microorganisms of the oral cavity develop the immune system (Dixon et al., 2004) by modulating adhesion molecule 1 (ICAM-1), E-selectin, and IL-8 (Dixon et al., 2000; Hasegawa et al., 2007; Cosseau et al., 2008). Kilian et al. in 2000 listed *Streptococcus mitis*, *Streptococcus oralis*, *Actinomyces naeslundii*, *Fusobacterium nucleatum*, *Haemophilus parainfluenzae*, *Eikenella corrodens* and some species of *Prevotella*. as 'true' oral commensal microorganisms

- *Microbiota Associated Oral Diseases*: The imbalance of resident microbiota is seen in caries and periodontitis. Different health conditions have different varieties and populations of microorganisms such as in caries, acidogenic or acid-tolerating species whereas obligately anaerobic and proteolytic bacteria which are predominantly Gram-negative are found in periodontal diseases. Mutans and non-mutant streptococci and *Lactobacilli*, *Bifidobacteria*, and species of *Actinomyces, Propionibacterium, Veillonella,* and *Atopobium* are found in caries (Aas et al., 2008; Tanner et al., 2002; Becker et al., 2002). Subgingival biofilms cause damage in periodontal disease, *Candida albicans* and other *Candida* species cause oral candidiasis and denture-associated stomatitis (Redding et al., 2004; Ramage et al., 2004) whereas Gram-negative anaerobes cause halitosis by the production of malodorous metabolic end products (especially volatile sulfur compounds) (Loesche and Kazor, 2000; Kazor et al., 2003).
- *Oral health and gut health alliance*: Oral health can be maintained by good gut health, which in turn can be maintained by providing specific substrates for the growth and metabolism of microorganisms that are beneficial for our health. These organisms maintain the composition of the gut microbiota and lead to the overall well-being of the host. Gastrointestinal microbiota is of utmost importance as it plays a significant role in immunity (Lenoir-Wijnkoop et al., 2007; Steidler et al., 2000). The prevalence of healthy microorganisms depends on pH, peristalsis, nutrient availability, oxidation-reduction potential within the tissue, age of the host, host health, bacterial adhesion, bacterial cooperation, mucin secretions containing Igs, bacterial antagonism, and transit time (Manley et al., 2007).

Different segments of the GI tract have different microbiota and contribute to maintaining health and preventing diseases. The first segment of the GI tract is the stomach, which has the lowest numbers (approximately 10^2 colony-forming units (CFU) per ml of contents) (Vandenbroucke et al., 2004) of microorganisms, namely *Lactobacilli*, *enterococci*, *catenabacteria,* and *bacilli* (Chang et al., 2003). The number of organisms changes with age, such as *Helicobacter pylori*, which causes acute or chronic gastritis, peptic ulcers, and gastric carcinoma, found in an average of 10% in young adulthood and 50% in old age (Manley et al., 2007). The next segment of the GI tract, the small intestine, has three sections: the duodenum, jejunum, and ileum. The bacterial count of the duodenum and jejunum is lower than that of the ileum, as the pH of the duodenum is lower (10^2–10^4 CFU per ml of contents) (Park and Kroll, 1993) than the ileum (10^6–10^8 CFU per ml of contents) (Manley et al., 2007). Most common species found in duodenum and jejunum are *lactobacilli*, *streptococci*, *veillonellae*, *staphylococci*, *actinobacilli* and yeasts (Manley et al., 2007). Small intestine is an important site of action for probiotics which become food for beneficial bacteria and interferes with the growth of diseases causes organisms namely *Salmonella* and *Escherichia coli*. Last segment of the GI tract is the large intestine, which is the highly populated site for organisms owing to an increase in undigested food availability, contact time period, microbial byproducts, and an optimum pH. It is often seen that with an increase in the colony of a particular organism, it can lead to toxin production. It is sometimes difficult to differentiate whether an organism would have a beneficial effect or not, as some of the resident organisms like clostridia and *Bacteroides* can be responsible for cancer, whereas some organisms form SCFA or help in vitamin synthesis. Their mechanism of action is quite interesting and differs from each other; for example, some have a lipopolysaccharide-deficient cell wall, some dominate on receptor sites and food, making it difficult for pathogens to grow. Based on these mechanisms, *bifidobacteria* and *lactobacilli* are classified as beneficial organisms, while *streptococci, enterococci, eubacteria,* and *bacteroides* have both beneficial or harmful effects. The effects of *Firmicutes*, Actinobacteria, and Bacteroidetes are still a mystery.

Microorganisms spend the most time in the colon, where conducive microorganisms exert their beneficial effects through their by-products, production of vitamins and antioxidants, and most importantly, some of them do not produce toxic metabolites as in the case of phytochemicals and prevent pathogens from invading and growing. The food for these beneficial organisms includes complex carbohydrates, fiber, undigested protein, and endogenous sources such as mucins, the main glycoprotein constituents of mucus (Roos et al., 2001). The strict anaerobes present in the colon grow on non-digestible carbohydrates and proteins (Hooper et al., 2000; Freitas et al., 2002). Carbohydrates are available in the initial segment of the colon, whereas proteins are available in the later segment of the colon (Cebra, 1999).

An average of 2–3 g/d of fermentable substrate is derived from mucins and chondroitin sulfate (Rakoff-Nahoum et al., 2004), whereas about 25 g of protein enters the colon daily (Tien et al., 2006). *Bacteroides, Bifidobacterium, Ruminococcus, Eubacterium, Lactobacillus,* and *Clostridium* form the colonic microflora, and *Bacteroides* and *Clostridium* are the main proteolytic species.

The SCFAs produced in the colon by fermentation of carbohydrates are mainly acetate, propionate, and butyrate (Collier-Hyams et al., 2005; Dixon et al., 2004; Dixon et al., 2000) that suppress the growth of pathogens (Cosseau et al., 2008). They also produce energy for the host (Collier-Hyams et al., 2005; Kroes et al., 1999; Diaz et al., 2006). Acetate helps in maintaining cholesterol levels as it is possibly a gluconeogenic substrate and regulates adipose tissue deposition (Kilian et al., 2006; Ledder et al., 2007). Butyrate, on the other hand, stimulates apoptosis and reduces colon cancer risk by regulating cell growth and differentiation (Dixon et al., 2004; Kumar et al., 2003). Rectally administered butyrate was also shown to relieve subjects from inflammatory bowel disease (IBD) symptoms (Stingu et al., 2008). However, some of the by-products of protein or amino acid metabolism produce toxic compounds such as ammonia, amines, and phenolic compounds (Tien et al., 2006) which pose a risk of colon cancer and IBD. Thus, efforts should be taken to prolong saccharolytic fermentation as it is considered beneficial for gut health.

4.5 PREBIOTICS AND ORAL HEALTH

The mouth is home to a diverse, exuberant, and diverse microbial community. Oral bacteria have developed strategies to recognize their surroundings.

Bacteria colonize the ecological niches exposed by the tooth surface and gingival epithelium. An extremely effective innate host defense system, on the other hand, constantly monitors bacterial colonization and keeps bacteria from invading local tissues. There is an imbalance between the bacteria in dental plaque and the innate host defense system. But over the past three decades, the equilibrium of the oral ecology has changed as a result of the increased use of antibiotics. Thus, in the early twentieth century, the concept of bacteriotherapy and the use of health-beneficial microorganisms to heal diseases or support immune function was introduced.

These include nutrients for microbial taxa that either inhibit acidogenic and aciduric microbes or improve pH recovery by producing alkali from these nutrients in terms of caries management. The two primary sources of alkali in the oral cavity are urea and arginine, which are metabolized by some oral bacteria, resulting in ammonia production and an increase in pH (Liu et al., 2012).

The existing evidence supports the health-related benefits of urea and arginine in caries management as discussed further below:

- ***Urea as Anti-Caries Prebiotic:*** Urea is essential for the body's metabolism of nitrogen-containing compounds. It is produced in the liver and excreted by the kidneys as nitrogen. The saliva of healthy people contains 3–10 mmol/L urea (Liu et al., 2012), whereas patients with kidney failure may have much higher levels. Urea can be converted to ammonia (NH_3) or ammonium (NH_4^+) and bicarbonate (HCO_3^-) ions by bacteria that have the enzyme urease. *Streptococcus salivarius* has been shown to have urease activity in the oral cavity. Studies have shown how urea metabolism helps to neutralize acids in plaque. Kesel demonstrated the effects of a mouthwash containing 3% urea and 5% dibasic ammonium phosphate in 1948; it decreased salivary *Lactobacillus* counts and lactic acid production in people with active dental caries and prevented the emergence of new caries lesions. Therefore, it was demonstrated that physiological salivary urea levels can cause a rise in pH in fasting plaque.

  Urea supplementation in oral care products has been discouraged further by reports on the role of urea in calculus formation. In a study, hemodialysis patients had significantly more calculus, higher salivary pH, buffer capacity, and salivary urea concentration than healthy controls. Calcium phosphate precipitation in plaque is caused by a quick pH rise brought on by microbial urea metabolism, according to *in vitro* mineralization studies. Acetohydroxamic acid, a urease inhibitor, was shown to be able to prevent the development of supragingival calculi by Son and Mühlemann in 1971.

Based on the findings of five studies, the European Food Safety Authority (EFSA) came to the conclusion that using sugar-free chewing gum with at least 20 mg of urea per piece for at least 20 minutes after eating or drinking can neutralize plaque acids. After discussing urea, it is obvious that the activity of the enzyme urease in dental plaque is inversely related to dental caries.

- ***Arginine as Anti-Caries Prebiotic:*** The amino acid arginine, which is naturally present in parotid saliva at concentrations of about 50 mol/L, is derived from protein-rich foods and salivary polypeptides. The bacterial arginine deiminase system (ADS) pathway hydrolyzes arginine and converts it to ammonia. Major oral microbiota players like *Streptococcus* sanguinis, *Streptococcus gordonii*, *Streptococcus parasanguinis*, *Streptococcus mitis*, as well as specific strains of *Lactobacilli* and *Actinomyces*, are among the bacteria that test positive for ADS. Additionally, some oral spirochetes that are part of the microbiota linked to periodontal disease have arginolytic activity. An increase in cytoplasmic and environmental pH is observed due to ammonia production driven by arginine metabolism, which provides ecological (pH increase) and bioenergetic benefits (e.g., ATP synthesis) to ADS-positive bacteria. Clinical studies using toothpaste with 8% arginine and fluoride for 4 weeks (Zheng et al., 2017) or 8 weeks (Koopman et al., 2017) improved the arginolytic potential and decreased acidogenicity, while toothpaste with 1.5% arginine and no fluoride increased ADS activity and ecologically advantageous microbial shifts in people with active caries.

Acid-alkali balance should be in equilibrium in a healthy oral ecosystem. Clinically healthy young adults who participated in a recent study that integrated the salivary microbiome, metabolome, and biochemical characteristics of their saliva were divided into five ecological types. The term "adaptive" was used to describe three of the five types (capable of adapting to both proteolytic and saccharolytic conditions and thus in a healthy balance with the host), while the term "specialized" was used to describe two other types (either a saccharolytic or a proteolytic ecological type). An early pro-inflammatory dysbiosis is suggested by high salivary pH and protein-degradation metabolites that are correlated with high microbial diversity and complexity, including proteolytic taxa, associated with gingival or periodontal inflammation. Saliva high in lipid metabolites and low in resting pH, on the other hand, was associated with low microbial diversity and enrichment of saccharolytic taxa, implying an early pro-caries dysbiosis in these individuals. The beneficial effect was seen in the latter group, as pH recovery was improved due to alkali produced from prebiotic supplements (such as urea or arginine). Additional arginine supplementation, on the other hand, had unfavorable effects in people who already had a highly proteolytic microbiome. Individuals who brushed their teeth with 8% arginine-containing toothpaste for 1 or 2 months had a significant increase in proteolytic taxa such as *Treponema*, *Porphyromonas*, and *Prevotella* (Koopman et al., 2017; Zheng et al., 2017).

Beneficial Effects of Prebiotics in Oral Health: Certain oligosaccharides have been shown to promote the growth of resident commensal gut bacteria, particularly *Bifidobacteria* and *Lactobacilli*. Thus, it is assumed that the primary mechanism of action of prebiotics is indirect, i.e., promoting growth of beneficial resident microflora which in turn would exert probiotic effect. Some prebiotics appear to have direct effects on the host, independent of their effects on resident bacterial populations. These include increased IL-10 and interferon gamma expression, IgA secretion, modulated inflammatory responses to pathogens, and stabilizing the gut mucosal barrier. Similarly, xylitol has been used to reduce acid production by mutans streptococci. Algal lectins and cranberry juice have been proposed to reduce oral streptococci adhesion. Cocoa polyphenols can reduce cariogenic bacteria viability and acid production.

There is some evidence supporting the positive effects of boosting *Bifidobacteria* and *Lactobacilli* populations in the gut that has been gathered over many years. According to Koll-Klais et al. (2005), homofermentative *Lactobacilli*, particularly *L. gasseri*, were more common in healthy than in periodontally diseased sites, and their presence was negatively correlated with both subgingival colonization by periodontal pathogens and clinical parameters related to chronic periodontitis. In addition, Hojo et al. (2007) discovered that *L. gasseri, L. salivarius*, and *Lactobacillus* fermentum were more common in healthy sites. The same study discovered that *Bifidobacterium* spp. were isolated from 80% of periodontally healthy subjects and *Bifidobacterium adolescentis* were isolated from 40% of healthy subjects, despite not being the most common in the mouth.

A 2016 study looked at the impact of *fructooligosaccharide* supplementation on school-aged children's oral hygiene as well as changes in their oral and gut microbiota. Children (8–12 years old) from the primary school in urban Vadodara were enrolled in the study and their oral hygiene status was assessed using the Oral Hygiene Index-Simplified (OHI-S). Sixty children with good oral hygiene were purposefully chosen and divided into experimental and control groups. A structured questionnaire was used to collect preliminary data on their basic oral hygiene practices. Standard methods were used to determine oral *Lactobacilli* in saliva and gut *Bifidobacterium* and *Lactobacilli*. The experimental group received FOS (7 g) in buttermilk (150 ml), while the control group received only buttermilk (150 ml). According to the study's findings, most children (41.9%) had fair to poor oral hygiene at the outset. The baseline OHI-S mean score was 2.003. The buttermilk and FOS-added buttermilk intervention improved the mean OHI score by 70% after one month. Gut *Bifidobacteria* and *Lactobacilli* increased as a result of the intervention by 11.5% and 25%, respectively, whereas *E. coli* decreased by 17%. The number of oral *Lactobacilli* rose by 16.03%. A daily dose of 7 g of FOS combined with buttermilk aids in improving clinical signs of oral hygiene in children and improves colonization of beneficial bacteria like oral *Lactobacilli*, gut *Bifidobacteria,* and *Lactobacilli* with a reduction in colonization of *E. coli* (Sheth et al., 2016). Thus, prebiotic therapies that promote the growth of certain *Bifidobacteria* and *Lactobacilli* are anticipated to enhance periodontal health.

KEYWORDS

- **bacteria**
- **cardiovascular diseases**
- **diabetes**
- **diseases**
- **eating disorders**
- **health differences**
- **health risks**
- **HIV infection**
- **proteins**
- **risk factors**

REFERENCES

Aas, J. A., Griffen, A. L., Dardis, S. R., Lee, A. M., Olsen, I., Dewhirst, F. E., et al. (2008). Bacteria of dental caries in primary and permanent teeth in children and young adults.

Journal of Clinical Microbiology, 46(4), 1407–1417. [In Devine, D. A., & Marsh, P. D. (2009). Prospects for the development of probiotics and prebiotics for oral applications. *Journal of Oral Microbiology*, *1*, 1949. https://doi.org/10.3402/jom.v1i0.1949].

Aas, J. A., Paster, B. J., Stokes, L. N., Olsen, I., & Dewhirst, F. E. (2005). Defining the normal bacterial flora of the oral cavity. *Journal of Clinical Microbiology, 43*(11), 5721–5732. [In Devine, D. A., & Marsh, P. D. (2009). Prospects for the development of probiotics and prebiotics for oral applications. *Journal of Oral Microbiology*, *1*, 1949. https://doi.org/10.3402/jom.v1i0.1949].

Andrade, M. R., Salazar, S. L., de Sá, L. F., Portela, M., Ferreira-Pereira, A., & Soares, R. M. (2015). Role of saliva in the caries experience and calculus formation of young patients undergoing hemodialysis. *Clinical Oral Investigations, 19*(8), 1973–1980. [In Zaura, E., & Twetman, S. (2019). Critical appraisal of oral pre- and probiotics for caries prevention and care. *Caries Research, 53*(5), 514–526. https://doi.org/10.1159/000499639].

Becker, M. R., Paster, B. J., Leys, E. J., Moeschberger, M. L., Kenyon, S. G., Galvin, J. L., et al. (2002). Molecular analysis of bacterial species associated with childhood caries. *Journal of Clinical Microbiology, 40*(3), 1001–1009. [In Devine, D. A., & Marsh, P. D. (2009). Prospects for the development of probiotics and prebiotics for oral applications. *Journal of Oral Microbiology*, *1*, 1949. https://doi.org/10.3402/jom.v1i0.1949].

Beighton, D., Rippon, H. R., & Thomas, H. E. C. (1987). The distribution of Streptococcus mutans serotypes and dental caries in a group of 5-year-old to 8-year-old Hampshire schoolchildren. *British Dental Journal, 162*(4), 103–106. [In Devine, D. A., & Marsh, P. D. (2009). Prospects for the development of probiotics and prebiotics for oral applications. *Journal of Oral Microbiology*, *1*, 1949. https://doi.org/10.3402/jom.v1i0.1949].

Beighton, D., Smith, K., & Hayday, H. (1986). The growth of bacteria and the production of exoglycosidic enzymes in the dental plaque of macaque monkeys. *Archives of Oral Biology, 31*(12), 829–835. [In Jiménez-Hernández, N., Serrano-Villar, S., Domingo, A., Pons, X., Artacho, A., Estrada, V., Moya, A., & Gosalbes, M. J. (2019). Modulation of saliva microbiota through prebiotic intervention in HIV-infected individuals. *Nutrients, 11*(6), 1346. https://doi.org/10.3390/nu11061346].

Bhardwaj, A., & Bhardwaj, S. V. (2012). Role of probiotics in dental caries and periodontal disease. *Archives of Clinical and Experimental Surgery, 1*(1), 45–49. [In Pandya, D. (2016). Benefits of probiotics in the oral cavity–A detailed review. *Annals of International Medical and Dental Research, 2*(5), 10–17].

Bhushan, J., & Chachra, S. (2010). Probiotics–Their role in prevention of dental caries. *Journal of Oral Health and Community Dentistry, 4*(3), 78–82. [In Pandya, D. (2016). Benefits of probiotics in the oral cavity–A detailed review. *Annals of International Medical and Dental Research, 2*(5), 10–17].

Burne, R. A., & Marquis, R. E. (2000). Alkali production by oral bacteria and protection against dental caries. *FEMS Microbiology Letters, 193*(1), 1–6. [In Zaura, E., & Twetman, S. (2019). Critical appraisal of oral pre- and probiotics for caries prevention and care. *Caries Research, 53*(5), 514–526. https://doi.org/10.1159/000499639].

Caglar, E., Kavaloglu, S. C., Kuscu, O. O., Sandalli, N., Holgerson, P. L., & Twetman, S. (2007). Effect of chewing gums containing xylitol or probiotic bacteria on salivary mutans streptococci and lactobacilli. *Clinical Oral Investigations, 11*(4), 425–429. [In Jiménez-Hernández, N., Serrano-Villar, S., Domingo, A., Pons, X., Artacho, A., Estrada, V., Moya, A., & Gosalbes, M. J. (2019). Modulation of saliva microbiota through prebiotic intervention in HIV-infected individuals. *Nutrients, 11*(6), 1346. https://doi.org/10.3390/nu11061346].

Cebra, J. J. (1999). Influences of microbiota on intestinal immune system development. *American Journal of Clinical Nutrition, 69*(5), 1046S–1051S. [In Devine, D. A., & Marsh, P. D. (2009). Prospects for the development of probiotics and prebiotics for oral applications. *Journal of Oral Microbiology, 1*, 1949. https://doi.org/10.3402/jom.v1i0.1949].

Chang, T. L. Y., Chang, C. H., Simpson, D. A., Xu, Q., Martin, P. K., Lagenaur, L. A., et al. (2003). Inhibition of HIV infectivity by a natural human isolate of Lactobacillus jensenii engineered to express functional two-domain CD4. *Proceedings of the National Academy of Sciences, 100*(20), 11672–11677. [In Devine, D. A., & Marsh, P. D. (2009). Prospects for the development of probiotics and prebiotics for oral applications. *Journal of Oral Microbiology, 1*, 1949. https://doi.org/10.3402/jom.v1i0.1949].

Collier-Hyams, L. S., Sloane, V., Batten, B. C., & Neish, A. S. (2005). Cutting edge: Bacterial modulation of epithelial signaling via changes in neddylation of cullin-1. *Journal of Immunology, 175*(8), 4194–4198. [In Devine, D. A., & Marsh, P. D. (2009). Prospects for the development of probiotics and prebiotics for oral applications. *Journal of Oral Microbiology, 1*, 1949. https://doi.org/10.3402/jom.v1i0.1949].

Cosseau, C., Devine, D. A., Dullaghan, E., Gardy, J. L., Chikatmarla, A., Gellatly, S., et al. (2008). The commensal Streptococcus salivarius down-regulates immune responses of human epithelial cells and promotes host-microbe homeostasis. *Infection and Immunity, 76*(9), 4163–4175. [In Devine, D. A., & Marsh, P. D. (2009). Prospects for the development of probiotics and prebiotics for oral applications. *Journal of Oral Microbiology, 1*, 1949. https://doi.org/10.3402/jom.v1i0.1949].

Devine, D. A., & Marsh, P. D. (2009). Prospects for the development of probiotics and prebiotics for oral applications. *Journal of Oral Microbiology, 1*(1), 1949. [In Jiménez-Hernández, N., Serrano-Villar, S., Domingo, A., Pons, X., Artacho, A., Estrada, V., Moya, A., & Gosalbes, M. J. (2019). Modulation of saliva microbiota through prebiotic intervention in HIV-infected individuals. *Nutrients, 11*(6), 1346. https://doi.org/10.3390/nu11061346].

Diaz, P. I., Chalmers, N. I., Rickard, A. H., Kong, C., Milburn, C. L., Palmer, R. J., Jr., et al. (2006). Molecular characterization of subject-specific oral microflora during initial colonization of enamel. *Applied and Environmental Microbiology, 72*(4), 2837–2848. [In Devine, D. A., & Marsh, P. D. (2009). Prospects for the development of probiotics and prebiotics for oral applications. *Journal of Oral Microbiology, 1*, 1949. https://doi.org/10.3402/jom.v1i0.1949].

Dixon, D. R., Bainbridge, B. W., & Darveau, R. P. (2004). Modulation of the innate immune response within the periodontium. *Periodontology 2000, 35*(1), 53–74. [In Devine, D. A., & Marsh, P. D. (2009). Prospects for the development of probiotics and prebiotics for oral applications. *Journal of Oral Microbiology, 1*, 1949. https://doi.org/10.3402/jom.v1i0.1949].

Dixon, D. R., Reife, R. A., Cebra, J. J., & Darveau, R. P. (2004). Commensal bacteria influence innate status within gingival tissues: A pilot study. *Journal of Periodontology, 75*(11), 1486–1492. [In Devine, D. A., & Marsh, P. D. (2009). Prospects for the development of probiotics and prebiotics for oral applications. *Journal of Oral Microbiology, 1*, 1949. https://doi.org/10.3402/jom.v1i0.1949].

Dixon, D. R., Reife, R. A., Cebra, J. J., & Darveau, R. P. (2004). Commensal bacteria influence innate status within gingival tissues: A pilot study. *Journal of Periodontology, 75*(11), 1486–1492. [In Devine, D. A., & Marsh, P. D. (2009). Prospects for the development of probiotics and prebiotics for oral applications. *Journal of Oral Microbiology, 1*, 1949. https://doi.org/10.3402/jom.v1i0.1949].

EFSA Panel on Dietetic Products, Nutrition and Allergies. (2011). Scientific opinion on the substantiation of health claims related to sugar-free chewing gum with carbamide and plaque acid neutralisation (ID 1153) pursuant to Article 13(1) of Regulation (EC) No 1924/2006. *EFSA Journal, 9*(4), 2071–2085. [In Zaura, E., & Twetman, S. (2019). Critical appraisal of oral pre- and probiotics for caries prevention and care. *Caries Research, 53*(5), 514–526. https://doi.org/10.1159/000499639].

FDI World Dental Federation. (2015). *The Challenge of Oral Disease–A Call for Global Action. The Oral Health Atlas* (2nd ed.). Geneva: FDI World Dental Federation.

Forchielli, M. L., & Walker, W. A. (2005). The role of gut-associated lymphoid tissues and mucosal defence. *British Journal of Nutrition, 93*(S1), S41–S48. [In Jiménez-Hernández, N., Serrano-Villar, S., Domingo, A., Pons, X., Artacho, A., Estrada, V., Moya, A., & Gosalbes, M. J. (2019). Modulation of saliva microbiota through prebiotic intervention in HIV-infected individuals. *Nutrients, 11*(6), 1346. https://doi.org/10.3390/nu11061346].

Freitas, M., Axelsson, L. G., Cayuela, C., Midtvedt, T., & Trugnan, G. (2002). Microbial-host interactions specifically control the glycosylation pattern in intestinal mouse mucosa. *Histochemistry and Cell Biology, 118*(2), 149–161. [In Devine, D. A., & Marsh, P. D. (2009). Prospects for the development of probiotics and prebiotics for oral applications. *Journal of Oral Microbiology, 1*, 1949. https://doi.org/10.3402/jom.v1i0.1949].

Gibbons, R. J. (1989). Bacterial adhesion to oral tissues–A model for infectious diseases. *Journal of Dental Research, 68*(5), 750–760. [In Devine, D. A., & Marsh, P. D. (2009). Prospects for the development of probiotics and prebiotics for oral applications. *Journal of Oral Microbiology, 1*, 1949. https://doi.org/10.3402/jom.v1i0.1949].

Gibson, G. R., McCartney, A. L., & Rastall, R. A. (2005). Prebiotics and resistance to gastrointestinal infections. *British Journal of Nutrition, 93*(S1), S31–S34. [In Jiménez-Hernández, N., Serrano-Villar, S., Domingo, A., Pons, X., Artacho, A., Estrada, V., Moya, A., & Gosalbes, M. J. (2019). Modulation of saliva microbiota through prebiotic intervention in HIV-infected individuals. *Nutrients, 11*(6), 1346. https://doi.org/10.3390/nu11061346].

Hasegawa, Y., Mans, J. J., Mao, S., Lopez, M. C., Baker, H. V., & Handfield, M., et al. (2007). Gingival epithelial cell transcriptional responses to commensal and opportunistic oral microbial species. *Infection and Immunity, 75*(5), 2540–2547. [In Devine, D. A., & Marsh, P. D. (2009). Prospects for the development of probiotics and prebiotics for oral applications. *Journal of Oral Microbiology, 1*, 1949. https://doi.org/10.3402/jom.v1i0.1949].

Hojo, K., Mizoguchi, C., Taketomo, N., Ohshima, T., Gomi, K., Arai, T., et al. (2007). Distribution of salivary Lactobacillus and Bifidobacterium species in periodontal health and disease. *Bioscience, Biotechnology, and Biochemistry, 71*(1), 152–157. [In Jiménez-Hernández, N., Serrano-Villar, S., Domingo, A., Pons, X., Artacho, A., Estrada, V., Moya, A., & Gosalbes, M. J. (2019). Modulation of saliva microbiota through prebiotic intervention in HIV-infected individuals. *Nutrients, 11*(6), 1346. https://doi.org/10.3390/nu11061346].

Hooper, L. V., Falk, P. G., & Gordon, J. I. (2000). Analyzing the molecular foundations of commensalism in the mouse intestine. *Current Opinion in Microbiology, 3*(1), 79–85. [In Devine, D. A., & Marsh, P. D. (2009). Prospects for the development of probiotics and prebiotics for oral applications. *Journal of Oral Microbiology, 1*, 1949. https://doi.org/10.3402/jom.v1i0.1949].

Kanapka, J. A., & Kleinberg, I. (1983). Catabolism of arginine by the mixed bacteria in human salivary sediment under conditions of low and high glucose concentration. *Archives of Oral Biology, 28*(11), 1007–1015. [In Zaura, E., & Twetman, S. (2019). Critical appraisal of oral pre- and probiotics for caries prevention and care. *Caries Research, 53*(5), 514–526. https://doi.org/10.1159/000499639].

Kazor, C. E., Mitchell, P. M., Lee, A. M., Stokes, L. N., Loesche, W. J., Dewhirst, F. E., et al. (2003). Diversity of bacterial populations on the tongue dorsa of patients with halitosis and healthy patients. *Journal of Clinical Microbiology, 41*(2), 558–563. [In Devine, D. A., & Marsh, P. D. (2009). Prospects for the development of probiotics and prebiotics for oral applications. *Journal of Oral Microbiology, 1*, 1949. https://doi.org/10.3402/jom.v1i0.1949].

Kesel, R. G. (1948). The effectiveness of dentifrices, mouthwashes, and ammonia-urea compounds in the control of dental caries. *Journal of Dental Research, 27*(2), 244–258. [In Zaura, E., & Twetman, S. (2019). Critical appraisal of oral pre- and probiotics for caries prevention and care. *Caries Research, 53*(5), 514–526. https://doi.org/10.1159/000499639].

Kilian, M., Frandsen, E. V. G., Haubek, D., & Poulsen, K. (2006). The etiology of periodontal disease revisited by population genetic analysis. *Periodontology 2000, 42*(1), 158–179. [In Devine, D. A., & Marsh, P. D. (2009). Prospects for the development of probiotics and prebiotics for oral applications. *Journal of Oral Microbiology, 1*, 1949. https://doi.org/10.3402/jom.v1i0.1949].

Kleessen, B., & Blaut, M. (2005). Modulation of gut mucosal biofilms. *British Journal of Nutrition, 93*(S1), S35–S40. [In Jiménez-Hernández, N., Serrano-Villar, S., Domingo, A., Pons, X., Artacho, A., Estrada, V., Moya, A., & Gosalbes, M. J. (2019). Modulation of saliva microbiota through prebiotic intervention in HIV-infected individuals. *Nutrients, 11*(6), 1346. https://doi.org/10.3390/nu11061346].

Kleinberg, I. (1967). Effect of urea concentration on human plaque pH levels in situ. *Archives of Oral Biology, 12*(12), 1475–1484. [In Zaura, E., & Twetman, S. (2019). Critical appraisal of oral pre- and probiotics for caries prevention and care. *Caries Research, 53*(5), 514–526. https://doi.org/10.1159/000499639].

Kleinberg, I. (1999). A new saliva-based anticaries composition. *Dentistry Today, 18*(2), 98–103. [In Zaura, E., & Twetman, S. (2019). Critical appraisal of oral pre- and probiotics for caries prevention and care. *Caries Research, 53*(5), 514–526. https://doi.org/10.1159/000499639].

Kolenbrander, P. E., Palmer, R. J., Rickard, A. H., Jakubovics, N. S., Chalmers, N. I., & Diaz, P. I. (2006). Bacterial interactions and successions during plaque development. *Periodontology 2000, 42*(1), 47–79. [In Devine, D. A., & Marsh, P. D. (2009). Prospects for the development of probiotics and prebiotics for oral applications. *Journal of Oral Microbiology, 1*, 1949. https://doi.org/10.3402/jom.v1i0.1949].

Koll-Klais, P., Mandar, R., Leibur, E., Marcotte, H., Hammarstrom, L., & Mikelsaar, M. (2005). Oral lactobacilli in chronic periodontitis and periodontal health: Species composition and antimicrobial activity. *Oral Microbiology and Immunology, 20*(6), 354–361. [In Jiménez-Hernández, N., Serrano-Villar, S., Domingo, A., Pons, X., Artacho, A., Estrada, V., Moya, A., & Gosalbes, M. J. (2019). Modulation of saliva microbiota through prebiotic intervention in HIV-infected individuals. *Nutrients, 11*(6), 1346. https://doi.org/10.3390/nu11061346].

Koopman, J. E., Hoogenkamp, M. A., Buijs, M. J., Brandt, B. W., Keijser, B. J., Crielaard, W., et al. (2017). Changes in the oral ecosystem induced by the use of 8% arginine toothpaste. *Archives of Oral Biology, 73*, 79–87. [In Zaura, E., & Twetman, S. (2019). Critical appraisal of oral pre- and probiotics for caries prevention and care. *Caries Research, 53*(5), 514–526. https://doi.org/10.1159/000499639].

Kroes, I., Lepp, P. W., & Relman, D. A. (1999). Bacterial diversity within the human subgingival crevice. *Proceedings of the National Academy of Sciences, 96*(25), 14547–14552. [In Devine, D. A., & Marsh, P. D. (2009). Prospects for the development of probiotics and prebiotics for oral applications. *Journal of Oral Microbiology, 1*, 1949. https://doi.org/10.3402/jom.v1i0.1949].

Kumar, P. S., Griffen, A. L., Barton, J. A., Paster, B. J., Moeschberger, M. L., & Leys, E. J. (2003). New bacterial species associated with chronic periodontitis. *Journal of Dental Research, 82*(5), 338–344. [In Devine, D. A., & Marsh, P. D. (2009). Prospects for the development of probiotics and prebiotics for oral applications. *Journal of Oral Microbiology, 1*, 1949. https://doi.org/10.3402/jom.v1i0.1949].

Ledder, R. G., Gilbert, P., Huws, S. A., Aarons, L., Ashley, M. P., Hull, P. S., et al. (2007). Molecular analysis of the subgingival microbiota in health and disease. *Applied and Environmental Microbiology, 73*(2), 516–523. [In Devine, D. A., & Marsh, P. D. (2009). Prospects for the development of probiotics and prebiotics for oral applications. *Journal of Oral Microbiology, 1*, 1949. https://doi.org/10.3402/jom.v1i0.1949].

Lenoir-Wijnkoop, I., Sanders, M. E., Cabana, M. D., Caglar, E., Corthier, G., Rayes, N., et al. (2007). Probiotic and prebiotic influence beyond the intestinal tract. *Nutrition Reviews, 65*(11), 469–489. [In Devine, D. A., & Marsh, P. D. (2009). Prospects for the development of probiotics and prebiotics for oral applications. *Journal of Oral Microbiology, 1*, 1949. https://doi.org/10.3402/jom.v1i0.1949].

Liu, Y. L., Nascimento, M. M., & Burne, R. A. (2012). Progress toward understanding the contribution of alkali generation in dental biofilms to inhibition of dental caries. *International Journal of Oral Science, 4*(3), 135–140. [In Zaura, E., & Twetman, S. (2019). Critical appraisal of oral pre- and probiotics for caries prevention and care. *Caries Research, 53*(5), 514–526. https://doi.org/10.1159/000499639].

Loesche, W. J., & Kazor, C. (2002). Microbiology and treatment of halitosis. *Periodontology 2000, 28*(1), 256–279. [In Devine, D. A., & Marsh, P. D. (2009). Prospects for the development of probiotics and prebiotics for oral applications. *Journal of Oral Microbiology, 1*, 1949. https://doi.org/10.3402/jom.v1i0.1949].

Manley, K. J., Fraenkel, M. B., Mayall, B. C., & Power, D. A. (2007). Probiotic treatment of vancomycin-resistant enterococci: A randomized controlled trial. *Medical Journal of Australia, 186*(9), 454–457. [In Devine, D. A., & Marsh, P. D. (2009). Prospects for the development of probiotics and prebiotics for oral applications. *Journal of Oral Microbiology, 1*, 1949. https://doi.org/10.3402/jom.v1i0.1949].

Marquis, R. E., Burne, R. A., Parsons, D. T., & Casiano-Colon, A. E. (1993). Arginine deiminase and alkali generation in plaque. In W. H. Bowen & L. A. Tabak (Eds.), *Cariology for the Nineties* (pp. 309–317). University of Rochester Press. [In Zaura, E., & Twetman, S. (2019). Critical appraisal of oral pre- and probiotics for caries prevention and care. *Caries Research, 53*(5), 514–526. https://doi.org/10.1159/000499639].

Marsh, P. D., Moter, A., & Devine, D. A. (2011). Dental plaque biofilms: Communities, conflict and control. *Periodontology 2000, 55*(1), 16–35. [In Zaura, E., & Twetman, S. (2019). Critical appraisal of oral pre- and probiotics for caries prevention and care. *Caries Research, 53*(5), 514–526. https://doi.org/10.1159/000499639].

Mead, G. C., & Barrow, P. A. (1990). Salmonella control in poultry by competitive-exclusion or immunization. *Letters in Applied Microbiology, 10*(6), 221–227. [In Devine, D. A., & Marsh, P. D. (2009). Prospects for the development of probiotics and prebiotics for oral applications. *Journal of Oral Microbiology, 1*, 1949. https://doi.org/10.3402/jom.v1i0.1949].

Munjal, S., & Munjal, S. (2014). Probiotics in oral health promotion: The new approach. *International Journal of Biomedical and Advance Research, 4*(1), 204–209. [In Pandya, D. (2016). Benefits of probiotics in oral cavity–A detailed review. *Annals of International Medical and Dental Research, 2*(5), 10–17].

Narwal, A. (2011). Probiotics in dentistry – A review. *Journal of Nutrition & Food Sciences, 1*(5), 1–4. [In Pandya, D. (2016). Benefits of probiotics in oral cavity–A detailed review. *Annals of International Medical and Dental Research, 2*(5), 10–17].

Nascimento, M. M. (2018). Potential uses of arginine in dentistry. *Advances in Dental Research, 29*(1), 98–103. In Zaura, E., & Twetman, S. (2019). Critical appraisal of oral pre- and probiotics for caries prevention and care. *Caries Research, 53*(5), 514–526. https://doi.org/10.1159/000499639].

Nascimento, M. M., Browngardt, C., Xiaohui, X., Klepac-Ceraj, V., Paster, B. J., & Burne, R. A. (2014). The effect of arginine on oral biofilm communities. *Molecular Oral Microbiology, 29*(1), 45–54. [In Zaura, E., & Twetman, S. (2019). Critical appraisal of oral pre- and probiotics for caries prevention and care. *Caries Research, 53*(5), 514–526. https://doi.org/10.1159/000499639].

Pandya, D. (2016). Benefits of probiotics in the oral cavity–A detailed review. *Annals of International Medical and Dental Research, 2*(5), 10–17.

Park, S. F., & Kroll, R. G. (1993). Expression of listeriolysin and phosphatidylinositol-specific phospholipase-C is repressed by the plant-derived molecule cellobiose in *Listeria monocytogenes*. *Molecular Microbiology, 8*, 653–661. [In Devine, D. A., & Marsh, P. D. (2009). Prospects for the development of probiotics and prebiotics for oral applications. *Journal of Oral Microbiology, 1*, 1949. https://doi.org/10.3402/jom.v1i0.1949].

Parwani, S. R., & Parwani, R. N. (2015). Nitric oxide and inflammatory periodontal disease. *General Dentistry, 63*(2), 34–40. [In Zaura, E., & Twetman, S. (2019). Critical appraisal of oral pre- and probiotics for caries prevention and care. *Caries Research, 53*(5), 514–526. https://doi.org/10.1159/000499639].

Percival, R. S., Devine, D. A., Duggal, M. S., Chartron, S., & Marsh, P. D. (2006). The effect of cocoa polyphenols on the growth, metabolism, and biofilm formation by *Streptococcus mutans* and *Streptococcus sanguinis*. *European Journal of Oral Sciences, 114*(5), 343–348. [In Jiménez-Hernández, N., Serrano-Villar, S., Domingo, A., Pons, X., Artacho, A., Estrada, V., Moya, A., & Gosalbes, M. J. (2019). Modulation of saliva microbiota through prebiotic intervention in HIV-infected individuals. *Nutrients, 11*(6), 1346. https://doi.org/10.3390/nu11061346].

Rakoff-Nahoum, S., Paglino, J., Eslami-Varzaneh, F., Edberg, S., & Medzhitov, R. (2004). Recognition of commensal microflora by toll-like receptors is required for intestinal homeostasis. *Cell, 118*(2), 229–241. [In Devine, D. A., & Marsh, P. D. (2009). Prospects for the development of probiotics and prebiotics for oral applications. *Journal of Oral Microbiology, 1*, 1949. https://doi.org/10.3402/jom.v1i0.1949].

Ramage, G., Tomsett, K., Wickes, B. L., Lopez-Ribot, J. L., & Redding, S. W. (2004). Denture stomatitis: a role for *Candida* biofilms. *Oral Surgery, Oral Medicine, Oral Pathology, Oral Radiology, and Endodontics, 98*(1), 53–59. [In Devine, D. A., & Marsh, P. D. (2009). Prospects for the development of probiotics and prebiotics for oral applications. *Journal of Oral Microbiology, 1*, 1949. https://doi.org/10.3402/jom.v1i0.1949].

Rangare, A. L., Umar, D., Basheer, B., & Baroudi, K. (2015). The magic of magic bugs in oral cavity: Probiotics. *Journal of Advanced Pharmaceutical Technology & Research, 6*(2), 43–47. In Pandya, D. (2016). Benefits of probiotics in oral cavity–A detailed review. *Annals of International Medical and Dental Research, 2*(5), 10–17].

Redding, S. W., Dahiya, M. C., Kirkpatrick, W. R., Coco, B. J., Patterson, T. F., & Fothergill, A. W. (2004). *Candida glabrata* is an emerging cause of oropharyngeal candidiasis in patients receiving radiation for head and neck cancer. *Oral Surgery, Oral Medicine, Oral Pathology, Oral Radiology, and Endodontics, 97*(1), 47–52. [In Devine, D. A., & Marsh, P. D. (2009). Prospects for the development of probiotics and prebiotics for oral applications. *Journal of Oral Microbiology, 1*, 1949. https://doi.org/10.3402/jom.v1i0.1949].

Roos, K., Hakansson, E. G., & Holm, S. (2001). Effect of recolonization with "interfering" alpha streptococci on recurrences of acute and secretory otitis media in children: Randomized placebo-controlled trial. *British Medical Journal, 322*(7280), 210–212. [In Devine, D. A., & Marsh, P. D. (2009). Prospects for the development of probiotics and prebiotics for oral applications. *Journal of Oral Microbiology, 1*, 1949. https://doi.org/10.3402/jom.v1i0.1949].

Rudney, J. D., Chen, R., & Sedgewick, G. J. (2005). *Actinobacillus actinomycetemcomitans, Porphyromonas gingivalis*, and *Tannerella forsythensis* are components of a polymicrobial intracellular flora within human buccal cells. *Journal of Dental Research, 84*(1), 59–63. In Devine, D. A., & Marsh, P. D. (2009). Prospects for the development of probiotics and prebiotics for oral applications. *Journal of Oral Microbiology, 1*, 1949. https://doi.org/10.3402/jom.v1i0.1949].

Rudney, J. D., Chen, R., & Zhang, G. (2005). Streptococci dominate the diverse flora within buccal cells. *Journal of Dental Research, 84*(12), 1165–1171. [In Devine, D. A., & Marsh, P. D. (2009). Prospects for the development of probiotics and prebiotics for oral applications. *Journal of Oral Microbiology, 1*, 1949. https://doi.org/10.3402/jom.v1i0.1949].

Sheth, M., & Hirdyani, H. (2016). Fructooligosaccharide intake affects the colonization of *Streptococcus mitis* establishment in oral cavity along with gut bacteria in young school children of urban Vadodara. *World Journal of Pharmaceutical Research, 5*(4), 1266–1276. https://doi.org/10.20959/wjpr20164-5941.

Sissons, C. H., Cutress, T. W., Hoffman, M. P., & Wakefield, J. S. (1991). A multi-station dental plaque microcosm (artificial mouth) for the study of plaque growth, metabolism, pH, and mineralization. *Journal of Dental Research, 70*(11), 1409–1416. [In Zaura, E., & Twetman, S. (2019). Critical appraisal of oral pre- and probiotics for caries prevention and care. *Caries Research, 53*(5), 514–526. https://doi.org/10.1159/000499639].

Socransky, S. S., & Haffajee, A. D. (2005). Periodontal microbial ecology. *Periodontology 2000, 38*(1), 135–187. [In Devine, D. A., & Marsh, P. D. (2009). Prospects for the development of probiotics and prebiotics for oral applications. *Journal of Oral Microbiology, 1*, 1949. https://doi.org/10.3402/jom.v1i0.1949].

Son, S., & Mühlemann, H. R. (1971). The effect of human supragingival calculus formation of acetohydroxamic acid. *Helvetica Odontologica Acta, 15*(2 Suppl 7), 158–159. [In Zaura, E., & Twetman, S. (2019). Critical appraisal of oral pre- and probiotics for caries prevention and care. *Caries Research, 53*(5), 514–526. https://doi.org/10.1159/000499639].

Steidler, L., Hans, W., Schotte, L., Neirynck, S., Obermeier, F., Falk, W., & Remaut, E. (2000). Treatment of murine colitis by *Lactococcus lactis* secreting interleukin-10. *Science, 289*(5483), 1352–1355. [In Devine, D. A., & Marsh, P. D. (2009). Prospects for the development of probiotics and prebiotics for oral applications. *Journal of Oral Microbiology, 1*, 1949. https://doi.org/10.3402/jom.v1i0.1949].

Stephan, R. M. (1943). The effect of urea in counteracting the influence of carbohydrates on the pH of dental plaques. *Journal of Dental Research, 22*(1), 63–71. [In Zaura, E., & Twetman, S. (2019). Critical appraisal of oral pre- and probiotics for caries prevention and care. *Caries Research, 53*(5), 514–526. https://doi.org/10.1159/000499639].

Stingu, C. S., Eschrich, K., Rodloff, A. C., Schaumann, R., & Jentsch, H. (2008). Periodontitis is associated with a loss of colonization by *Streptococcus sanguinis*. *Journal of Medical Microbiology, 57*(4), 495–499. [In Devine, D. A., & Marsh, P. D. (2009). Prospects for the development of probiotics and prebiotics for oral applications. *Journal of Oral Microbiology, 1*, 1949. https://doi.org/10.3402/jom.v1i0.1949].

Tanner, A. C. R., Milgrom, P. M., Kent, R., Mokeem, S. A., Page, R. C., Riedy, C. A., ... & Stoltenberg, J. L. (2002). The microbiota of young children from tooth and tongue samples. *Journal of Dental Research, 81*(1), 53–57. [In Devine, D. A., & Marsh, P. D. (2009). Prospects for the development of probiotics and prebiotics for oral applications. *Journal of Oral Microbiology, 1*, 1949. https://doi.org/10.3402/jom.v1i0.1949].

Teixeira, E. H., Napimoga, M. H., Carneiro, V. A., de Oliveira, T. M., Nascimento, K. S., Nagano, C. S., & Cavada, B. S. (2007). In vitro inhibition of oral streptococci binding to the acquired pellicle by algal lectins. *Journal of Applied Microbiology, 103*(3), 1001–1006. [In Jiménez-Hernández, N., Serrano-Villar, S., Domingo, A., Pons, X., Artacho, A., Estrada, V., Moya, A., & Gosalbes, M. J. (2019). Modulation of saliva microbiota through prebiotic intervention in HIV-infected individuals. *Nutrients, 11*(6), 1346. https://doi.org/10.3390/nu11061346].

Tien, M. T., Girardin, S. E., Regnault, B., Le Bourhis, L., Dillies, M. A., Coppee, J. Y., & Sansonetti, P. J. (2006). Anti-inflammatory effect of *Lactobacillus casei* on *Shigella*-infected human intestinal epithelial cells. *Journal of Immunology, 176*(2), 1228–1237. [In Devine, D. A., & Marsh, P. D. (2009). Prospects for the development of probiotics and prebiotics for oral applications. *Journal of Oral Microbiology, 1*, 1949. https://doi.org/10.3402/jom.v1i0.1949].

Vandenbroucke, K., Hans, W., Van Huysse, J., Neirynck, S., Demetter, P., Remaut, E., & Rottiers, P. (2004). Active delivery of trefoil factors by genetically modified *Lactococcus lactis* prevents and heals acute colitis in mice. *Gastroenterology, 127*(2), 502–513. [In Devine, D. A., & Marsh, P. D. (2009). Prospects for the development of probiotics and prebiotics for oral applications. *Journal of Oral Microbiology, 1*, 1949. https://doi.org/10.3402/jom.v1i0.1949].

Wong, L., Sissons, C. H., Pearce, E. I. F., & Cutress, T. W. (2002). Calcium phosphate deposition in human dental plaque microcosm biofilms induced by a ureolytic pH-rise procedure. *Archives of Oral Biology, 47*(11), 779–790. [In Zaura, E., & Twetman, S. (2019). Critical appraisal of oral pre- and probiotics for caries prevention and care. *Caries Research, 53*(5), 514–526. https://doi.org/10.1159/000499639].

Yamanaka, A., Kimizuka, R., Kato, T., & Okuda, K. (2004). Inhibitory effects of cranberry juice on attachment of oral streptococci and biofilm formation. *Oral Microbiology and Immunology, 19*(3), 150–154. [In Jiménez-Hernández, N., Serrano-Villar, S., Domingo, A., Pons, X., Artacho, A., Estrada, V., Moya, A., & Gosalbes, M. J. (2019). Modulation of saliva microbiota through prebiotic intervention in HIV-infected individuals. *Nutrients, 11*(6), 1346. https://doi.org/10.3390/nu11061346].

Zaura, E., Brandt, B. W., Prodan, A., Teixeira de Mattos, M. J., Imangaliyev, S., Kool, J., & Herpers, B. (2017). On the ecosystemic network of saliva in healthy young adults. *ISME Journal, 11*(5), 1218–1231. [In Zaura, E., & Twetman, S. (2019). Critical appraisal of oral pre- and probiotics for caries prevention and care. *Caries Research, 53*(5), 514–526. https://doi.org/10.1159/000499639].

Zheng, X., He, J., Wang, L., Zhou, S., Peng, X., Huang, S., & Xu, X. (2017). Ecological effect of arginine on oral microbiota. *Scientific Reports, 7*(1), 7206. [In Zaura, E., & Twetman, S. (2019). Critical appraisal of oral pre- and probiotics for caries prevention and care. *Caries Research, 53*(5), 514–526. https://doi.org/10.1159/000499639].

CHAPTER 5

Role of Prebiotics in Gastrointestinal Health: Constipation

KANKONA DEY

Department of Foods and Nutrition, The Maharaja Sayajirao University of Baroda, Gujarat, India

Constipation is a common problematic condition, representing a large socioeconomic burden. Despite the fact that the majority of patients with chronic constipation do not seek medical attention for their symptom, it is nevertheless identified in more than one million consultations with general practitioners and 63,000 admissions to hospitals in the UK each year (Science Daily, 2019).

5.1 INTRODUCTION

Constipation generally refers to three or fewer bowel movements per week accompanied by small, dry, and/or hard defecation and difficulty with discharge. Studies have estimated that constipation varies between 3% and 35% among the population. Almost 60% of patients taking prescriptions for constipation report being dissatisfied with their results (Science Daily, 2019).

Unfortunately, many commonly used medications for treating constipation are unable to cure and sometimes even worsen gastrointestinal symptoms (Portalatin and Winstead, 2012). Chronic constipation can be classified into three broad categories based on the Rome IV criteria for adults and more

Prebiotic Foods for a Healthy Gut: Scientific Studies to Manage Metabolic Disorders.
Mini K. Sheth and Shruti Dwivedi (Eds.)

recent criteria for children: normal-transit constipation, slow-transit constipation (STC), and defecatory or rectal evacuation disorders. Constipation is a functional GI disorder that can have a negative impact on both quality of life and healthcare costs. It does not only affect the elderly but also middle-aged and young adults, and sedentary lifestyles and physical inactivity are major contributors (Lacy et al., 2016) (Figure 5.1).

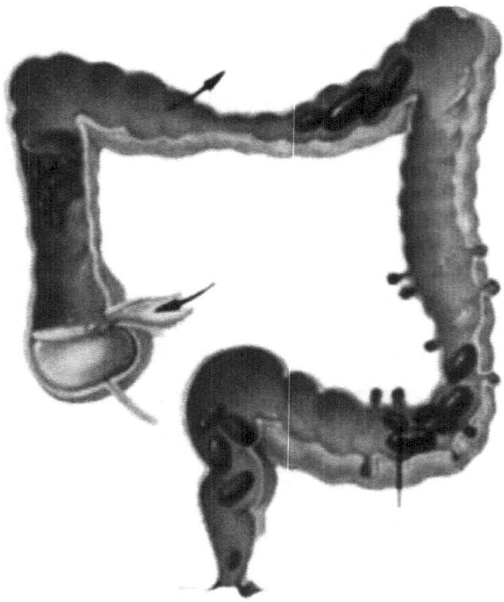

FIGURE 5.1 Constipation.

Depending on the definition used, it can be challenging to estimate the prevalence of constipation. Patients and clinicians typically have different definitions of constipation; while clinicians tend to focus on bowel frequency, patients tend to focus on other symptoms. Based on the published literature, the statistics on the epidemiology of constipation prevalence rates range from 0.7% to 29.6% in children and from 2.5% to 79% in adults worldwide.

5.2 EPIDEMIOLOGY OF CONSTIPATION

In the community, functional constipation is a common bowel disorder with a variable prevalence across cross-sectional surveys (Brigida et al., 2021). Up to one fifth of the general population suffers from chronic constipation once in their lifetime (Table 5.1).

TABLE 5.1 Global Population of Constipation

Year	Author	Place	Statistics
2021	Brigida et al	Worldwide	10·1%
2020	Katsuhisa Omagari et al	Japan	9.2%
2020	Palsson, O. S et al	US; UK	6.9%; 8.6%
2019	Russo et al	Worldwide	18.2%
2017	Wald A et al	Worldwide	0.7–79%
2002	Barnes PM, Maddocks A	Iran	3.1–28%

Despite the lack of information, studies show that constipation is a common health issue in India. This goes against the common belief that constipation is uncommon due to vegetarianism, high-fiber diets, and higher bowel movement frequency, and raises the possibility that constipation is actually underreported. The National Health Portal reports that the prevalence of constipation ranges from 24 to 37% (Table 5.2).

TABLE 5.2 Indian Population of Constipation

Author	Place	Sample Size	Criteria for Diagnosis	Prevalence for Constipation
Goyal et al., 2021	North India	1309	Rome IV	2.1
Ghoshal and Singh, 2017	Rural UP	2774	Rome III	2.4
Shah et al., 2014	Western India	128	Rome III	58%
Rajput and Saini, 2014	Chandigarh	505	Rome II	16.8
			Self-perception	24.8
Vidyut Bhatia, 2016	Delhi	1115	Rome II	0.5%
Panigrahi et al., 2013	Odisha	2400	<3 stools/week	2.6
Makharia et al., 2011	Rural Haryana	4767	Self-perception	11.6
			Rome III	0.3
Ghoshal et al., 2008	Multicentric (22 centers)	2785	Self-perception	53

5.3 CAUSES OF CONSTIPATION

Multiple pathogenic factors, including irregular intestinal wall contractility, visceral sensation dysfunction, unsatisfactory transit, and expulsion of endoluminal contents, contribute to functional constipation. There is currently no effective treatment for functional constipation due to the complex pathophysiological mechanisms (Bazzocchi et al., 2014). Functional constipation can cause a variety of discomforts, including abdominal distension, abdominal pain, headaches, dizziness, and poor appetite. These symptoms can significantly reduce a patient's quality of life (Valerio et al., 2010). The pathophysiology of functional constipation in children is heavily influenced by withholding behavior, whereas adults frequently have dysynergic defecation patterns. Functional constipation can also be brought on by lifestyle factors, psychological factors, and behavioral factors. Constipation is becoming more common in modern society as a result of the consumption of high-protein foods, animal-based and instant foods, as well as unbalanced dietary practices. Lifestyle factors reported to cause functional constipation in epidemiological surveys include inadequate dietary fiber and fluid intake, irregular and inadequate toilet use, sedentary lifestyle, prolonged bed rest, systemic illnesses, and chronic use of drugs that cause constipation (Ghoshal, 2007).

5.4 SYMPTOMS OF CONSTIPATION

The main symptoms of constipation in children seeking Brazilian primary health care settings refer to hard stools and large intervals between bowel movements. Less than 15% of patients reported fecal retentive incontinence. Primary care treatment prescribed for early constipation symptoms in children varies considerably, and usually less than 50% of these patients are successfully treated (Bruno-Barcena and Azcarate-Peril, 2015).

5.5 HOW TO DIAGNOSE CONSTIPATION

Accurately identifying a condition is at the root of good care. Currently, prescription medication for constipation fails in nearly 60% of patients, and almost half report not being satisfied with their treatment. The methods used to detect it differ greatly. Some patients self-diagnose, some doctors make a logical diagnosis, and some people rely on formal criteria (like the Rome IV criteria), which list combinations of symptoms that must be present over a certain amount of time (Science Daily, 2019).

In accordance with the most recent version of the Rome IV criteria, a patient must have at least two of the following symptoms for a period of at least three months, with the onset of symptoms occurring at least six months before the diagnosis: straining during 25% of bowel movements; lumpy or hard stools in 25% of bowel movements; feeling of incomplete evacuation for 25% of bowel movements; feeling of anorectal obstruction/blockage for 25% of bowel movements; manual maneuvers to facilitate 25% of bowel movements; and having less than 3 bowel movements per week (Imran et al., 2020).

5.6 TREATING CONSTIPATION

Treatment should aim to alleviate the symptoms while also addressing the underlying pathophysiology. Prior to any pharmacological interventions, conservative measures like dietary and lifestyle changes are typically advised; however, if these measures are unsuccessful in alleviating symptoms to the satisfaction of the patient, laxatives such as bulk-forming laxatives, stool softeners, osmotic laxatives (both sugar-based and non-sugar-based), and stimulant laxatives are frequently used. Diet modification to increase fiber consumption is considered an important component in the management of constipation. Therefore, there is a need for more efficient and safe treatments.

Exploring prebiotics with the potential to control gut flora and treat constipation while having few side effects has recently attracted increasing interest. Functional foods containing probiotics or prebiotics have been identified as useful for regulating bowel habits in children.

5.7 RELATION BETWEEN GUT MICROBIOTA AND CONSTIPATION

The gut microbiota differs in both structure and function between constipated and non-constipated individuals. The pathology of constipation is significantly influenced by the gut microbiota, which can have an impact on a variety of gut functions (Zhu et al., 2014). Intestinal motor functions are either directly or indirectly affected by the gut microbiota (Choi and Chang, 2015). Constipation patients' stools have been shown to have an unbalanced microbiota composition (Bazzocchi et al., 2014) with a decrease in obligate bacteria and an increase in potentially pathogenic bacteria. In this regard, new therapeutic methods for treating constipation are based on altering the intestinal microbiota, which can affect the colon's peristalsis (Guarner and Malagelada, 2003). Gut microbiota such as *Bifidobacterium*, *Clostridium*, Bacteroidetes, and *Lactobacilli* have been demonstrated in functional gastrointestinal disorders and prebiotics to play a role in augmenting their presence.

A causal relationship between the altered composition of commensal microbes and constipation has been demonstrated. Gnotobiotic animals (germ-free animals re-colonized with human fecal material) receiving fecal microbiota from patients with constipation showed reduced intestinal peristalsis and abnormal defecation parameters. Commensal bacteria play a key role in the development and maintenance of intestinal immune, vascular, sensory, and motor functions, as demonstrated in germ-free mice that exhibit significantly impaired gut development, delayed gastric emptying, and prolonged gastrointestinal transit time compared with conventionally raised animals.

The distal gastrointestinal tract harbors many different microbial species, and imbalances in the composition and function of these intestinal microbes are known as gut microbiota dysbiosis. A new scientifically based approach has re-examined the possibility of targeting the composition of intestinal bacterial populations for the treatment of functional constipation, considering the growing understanding of the role of gut microbes in regulating gut motility and stool consistency. Molecular methods based on the 16S rRNA gene sequence have replaced culture-based methods in the analysis of gut microbiota. Advancements in culture-independent technologies have demonstrated the enormous diversity, versatility, and age-related dynamics of the human microbiome.

5.8 PREBIOTICS AND SYNBIOTIC AS TREATMENT TO CONSTIPATION

The mechanisms of action of prebiotics in constipation have been investigated in a large number of human intervention studies. A dietary prebiotic is a substance that is fermented selectively and induces specific changes to the makeup or functioning of the gut's microbiota, with health benefits.

Prebiotics have recently started to take the place of probiotics due to increased resistance to digestive barriers, lower costs, lower risks, and simpler diet integration. In addition to prebiotics such as inulin, fructo-oligosaccharide (FOS), and galacto-oligosaccharide (GOS), isomalto-oligosaccharide, xylo-oligosaccharides, resistant starch, and soybean oligosaccharides are used as prebiotics (Wang et al., 2017). The most widely studied prebiotics are fructo-oligosaccharides (FOS), galacto-oligosaccharides (GOS), and inulin. A dietary synbiotic is a product that contains both prebiotics and probiotics.

These are living microorganisms, which if given in appropriate quantities, can be beneficial to the health of the host.

Prebiotics and synbiotics are used to modulate gut microbiota with potential benefits for patients with constipation symptoms. The benefits include lowering of the pH of the bowel, increasing fecal mass, and causing an osmotic effect leading to increased water in the large bowel. Many studies have reported the effects of prebiotics and synbiotics on functional constipation. Fructooligosaccharide and galactooligosaccharides soften fecal bolus and increase the frequency of bowel movements when added to infant formula. Studies suggest that prebiotics, even at a low dose of 5–8 g/day, significantly amplify fecal bacteria like *Bifidobacteria* and *Lactobacillus*. An alternative among nutritional therapies for constipation is the use of prebiotics, including galacto-oligosaccharides (GOS) and fructo-oligosaccharides (FOS) (Davis et al., 2010). Various studies have shown that feeding FOS to adults increases the colonization of *Bifidobacteria* and *Lactobacillus* and reduces the colonization of pathogenic bacteria like *E. coli* and *Clostridium* (Sheth et al., 2017; Mahendra and Sheth, 2015).

Human breast milk is the natural source of GOS for infants and contains approximately 100 times more GOS than cow's milk (Oozer et al., 2013). The infants fed with human breast milk have *Bifidobacteria* as the predominant microflora. For this reason, manufactured GOS are incorporated as the most common component of infant formulas to strengthen functional attributes associated with breast milk (Coulet et al., 2014; Knol et al., 2005). GOS are globally used in various commercial products such as infant formulas, various dairy products, breakfast cereals, beverages, and products related to bakery items.

One study indicated that in loperamide-induced constipated rats, oral GOS administration increases the moisture content of feces, maintains the integrity of intestinal mucosal cells, crypt cells, and ICC associated with intestinal motility, and increases intestinal LAB. Because of the intake of prebiotics and resistant starch, GOS consumption was linked to higher levels of *Bifidobacteria* and *Lactobacilli*, as well as a group of intestinal microorganisms that continuously selectively metabolize specific prebiotic/carbohydrate-type substances. In particular, GOS consumption improved constipation by causing the proliferation of LAB in the intestinal flora, as well as increasing SCFA production.

One study shows that GOS can promote small bowel propulsion in constipated mice, shorten the first defecation time of constipated mice, and increase the fecal weight and the number of feces particles. According to the relevant provisions of the health food functional evaluation procedures and testing methods, these results show that GOS has the effect of relieving constipation and promoting fecal excretion (Jin, 2020).

Randomized clinical trials of the probiotics *Lactobacillus casei, Lactobacillus rhamnosus, Streptococcus thermophiles, Bifidobacterium breve, Lactobacillus acidophilus, Bifidobacterium infantis,* and *Lactobacillus bulgaricus* were conducted by Khodadad et al. (2010) and Sadeghzadeh et al. (2014).

According to the literature, the addition of 12 g/day of GOS was shown to be effective and well tolerated, resulting in increased fecal *Bifidobacteria* in healthy individuals. It was shown that the estimated safe adult dose ranges from 0.3 to 0.4 g/kg/day and 2.5 g of GOS is sufficient to induce a bifidogenic effect. An excessive amount of GOS could promote adverse osmotic effects leading to osmotic diarrhea, symptoms not reported by the present patients during GOS ingestion. GOS ingestion promoted relief of complaints during its ingestion, regardless of other adjuvant recommendations. The prebiotic GOS ingested by constipated children was effective at increasing bowel movement frequency, relieving pain or discomfort during stool passage, and softening stool consistency. No adverse effects were reported. Probiotics, prebiotics, and synbiotics may be effective treatment options for constipation.

Table 5.3 shows the association between the defecation profile of obese bank employees before and after intervention with FOS as observed in a clinical study by Assudani and Sheth. As depicted, post FOS consumption after 90 days, a significant improvement in the overall defecation profile was observed in the experimental group. Reduction in constipation was reported in 14% of obese employees. The frequency of passing stool twice increased in 15% of them and improved by clearing bowels completely. The quantity of stool [bulk] increased in 8%, hardness reduced in 17%, foulness in stool odor reduced in 8%, and 17% of obese employees felt much better after defecation with FOS intervention.

There is a common misconception that constipation may go away on its own or that it is simple to treat, but this is not always the case. If stigma and embarrassment are not addressed concurrently with efforts to alter public perceptions of constipation and its treatment, the burden that these patients bear will not lessen. Greater openness and awareness will enable symptoms to be treated earlier, lowering the costs of constipation—both monetary and quality-of-life-related (Lancet Editorial, 2019).

TABLE 5.3 Association Between Defecation Profile of Obese Bank Employees Before and After Intervention

Defecation Profile		Placebo Group N = 70 [%]	Experiment Group N = 72 (%)	χ^2 Value
		Constipation (Subjects Perception)		
Pre	Present	34 [23.94]	27.[19.01]	**1.78 NS**
	Absent	36 [25.35]	45 [31.69]	**p-0.18**
Post	Present	24 [16.90]	7 [4.93]	**12.55**
	Absent	46.[32.39]	65 [45.77]	**p-0.000**
	χ^2 Value	**2.95 NS; p-0.08**	**15.40***; p-0.000**	
		Frequency (time/day)		
Pre	1	50 [35.21]	38 [26.76]	**5.24 NS**
	2	20 [14.08]	34 [23.94]	**P-0.02**
Post	1	43 [30.28]	17 [11.97]	**20.8*****
	2	27 [19.01]	55 [38.73]	**p-0.000**
	χ^2 Value	**1.57 NS; p-020**	**12.97***; p-0.000**	
		Quantity of stool		
Pre	Small	23 [16.20]	16 [11.27]	**2.02 NS**
	Middle to large	47[33.09]	56 [39.44]	**p-0.16**
Post	Small	18 [12.68]	4 [2.82]	**11.02*****
	Middle to large	52 [36.62]	68 [47.89]	**p-0.0009**
	χ^2 Value	**0.86 NS; p-0.35**	**8.36**; p-0.0038**	
		Hardness of stool		
Pre	Very hard to hard	34 [23.94]	34 [23.94]	**0.02 NS**
	Medium to soft	36[25.35]	38 [26.76]	**p-0.87**
Post	Very hard to hard	27[19.01]	10 [7.04]	**11.23*****
	Middle to large	43 [30.28]	62. [43.66]	**p-0.0008**
	χ^2 Value	**1.42 NS; p-0.23**	**18.85***; p-000**	
		Color of Stool		
Pre	BLack to middle	59 [41.55]	70 [49.29]	**714****
	Yellowish	11 [7.75]	2 [1.41]	**p-0.007**
Post	BLack to middle	61[42.96]	69 [48.59}	**3.46 NS**
	Yellowish	9 [6.34]	3 [2.11]	**p-0.064**
	χ^2 Value	**0.233 NS,p- 0.63**	**0.21 NS, p- 0.65**	
		Odor of Stool		
Pre	Strong	17 [11.97]	17 [11.97]	**0.01 NS**
	Middle to Weak	53 [37.32]	55 [38.73]	**p-0.92**
Post	Strong	12 [8.45]	5 [3.52]	**3.50 NS**
	Middle to Weak	58 [40.84]	67 [47.18]	**p-0.06**
	χ^2 Value	**1.09 NS, P-0.29**	**7.73**,P-0.005**	

NOTE: Figure in parenthesis represent present of Subject, NS -non- significant , p < 0.05.* , p < 0.01.**, p < 0.01.***

KEYWORDS

- **bacteria**
- **constipation**
- **diseases**
- **drug use**
- **health benefits**
- **intestines**
- **obesity**
- **patients**
- **symptoms**
- **young adults**

REFERENCES

Barberio, B., Judge, C., Savarino, E. V., & Ford, A. C. (2021). Global prevalence of functional constipation according to the Rome criteria: A systematic review and meta-analysis. *The Lancet Gastroenterology & Hepatology, 6*(8), 638–648. https://doi.org/10.1016/S2468-1253(21)00084-8.

Bazzocchi, G., Giovannini, T., Giussani, C., Brigidi, P., & Turroni, S. (2014). Effect of a new synbiotic supplement on symptoms, stool consistency, intestinal transit time and gut microbiota in patients with severe functional constipation: A pilot randomized double-blind, controlled trial. *Techniques in Coloproctology, 18*(10), 945–953. https://doi.org/10.1007/s10151-014-1194-4.

Bazzocchi, G., Giovannini, T., Giussani, C., Brigidi, P., & Turroni, S. (2014). Effect of a new synbiotic supplement on symptoms, stool consistency, intestinal transit time and gut microbiota in patients with severe functional constipation: A pilot randomized double-blind, controlled trial. *Techniques in Coloproctology, 18*(10), 945–953. https://doi.org/10.1007/s10151-014-1194-4.

Beleli, C. A. V., Antonio, M. A. R. G. M., dos Santos, R., Pastore, G. M., & Lomazi, E. A. (2015). Effect of 4′galactooligosaccharide on constipation symptoms. *Jornal de Pediatria, 91*(6), 567–573. https://doi.org/10.1016/j.jped.2015.01.007.

Bruno-Barcena, J. M., & Azcarate-Peril, M. A. (2015). Galacto-oligosaccharides and colorectal cancer: Feeding our intestinal probiome. *Journal of Functional Foods, 12*, 92–108. https://doi.org/10.1016/j.jff.2014.11.011.

Ceresola, E. R., Ferrarese, R., Preti, A., & Canducci, F. (2018). Targeting patients' microbiota with probiotics and natural fibers in adults and children with constipation. *European Review for Medical and Pharmacological Sciences, 22*, 7045–7057. https://doi.org/10.26355/eurrev_201812_16593.

Choi, C. H., & Chang, S. K. (2015). Alteration of gut microbiota and efficacy of probiotics in functional constipation. *Journal of Neurogastroenterology and Motility, 21*(1), 4–7. https://doi.org/10.5056/jnm14097.

Coulet, M., Phothirath, P., Allais, L., et al. (2014). Pre-clinical safety evaluation of the synthetic human milk, nature-identical, oligosaccharide 2′-O-fucosyllactose (2′ FL). *Regulatory Toxicology and Pharmacology, 68*(1), 59–69. https://doi.org/10.1016/j.yrtph.2013.10.003.

Emmanuel, A., Quigley, E. M., Simrén, M., Feng, Y., Müller-Lissner, S., Urbain, D., & Solomon, D. (2013). Factors affecting satisfaction with treatment in European women with chronic constipation: An internet survey. *United European Gastroenterology Journal, 1*(5), 375–384. https://doi.org/10.1177/2050640613501814.

Ghoshal, U. C. (2007). Review of pathogenesis and management of constipation. *Tropical Gastroenterology, 28*(3), 91–95.

Ghoshal, U. C., Abraham, P., Bhatia, S. J., Choudhuri, G., Misra, S. P., Issar, S. K., & Jayanthi, V. (2018). The Indian Society of Gastroenterology consensus on chronic constipation in India. *Indian Journal of Gastroenterology, 37*(6), 526–544. https://doi.org/10.1007/s12664-018-0894-y.

Ghoshal, U. C., Abraham, P., Bhatia, S. J., Choudhuri, G., Misra, S. P., Issar, S. K., & Jayanthi, V. (2017). The Indian Society of Gastroenterology consensus on chronic constipation in India. *Indian Journal of Gastroenterology, 36*(6), 508–528. https://doi.org/10.1007/s12664-017-0810-7.

Gopala, P. K., Sullivan, P. A., & Smart, J. B. (2001). Utilization of galacto-oligosaccharides as selective substrates for growth by lactic acid bacteria including Bifidobacterium lactis DR10 and Lactobacillus rhamnosus DR20. *International Dairy Journal, 11*(1–2), 19–25. https://doi.org/10.1016/S0958-6946(01)00098-3.

Goyal, O., Nohria, S., Dhaliwal, A. S., Goyal, P., Soni, R. K., Chhina, R. S., & Sood, A. (2021). Prevalence, overlap, and risk factors for Rome IV functional gastrointestinal disorders among college students in northern India. *Indian Journal of Gastroenterology, 40*(2), 144–153. https://doi.org/10.1007/s12664-020-01091-w.

Guarner, F., & Malagelada, J. R. (2003). Gut flora in health and disease. *The Lancet, 361*(9356), 512–519. https://doi.org/10.1016/S0140-6736(03)12489-0.

Imran, A., Whitehead, W. E., Palsson, O. S., Törnblom, H., & Simrén, M. (2020). An approach to the diagnosis and management of Rome IV functional disorders of chronic constipation. *Expert Review of Gastroenterology & Hepatology, 14*(9), 739–746. https://doi.org/10.1080/17474124.2020.1794994.

Jain, M., Verma, M., Jaiswal, N., Yadav, N. K., & Singh, K. (2018). Juvenile toxicity study of Gossence™ (galactooligosaccharides) in Sprague Dawley rats. *Toxicology Research and Application, 4*, 1–3. https://doi.org/10.1177/2397847318763248.

Khodadad, A., & Sabbaghian, M. (2010). Role of symbiotic in the treatment of childhood constipation: A double-blind randomized placebo-controlled trial. *Iranian Journal of Pediatrics, 20*(4), 387–392.

Kim, M. G., Jo, K., Chang, Y. B., Suh, H. J., & Hong, K. B. (2020). Changes in the gut microbiome after galacto-oligosaccharide administration in loperamide-induced constipation. *Journal of Personalized Medicine, 10*(4), 161. https://doi.org/10.3390/jpm10040161.

Knol, J., Scholtens, P., Kafka, C., et al. (2005). Colon microflora in infants fed formula with galacto- and fructo-oligosaccharides: More like breast-fed infants.

Journal of Pediatric Gastroenterology and Nutrition, 40(1), 36–42. https://doi.org/10.1097/00005176-200501000-00010.

Li, T., Lu, X., & Yang, X. (2013). Stachyose-enriched α-galacto-oligosaccharides regulate gut microbiota and relieve constipation in mice. *Journal of Agricultural and Food Chemistry, 61*(50), 11825–11831. https://doi.org/10.1021/jf403802a.

Li, T., Lu, X., & Yang, X. (2017). Evaluation of clinical safety and beneficial effects of stachyose-enriched α-galacto-oligosaccharides on gut microbiota and bowel function in humans. *Food & Function, 8*(1), 262–269. https://doi.org/10.1039/C6FO01308A.

Lim, Y. J., Jamaluddin, R., Hazizi, A. S., & Chieng, J. Y. (2018). Effects of symbiotic among constipated adults in Serdang, Selangor, Malaysia: A randomized, double-blind, placebo-controlled trial. *Nutrients, 10*(7), 810. https://doi.org/10.3390/nu10070810.

Manning, T. S., & Gibson, G. R. (2004). Prebiotics. *Best Practice & Research: Clinical Gastroenterology, 18*(2), 287–298. https://doi.org/10.1016/j.bpg.2003.10.008.

Marín-Manzano, M. C., Abecia, L., Hernández-Hernández, O., Sanz, M. L., Montilla, A., & Olano, A., et al. (2013). Galacto-oligosaccharides derived from lactulose exert a selective stimulation on the growth of Bifidobacterium animalis in the large intestine of growing rats. *Journal of Agricultural and Food Chemistry, 61*(32), 7560–7567. https://doi.org/10.1021/jf401407z.

McRae, M. P. (2020). Effectiveness of fiber supplementation for constipation, weight loss, and supporting gastrointestinal function: A narrative review of meta-analyses. *Journal of Chiropractic Medicine, 19*(2), 101–107. https://doi.org/10.1016/j.jcm.2020.03.001.

Mussatto, S. I., & Mancilha, I. M. (2007). Non-digestible oligosaccharides: A review. *Carbohydrate Polymers, 68*(3), 587–597. https://doi.org/10.1016/j.carbpol.2006.12.011.

Niittynen, L., Kajander, K., & Korpela, R. (2007). Galacto-oligosaccharides and bowel function. *Scandinavian Journal of Food and Nutrition, 51*(2), 62–66. https://doi.org/10.1080/17482970701380082.

Ohkusa, T., Koido, S., Nishikawa, Y., & Sato, N. (2019). Gut microbiota and chronic constipation: A review and update. *Frontiers in Medicine, 6*, 19. https://doi.org/10.3389/fmed.2019.00019.

Oozeer, R., van Limpt, K., Ludwig, T., et al. (2013). Intestinal microbiology in early life: Specific prebiotics can have similar functionalities as human-milk oligosaccharides. *American Journal of Clinical Nutrition, 98*(2), 561S–571S. https://doi.org/10.3945/ajcn.112.038893.

Palsson, O. S., Whitehead, W., Törnblom, H., Sperber, A. D., & Simrén, M. (2020). Prevalence of Rome IV functional bowel disorders among adults in the United States, Canada, and the United Kingdom. *Gastroenterology, 158*(5), 1262–1273. https://doi.org/10.1053/j.gastro.2019.12.021.

Portalatin, M., & Winstead, N. (2012). Medical management of constipation. *Clinics in Colon and Rectal Surgery, 25*(1), 12–19. https://doi.org/10.1055/s-0032-1301754.

Sadeghzadeh, M., Rabieefar, A., Khoshnevisasl, P., Mousavinasab, N., & Eftekhari, K. (2014). The effect of probiotics on childhood constipation: A randomized controlled double-blind clinical trial. *International Journal of Pediatrics, 2014*, Article ID 937212. https://doi.org/10.1155/2014/937212.

Science Daily. (2019). Research defines constipation. Retrieved from: https://www.sciencedaily.com/releases/2019/06/190605201123.htm (accessed on 19 July 2024).

Tabbers, M. M., Chmielewska, A., Roseboom, M. G., Crastes, N., Perrin, C., Reitsma, J. B., & Benninga, M. A. (2011). Fermented milk containing Bifidobacterium lactis DN-173 010 in childhood constipation: A randomized, double-blind, controlled trial. *Pediatrics, 127*(6), e1392–e1399. https://doi.org/10.1542/peds.2010-2590.

Tanaka, Y., Takami, K., Nishijima, T., Aoki, R., Mawatari, T., & Ikeda, T. (2015). Short- and long-term dynamics in the intestinal microbiota following ingestion of Bifidobacterium animalis subsp. lactis GCL2505. *Bioscience of Microbiota, Food and Health, 34*(2), 77–85. https://doi.org/10.12938/bmfh.2015-002.

Tomomatsu, H. (1994). Health effects of oligosaccharides. *Food Technology, 48*(10), 61–65.

Valerio, F., Russo, F., de Candia, S., Riezzo, G., Orlando, A., & Lonigro, S. L., et al. (2010). Effects of probiotic Lactobacillus paracasei-enriched artichokes on constipated patients: A pilot study. *Journal of Clinical Gastroenterology, 44*(Supplement 1), S49–S53. https://doi.org/10.1097/MCG.0b013e3181dffb25.

Vriesman, M. H., Koppen, I. J. N., Camilleri, M., Di Lorenzo, C., & Benninga, M. A. (2020). Management of functional constipation in children and adults. *Nature Reviews Gastroenterology & Hepatology, 17*(1), 21–39. https://doi.org/10.1038/s41575-019-0222-y.

Wang, J., Du, P., & Wang, Y. (2020). Study on the effect of galacto-oligosaccharide (GOS) in relieving constipation and defecating feces excretion. *IOP Conference Series: Materials Science and Engineering, 730*(1), 012011. https://doi.org/10.1088/1757-899X/730/1/012011.

Wang, L., Hu, L., Yan, S., Jiang, T., Fang, S., Wang, G., Zhao, J., Zhang, H., & Chen, W. (2017). Effects of different oligosaccharides at various dosages on the composition of gut microbiota and short-chain fatty acids in mice with constipation. *Food & Function, 8*(5), 1966–1978. https://doi.org/10.1039/C7FO00278H.

Xu, Q., Chao, Y., & Wan, Q. (2009). Health benefit application of functional oligosaccharides. *Carbohydrate Polymers, 77*(3), 435–441. https://doi.org/10.1016/j.carbpol.2009.01.020.

Zhu, L., Liu, W., Alkhouri, R., Baker, R. D., Bard, J. E., & Quigley, E. M., et al. (2014). Structural changes in the gut microbiome of constipated patients. *Physiological Genomics, 46*(18), 679–686. https://doi.org/10.1152/physiolgenomics.00082.2014.

CHAPTER 6

Prebiotics as a Symptom Reliever in Irritable Bowel Syndrome

ARPI SHAH

International Institute of Sports and Management, Mumbai, Maharashtra, India

Irritable Bowel Syndrome (IBS) is characterized by abdominal discomfort, followed by gas, bloating constipation and diarrhea. Patients with IBS have generally low attendance in school or at work, decreased productivity, increased healthcare cost and low quality of life.

6.1 INTRODUCTION

Irritable bowel disorder is a gastrointestinal disorder associated with chronic or acute abdominal pain. The main symptoms also include the presence of constipation, diarrhea, or a mixed stool pattern (both constipation and diarrhea) in patients. Traditionally, IBS was considered a disease with an unknown pathophysiology, but with advances in research, we have found that it is mainly associated with abnormalities in the serotonin mechanism, other GI diseases, ulcers, genetic factors, and the gut-brain axis. It is now widely accepted that IBS is an unexplained gut-brain disorder (Gerald et al., 2016).

It is a long-term GI tract illness that causes chronic pain. This disease affects 6 to 18% of the world's population, primarily women. IBS is not

Prebiotic Foods for a Healthy Gut: Scientific Studies to Manage Metabolic Disorders.
Mini K. Sheth and Shruti Dwivedi (Eds.)

a life-threatening condition, and it does not put you at risk for other colon diseases like colon cancer, ulcerative colitis, or Crohn's disease. It is more likely to interfere with a person's daily activities. IBS can be classified into three types: IBS-C, IBS-D, and IBS-M/IBS-A. One of the most common types of IBS is IBS-C. This disorder causes the patient to have irregular bowel movements, which include hard and lumpy stools. IBS-D, often known as diarrhea-related IBS, causes patients to have frequent stools and abdominal bloating.

6.2 SYMPTOMS OF IRRITABLE BOWEL DISEASE

Constipation, diarrhea, nausea, abdominal pains and cramps, abdominal bloating, unusual stool consistency, sudden need to use the restroom, changing pattern of constipation and diarrhea, mucus passing from the rectal area, food intolerance are some of the most common symptoms of IBS. These symptoms usually change over a few days to a few weeks.

Some unusual symptoms include frequent headaches, fatigue, a sudden need to urinate, joint or muscle aches, irregular menstrual cycles, dyspareunia, and halitosis. Anxiety and depression are also reported in some patients.

6.3 CAUSES OF IRRITABLE BOWEL SYNDROME

IBS's exact etiology is still unknown, but there may be a number of contributing factors. Alterations in the gastrointestinal tract, food intolerances, abnormal nervous system signals, and hypersensitivity to pain are among the factors that can cause the development of IBS. Unusual levels of serotonin in the colon cause irregular bowel movements and motility. Mild celiac disease can also contribute to the development of IBS symptoms.

6.4 MAIN CAUSES AND TRIGGERS

Some of the main symptoms that are found to cause IBS are mentioned below:

- Sensitivity to pain caused by gas or a full bowel
- Small intestinal bacterial overgrowth (SIBO)
- External elements, such as stress
- Food intolerances, possibly as a result of a poor ability to absorb acids or sugars from food
- Abnormal nervous system signals or reproductive hormone levels

- A person's psychological and emotional state (depression or anxiety) can lead to the development of IBS.
- Gastroenteritis (a viral stomach and gut infection caused by the presence of harmful bacteria)

6.5 PREVALENCE OF IRRITABLE BOWEL SYNDROME

6.5.1 GLOBAL PREVALENCE

A meta-analysis of 390 cross-sectional studies containing 260,960 subjects worldwide reported an overall 11.2% prevalence of IBS. The prevalence was higher in women compared to men and lower in individuals over 50 years of age. There was no significant association between socioeconomic status and the prevalence of IBS. The lowest prevalence of IBS was in Southeast Asia (7.0%) and the highest was in South America (21%) (Rebecca et al., 2012).

A systematic review of 184 studies on the prevalence of IBS from January 2006 to April 2019 stated that the pooled prevalence of IBS was 9.2%, out of which 33.8% of the population had IBS with mixed bowel habits. The prevalence of IBS was higher in women compared to men (Priya et al., 2020).

Another systematic review of 27 studies that enrolled a total of 3,169 patients aged 16 years and above concluded that the prevalence of IBS was 32.5%, and the prevalence was higher in patients with Crohn's disease (Keeley et al., 2020).

World Gastroenterology Report suggests that the prevalence of IBS is common in developing countries. It mainly occurs in the age range of 15 to 65 years, and the incidence decreases in older individuals (World Gastroenterology Organization Global Guideline).

A study from Norway checked the prevalence of IBS in 2001 on 11,078 subjects. The results showed that 42% of subjects had IBS symptoms, out of which 8.2% of subjects had confirmed IBS with a female predominance. The results also suggested that subjects with IBS frequently had somatic and psychiatric comorbidities (Per et al., 2009).

6.5.2 NATIONAL PREVALENCE

A study was conducted in northern India to study the prevalence of IBS from 2008 to 2009. A total of 4767 participants were enrolled. The results stated that 4% of subjects had confirmed IBS. The prevalence of constipation-predominant IBS was 0.3%, diarrhea-predominant IBS was 1.5%, and mixed IBS was 1.7%. The data also suggested a greater prevalence of IBS in females compared to males (Govind Makharia et al., 2011).

Prevalence of IBS in Asian countries is less than 5%. It is found to be more prevalent in affluent cities like Singapore and Tokyo with the highest prevalence of 8.6% and 9.8% respectively, while India (4.2%) has the lowest prevalence (Kok et al., 2009).

6.6 EFFECT OF IBS ON QUALITY OF LIFE

A study conducted on 253 French patients with questionnaires on quality of life showed that quality of life scores were significantly lower in subjects with IBS as compared to the general population ($p<0.05$). The results also showed that women had significantly more deteriorated QOL compared to men. Data showed a positive correlation between low QOL and increased intensity of pain or discomfort (Michel et al., 2006).

A study evaluated the gender-related differences in 714 subjects for extracolonic symptoms, psychological symptoms, and quality of life. The results revealed that abdominal distension with a sensation of bloating was more common in female patients. Females also reported nausea, alteration of taste and smell, greater food sensitivity, and side effects from medications. Females (40%) reported irregular menstrual cycle-related symptoms. Hence, the study concluded that females have higher levels of intestinal and non-intestinal symptoms of IBS compared to males with the same severity of the disease (Young et al., 2001). A UK study on 209 IBS patients showed that serious consequences of the disease were related to lower quality of life, lower health satisfaction, and higher depression and anxiety scores ($p<0.01$). Subjects with lower illness scores were associated with better health satisfaction, but it didn't improve their quality of life (Claire et al., 2002).

Studies have concluded that patients with IBS have three times more chances of anxiety and depressive disorders. An Indian study stated that the prevalence of depression was 37.1% and that anxiety was 31.4% among subjects with IBS (Neerja et al., 2013). A systematic review incorporated 73 research papers with the aim of investigating the prevalence of anxiety and depression in IBS patients. The results revealed that the prevalence of anxiety symptoms was 39.1% and the prevalence of depressive disorder was 28.8% (Mohammad et al., 2019).

An article in the Journal of Clinical Gastroenterology showed that patients with IBS have a worse quality of life compared to patients with other diseases. The severity of IBS symptoms affects the health-related quality of life (HRQoL) of patients (Mönnikes et al., 2011).

A study was conducted to see the association between early adverse life events (EALs) and IBS, evaluating 729 women out of which 294 had IBS and 435 were control subjects. Data was collected on EALs which provide information on physical, emotional, general trauma, and sexual abuse. The results revealed that compared to controls, IBS patients reported a higher prevalence of general trauma, physical punishment, and sexual events ($p < 0.001$), concluding that there is a strong association between trauma and the development of IBS (Kara et al., 2012).

A case-control study conducted in the UK to assess the association between Health-Related Quality of Life (HRQoL) and the cost impact of IBS. The research evaluated QoL, time off work, and 12 months of hospital records of the selected subjects. The results showed that patients with IBS had a lower HRQoL than controls and also had worse scores in all components of the HRQoL measure (Ron et al., 2000).

A study from Sweden assessed Gastrointestinal-specific anxiety (GSA) and quality of life in patients with irritable bowel syndrome (IBS). The study selected 60 healthy subjects and 306 patients fulfilling Rome II criteria for IBS. The results showed that patients with IBS had a more severe GSA score. Specifically, patients who had more severe symptoms had a more prevalent GSA, general anxiety, and depression. GSA, along with anxiety and depression (Jerndal et al., 2010).

IBS patients (227) were assessed for quality of life. The results showed that patients had significantly higher scores for anxiety and depression. Food avoidance had the highest score for a decrease in quality of life. Food avoidance and anxiety were higher in females compared to males (Liming et al., 2015).

Another study was conducted in Taiwan on 1894 female university students to see the prevalence of IBS and the impact of stress markers and quality of life on its etiology. The study resulted in 10.1% of females suffering from IBS. Students with IBS had higher stress levels and lower quality of life compared to those who didn't have IBS (Huan-Hwa et al., 2021).

A population-based cohort study was conducted in Germany from 2008 to 2012 to observe the impact of mental health and quality of life in patients with IBS. The study enrolled 4194 subjects, out of which 3.5% had IBS. The results concluded that there was a significant association between the prevalence of IBS and underlying complex somatic symptom disorder (Birgit Schauer et al., 2018).

6.7 MODES OF TREATMENT IN IRRITABLE BOWEL SYNDROME (IBS)

Dietary and lifestyle changes are the primary treatment option for IBS. If dietary and lifestyle changes are not adequate medical treatments are given for relief. Some of the lifestyle changes includes:

- Probiotics consumption
- Quit intake of alcohol and smoking
- Eat smaller, more frequent meals
- Avoid caffeine and beverages
- Exercise regularly
- Mindfulness training
- Cognitive behavioral therapy (CBT) or psychotherapy
- Gut-directed hypnosis
- Getting proper sleep
- Applying pain management techniques
- Stress management and relaxation techniques
- Walking daily and practicing yoga
- Avoiding laxatives

In most cases of IBS, surgery is not required and can be treated through non-surgical measures.

A cohort study conducted in the years 2000/2001 on 645 subjects with IBS in Birmingham. Data was collected on various treatment methods for IBS and the acceptance of subjects to these modes of treatment. The data revealed that tablets were the most accepted form of treatment for IBS (84%), followed by lifestyle changes (82%) and yoga (77%). Acupuncture (59%) and suppositories (57%) were less accepted methods for IBS treatment. The study concluded that most of the patients were willing to accept various forms of treatment but fears and misconceptions were stopping them from trying new methods (Lynsey and Lesley, 2008).

In January 2002, 668 subjects were selected for a survey on medications for IBS and their side effects. Out of all the subjects, 504 had constipation-predominant IBS and 88% were females. The results showed that nearly three-quarters of subjects discontinued treatment because of side effects as it led to missed school or work. Side effects of treatments negatively affected patients with IBS (Anthony Lembo, 2004).

6.7.1 ROLE OF PREBIOTIC, PROBIOTIC AND SYNBIOTIC IN IRRITABLE BOWEL SYNDROME

A review article stated that there are very few known treatments for IBS. Prebiotics are among the effective treatment methods for IBS. Prebiotics are the metabolic fuel that gives energy to good bacteria for their expansion and growth. Probiotics have also shown an encouraging effect in the treatment of IBS. In a meta-analysis of nine randomized controlled trials, children with IBS who were intervened with probiotics *Lactobacillus* GG, *Lactobacillus reuteri* DSM 17938, and the probiotic mixture VSL#3 showed improvement in abdominal pain. The patients who benefited from probiotics mainly had diarrhea-predominant IBS (Hui Juan et al., 2020).

Some of the beneficial effects of Prebiotics are:

1. Prevention and reduction of the duration and severity of rotavirus-induced or antibiotic-associated diarrhea, as well as relief of lactose intolerance complaints.
2. Cancer-promoting enzymes and putrefactive (bacterial) metabolites in the gut are reduced.
3. Prevention and treatment of irregular gastrointestinal complaints in healthy individuals.
4. Beneficial effects on microbial abnormalities, inflammation, and other complaints associated with: inflammatory gastrointestinal diseases, *Helicobacter pylori* infection, or bacterial overgrowth.
5. Ease constipation and irritable colon.
6. Allergies and atopic diseases in infants can be prevented or alleviated.
7. Treatment of urogenital infections as well as prevention of respiratory tract infections (common cold, influenza) and other infectious diseases (Michael and Schrezenmeir, 2008).

A review conducted in 2020 on the role of pre/probiotics in celiac disease. The results showed that supplementation of GFD with probiotics, such as *Bifidobacterium* and *Lactobacilli*, reported a potential to restore gut microbiota composition and to pre-digest gluten in the intestinal lumen, reducing the inflammation associated with gluten intake, the intestinal permeability, and the cytokine and antibody production (Giovanni Marasco et al., 2020).

A meta-analysis conducted on forty-three randomized control trials to showcase the effect of prebiotics, probiotics, and synbiotics in IBD revealed that probiotics appeared to have a beneficial effect on symptoms of IBS like abdominal pain, flatulence, bloating, etc., and CIC (mean increase in the number of stools per week). Supplementation of synbiotics also proved to be beneficial (Ford et al., 2014).

The microbial alterations due to disease checked in 37 IBS female patients revealed that there were vast changes in microbiota composition of the gut. It was characterized by an increase in *Firmicutes*-associated taxa and depletion of Bacteroidetes-related taxa (Ian et al., 2012).

Studies have concluded that multispecies probiotics are good and safe options to treat symptomatic IBS (Kajander, et al., 2007). A 5-month clinical trial conducted on 86 patients where they received multispecies probiotic supplementation daily or placebo control group showed that IBS symptoms score decreased by 14 points in the treatment group. Distension and abdominal pain showed the most significant decline. Stability of microbial content was also seen. Another placebo-controlled study carried out on 55 IBS patients over a period of 6 months showed that prebiotic strain increased in fecal samples of the treatment group (Kajander et al., 2007).

In a systematic review, forty-three studies were selected for inclusion which recruited adults with either IBS or CIC. The results showed that probiotic consumption had a global impact on IBS. Improvement was also seen in abdominal pain, bloating, and flatulence scores in patients. Synbiotic consumption also showed a desirable effect (Ford, 2014). A total of 33 randomized controlled trials were taken into account which enrolled 4321 subjects. Factors under consideration were global IBS symptoms, abdominal pain, secondary symptoms, and frequency of adverse events. The results showed that overall probiotics significantly increased the improvement in IBS symptoms in the treatment group as compared to the placebo group. Both single and multi-strain probiotics showed improvement in IBS symptoms, symbiotic supplementation showed a similar effect (Mohammad and Sundos, 2020).

A review was conducted on the effect of two main kinds of prebiotics: inulin-fructans and galacto-oligosaccharides in gastrointestinal disorders. The results showed that prebiotic supplementation leads to an increase in good bacteria in the gut (e.g., *Bifidobacterium*). There has been evidence of modulation of the gut bacteria in lowering IBS symptoms after supplementation of low doses of prebiotics (Bridgette and Kevin, 2017).

KEYWORDS

- **celiac disease**
- **chronic pain**
- **intestines**
- **irritability**
- **lifestyle**
- **pain**
- **patients**
- **respiratory tract diseases**
- **socioeconomic status**
- **symptoms**

REFERENCES

Akehurst, R. L., Kaltenthaler, E., Brazier, J., Morgan, A., Platts, M., O'Keeffe, C., & Mathers, N. (2002). Health-related quality of life and cost impact of irritable bowel syndrome in a UK primary care setting. *PharmacoEconomics, 20*(7), 455–462. https://doi.org/10.2165/00019053-200220070-00004.

Amouretti, M., Arnould, B., El Hasnaoui, A., Fauconnier, J., Gensac, N., Ziegler, K., & Lançon, C. (2006). Impact of irritable bowel syndrome (IBS) on health-related quality of life (HRQOL). *Gastroentérologie Clinique et Biologique, 30*(2), 241–246. https://doi.org/10.1016/S0399-8320(06)73150-3.

Asha, M. Z., & Khalil, S. F. H. (2020). Efficacy and safety of probiotics, prebiotics, and symbiotic in the treatment of irritable bowel syndrome. *Sultan Qaboos University Medical Journal, 20*(4), e13–e24. https://doi.org/10.18295/squmj.2020.20.04.004.

Bolge, S. C., Doan, J. F., Kannan, H., & Baran, R. (2010). Association of health-related quality of life with work productivity and indirect costs among patients with irritable bowel syndrome with diarrhea. *Health and Quality of Life Outcomes, 8*, 90. https://doi.org/10.1186/1477-7525-8-90.

Bradford, K., Shih, W., Videlock, E. J., Presson, A. P., Naliboff, B. D., & Mayer, E. A., & Chang, L. (2012). Association between early adverse life events and irritable bowel syndrome. *Clinical Gastroenterology and Hepatology, 10*(4), 385–390. https://doi.org/10.1016/j.cgh.2011.12.018.

Chang, L., Lee, O. Y., Naliboff, B., Schmulson, M., Mayer, E. A., & Chang, L. (2001). Gender-related differences in IBS symptoms. *The American Journal of Gastroenterology, 96*(7), 2184–2193. https://doi.org/10.1111/j.1572-0241.2001.04027.x.

Chao, G. Q., & Zhang, S. (2014). Effectiveness of acupuncture to treat irritable bowel syndrome: A meta-analysis. *World Journal of Gastroenterology, 20*(7), 1871–1877. https://doi.org/10.3748/wjg.v20.i7.1871.

Chen, H. H., Hung, C. H., Kao, A. W., & Hsieh, H. F. (2021). Exploring quality of life, stress, and risk factors associated with irritable bowel syndrome for female university students in Taiwan. *International Journal of Environmental Research and Public Health, 18*(8), 3936. https://doi.org/10.3390/ijerph18083936.

Choghakhori, R., Abbasnezhad, A., Amani, R., & Alipour, M. (2017). Sex-related differences in clinical symptoms, quality of life, and biochemical factors in irritable bowel syndrome. *Digestive Diseases and Sciences, 62*(6), 1550–1560. https://doi.org/10.1007/s10620-017-4541-5.

de Vrese, M., & Schrezenmeir, J. (2008). Probiotics, prebiotics, and symbiotic. *Food Biotechnology, 3*(4), 93–117. https://doi.org/10.1007/978-3-540-79011-2_3.

Ford, C., Quigley, E. M., Lacy, B., Lembo, A., Schiller, L., Soffer, E., Moayyedi, P., & Brennan, M. (2014). Efficacy of prebiotics, probiotics, and symbiotic in irritable bowel syndrome and chronic idiopathic constipation: Systematic review and meta-analysis. *The American Journal of Gastroenterology, 109*(10), 1547–1561. https://doi.org/10.1038/ajg.2014.202.

Gwee, K. A., Chua, A. S. B., & Goh, K. L. (2009). Epidemiology of irritable bowel syndrome in Asia: Something old, something new, something borrowed. *Journal of Gastroenterology and Hepatology, 24*(10), 1601–1607. https://doi.org/10.1111/j.1440-1746.2009.05984.x.

Holtmann, G. J., Ford, A. C., & Talley, N. J. (2016). Pathophysiology of irritable bowel syndrome. *The Lancet Gastroenterology & Hepatology, 1*(2), 133–146. https://doi.org/10.1016/S2468-1253(16)30023-1.

Jeffery, I. B., O'Toole, P. W., Öhman, L., Claesson, M. J., Deane, J., Quigley, E. M. M., & Simrén, M. (2012). An irritable bowel syndrome subtype defined by species-specific alterations in fecal microbiota. *Gut, 61*(7), 997–1006. https://doi.org/10.1136/gutjnl-2011-301501.

Jerndal, P., Ringström, G., Agerforz, P., Karpefors, M., Akkermans, L. M., Bayati, S., & Simrén, M. (2010). Gastrointestinal-specific anxiety: An important factor for severity of GI symptoms and quality of life in IBS. *Neurogastroenterology and Motility, 22*(6), 646–e179. https://doi.org/10.1111/j.1365-2982.2010.01472.x.

Jing, H. J., Wang, Q. H., Wang, L. S., Wang, Y. H., & Zhang, F. (2021). Probiotics, prebiotics, and symbiotic for the improvement of metabolic profiles in patients with chronic kidney disease: A systematic review and meta-analysis of randomized controlled trials. *Critical Reviews in Food Science and Nutrition, 61*(4), 577–598. https://doi.org/10.1080/10408398.2020.1748515.

Kajander, K., Krogius-Kurikka, L., Rinttilä, T., Karjalainen, H., Palva, A., & Korpela, R. (2007). Effects of multispecies probiotic supplementation on intestinal microbiota in irritable bowel syndrome. *Alimentary Pharmacology & Therapeutics, 26*(3), 463–473. https://doi.org/10.1111/j.1365-2036.2007.03362.x.

Kajander, K., Myllyluoma, E., Rajilić-Stojanović, M., Kyrönpalo, S., Rasmussen, M., Järvenpää, S., Zoetendal, E. G., De Vos, W. M., Vapaatalo, H., & Korpela, R. (2008). Clinical trial: Multispecies probiotic supplementation alleviates the symptoms of irritable bowel syndrome and stabilizes intestinal microbiota. *Alimentary Pharmacology & Therapeutics, 27*(1), 48–57. https://doi.org/10.1111/j.1365-2036.2007.03542.x.

Keeley, B. R., Creedon, D., Battaglia, T., Lacy, B. E., & Ford, A. C. (2020). Prevalence of irritable bowel syndrome-type symptoms in patients with inflammatory bowel disease in remission: A systematic review and meta-analysis. *The Lancet Gastroenterology & Hepatology, 5*(12), 1053–1062. https://doi.org/10.1016/S2468-1253(20)30244-7.

Khoshnood, N., & Niaz, A. (2013). Prevalence of depression and anxiety in irritable bowel syndrome: A clinic-based study from India. *Indian Journal of Psychiatry, 55*(1), 77–80. https://doi.org/10.4103/0019-5545.105508.

Krause, M. V., & Mahan, L. K. (Eds.). (2016). *Krause's Food & the Nutrition Care Process* (14th ed.). Elsevier.

Le, C., & Rhodes, D. (2002). Illness representation, coping, and outcome in irritable bowel syndrome (IBS). *British Journal of Health Psychology, 7*(4), 377–391. https://doi.org/10.1348/135910702320645381.

Lembo, A. (2004). Irritable bowel syndrome medications side effects survey. *Journal of Clinical Gastroenterology, 38*(9), 776–781. https://doi.org/10.1097/01.mcg.0000139052.56956.83.

Lovell, R. M., & Ford, A. C. (2012). Global prevalence of and risk factors for irritable bowel syndrome: A meta-analysis. *Clinical Gastroenterology and Hepatology, 10*(7), 712–721. https://doi.org/10.1016/j.cgh.2012.02.029.

Makharia, G. K., Verma, A. K., Amarchand, R., Goswami, A., Singh, P., Agnihotri, A., Suhail, F., & Krishnan, A. (2011). Prevalence of irritable bowel syndrome: A community-based study from northern India. *Journal of Neurogastroenterology and Motility, 17*(1), 82–87. https://doi.org/10.5056/jnm.2011.17.1.82.

Marasco, G., Cirota, G. G., Rossini, B., Lungaro, L., Di Biase, A. R., Colecchia, A., Volta, U., De Giorgio, R., Fest, D., & Caio, G. (2020). Probiotics, prebiotics, and other dietary supplements for gut microbiota modulation in celiac disease patients. *Nutrients, 12*(9), 2674. https://doi.org/10.3390/nu12092674.

Mönnikes, H. (2011). Quality of life in patients with irritable bowel syndrome. *Journal of Clinical Gastroenterology, 45*(Suppl), S98–S101. https://doi.org/10.1097/MCG.0b013e31822ff53d.

National Institute of Diabetes and Digestive and Kidney Diseases. (n.d.). Treatment for irritable bowel syndrome (IBS). Retrieved from: https://www.niddk.nih.gov/health-information/digestive-diseases/irritable-bowel-syndrome/treatment (accessed on 19 July 2024).

Oka, P., Parr, H., Barberio, B., Black, C. J., Savarino, E. V., & Ford, A. C. (2020). Global prevalence of irritable bowel syndrome according to Rome III or IV criteria: A systematic review and meta-analysis. *The Lancet Gastroenterology & Hepatology, 5*(10), 908–917. https://doi.org/10.1016/S2468-1253(20)30217-X.

Oniszczuk, A., Oniszczuk, T., Gancarz, M., & Szymańska, J. (2021). Role of gut microbiota, probiotics, and prebiotics in cardiovascular diseases. *Molecules, 26*(4), 1172. https://doi.org/10.3390/molecules26041172.

Pletikosić, S., & Tkalčić, M. (2017). A measure of suffering in relation to anxiety and quality of life in IBS patients: Preliminary results. *BioMed Research International, 2017*, 1–8. https://doi.org/10.1155/2017/3129594.

Rimmer, L. R., & Rimmer, L. (2008). Treatments for irritable bowel syndrome: Patients' attitudes and acceptability. *BMC Complementary Medicine and Therapies, 8*, 65. https://doi.org/10.1186/1472-6882-8-65.

Schauer, B., Grabe, H. J., Ittermann, T., Lerch, M. M., Weiss, F. U., Mönnikes, H., Völzke, H., Enck, P., & Schwille-Kiuntke, J. (2018). Irritable bowel syndrome, mental health, and quality of life: Data from a population-based survey in Germany. *Neurogastroenterology and Motility, 31*(3), e13348. https://doi.org/10.1111/nmo.13348.

Spiegel, B. M. R., DeRosa, V. P., Gralnek, I. M., Chang, L., & Dulai, G. S. (2004). The impact of irritable bowel syndrome on health-related quality of life. *Gastroenterology, 126*(1), 210–219. https://doi.org/10.1053/j.gastro.2003.10.049.

Vandvik, P. O., Lydersen, S., & Farup, P. G. (2006). Prevalence, comorbidity and impact of irritable bowel syndrome in Norway. *Scandinavian Journal of Gastroenterology, 41*(6), 650–656. https://doi.org/10.1080/00365520500441264.

Wilson, B., & Whelan, K. (2017). Prebiotic inulin-type fructans and galacto-oligosaccharides: Definition, specificity, function, and application in gastrointestinal disorders. *Journal of Gastroenterology and Hepatology, 32*(3), 416–427. https://doi.org/10.1111/jgh.13592.

World Gastroenterology Organization Global Guideline. (2015). Irritable bowel syndrome: A global perspective. Retrieved from: https://www.worldgastroenterology.org/guidelines/global-guidelines/irritable-bowel-syndrome (accessed on 19 July 2024).

Zamani, M., Alizadeh-Tabari, S., & Zamani, V. (2019). Systematic review with meta-analysis: The prevalence of anxiety and depression in patients with irritable bowel syndrome. *Alimentary Pharmacology & Therapeutics, 50*(2), 132–143. https://doi.org/10.1111/apt.15310.

Zhu, L., Huang, D., Shi, L., Xu, T., Xu, T., Chen, W., Wu, D., Zhang, F., & Fang, X. (2015). Intestinal symptoms and psychological factors jointly affect quality of life of patients with irritable bowel syndrome with diarrhea. *Health and Quality of Life Outcomes, 13*(1), 49. https://doi.org/10.1186/s12955-015-0232-4.

CHAPTER 7

Role of Prebiotic and Fermented Beverages in Mental Health – Depression

SHRUTI DWIVEDI

Department of Foods and Nutrition, The Maharaja Sayajirao University of Baroda, Gujarat, India

Mental health statistics reflects the crux of integrated view of any society's overall health. Still, it is often underestimated. The 2017 data states about 197.3 million people are victim of mental disorder in the country. Anxiety disorders affect 44.9 million people, followed by depressive disorders (45 7 million). Looking at the rising trend there is urgent need to address this problem. One probable solution is psych biotics.

7.1 INTRODUCTION

Mental health should be treasured for societal well-being and human capital. To think and thrive, good psychological status is a necessity. It's vital not only to meet the needs of people with special mental needs but also to preserve, protect, and promote the psychological health of the population in general. Mental disabilities affect more than 450 million people all over the world (Updated factsheet 2016 by the World Health Organization). Individual traits, social, and environmental factors influence the mind's ability. These determinants work in a complex way, and they can either threaten

Prebiotic Foods for a Healthy Gut: Scientific Studies to Manage Metabolic Disorders.
Mini K. Sheth and Shruti Dwivedi (Eds.)

or protect a person's mental status. According to data from the Institute for Health Metrics and Evaluation from 2016 (IHME), 121.24 million people reported suffering from mental and substance use disorders, followed by 109.46 million people affected by neurological disorders. Data from 2017 points toward an increase in this graph peak, indicating a rise in the growth of mental disorders to 792 million. According to WHO estimates, depression affects more than 300 million people globally. Nearly half of the world's depressed people live in the South-East Asia and Western Pacific region, which includes India. By 2030, depression and anxiety disorders are expected to contribute the most to the global burden of disease, respectively. Anxiety disorders are diagnosed in nearly half of all people who are diagnosed with depression (Mathers and Loncar, 2006).

7.2 MENTAL DISORDERS' CONTRIBUTION TO DISABILITY ADJUSTED LIFE YEARS

Disability Adjusted Life Years is the measure of the overall burden of disease. Mental disorders' contribution to disability adjusted life years and the mean age group affected by the condition across different Indian states have been clubbed in a 2020 study. Depressive disorders are more prevalent (33.8%, 29.5–38.5), followed by anxiety disorders (19.0%, 15.9–22.4), idiopathic developmental intellectual disability (10.8%, 6.3–15.9), schizophrenia (9.8%, 7.7–12.4), bipolar disorder (6.9%, 4.9–9.6), conduct disorder (5.9%, 4.0), autism spectrum disorders (3.2%, 2.7–3.8), eating disorders (2.2%, 1.7–2.8), and attention-deficit hyperactivity disorder (0.3%, 0.2–0.5) (Sagar et al., 2020). Looking at the 27-year graph from 1990 to 2017, the share of mental disorders to DALYs in India almost doubled.

Globally, more years are lost due to disability due to depression than any other ailment. The DSM has undergone five revisions between 1952 and 2013. Some of the following symptoms are common to all depressive disorders:

- Suffering from a sad or depressed mood
- Loss of enjoyment or interest in once-enjoyed activities
- Appetite changes, weight gain or reduction unrelated to diets
- Sleep issues or excessive sleep
- A decrease in energy or increased tiredness
- A rise in pointless movement (such as pacing or hand-wringing) or slower speech and movement (actions observable by others)
- Feeling guilty or unworthy

- Difficulty focusing, thinking, or making choices
- Dismal and gloomy predictions about the future
- Decreased confidence and self-esteem
- Suicidal or self-destructive ideas or behaviors

Symptoms should continue for at least two weeks for the diagnosis of depression. Table 7.1 depicts share of global population of various mental disorders.

TABLE 7.1 Share of Global Population of Various Mental Disorders

Disorder	Share of Global Population with Disorder (2017) [Difference Across Countries]	Number of People with the Disorder (2017)	Share of Males: Females with Disorder (2017)
Mental health disorder (Any)	10.7%	792 million	9.3% males 11.9% females
Depression	3.4% [2–6%]	264 million	2.7% males 4.1% females
Anxiety disorders	3.8% [2.5–7%]	284 million	2.8% males 4.7% females
Bipolar disorder	0.6% [0.3–1.2%]	46 million	0.55% males 0.65% females
Eating disorders (clinical anorexia & bulimia)	0.2% [0.1–1%]	16 million	0.13% males 0.29% females
Schizophrenia	0.3% [0.2–0.4%]	20 million	0.26% males 0.25% females
Any mental or substance use disorder	13% [11–18%]	970 million	12.6% males 13.3% females
Alcohol use disorder	1.4% [0.5–5%]	107 million	2% males 0.8% females

Source: IHME, 2017.

7.3 DEPRESSION IN INDIA

As per 2017 data, 197.3 million people reported being victims of mental disorders in the country, which is equivalent to one in seven Indians, with 44.9 million anxiety disorders followed by 45.7 million depressive disorders. The national mental health survey conducted across 12 Indian states in 2015–16 reported that in India, one in every twenty (5.25%) people over the age of 18 has suffered from depression at least once in their lives, amounting to over 45 million people. When compared to males (4.75%), females had a higher lifetime prevalence of depressive disorders (5.72%).

The WHO in 2008 reported India as one of the world's most depressed countries, with 9% of people having depression and 36% experiencing major depressive episodes (Smith, 2014). It is a serious health condition that may even start at a young age and may last long; once treated, it may recur. Every year, nearly eight lakh people commit suicide, making it the leading cause of death among people aged 15 to 29 according to the WHO factsheet from 2020 (Figure 7.1).

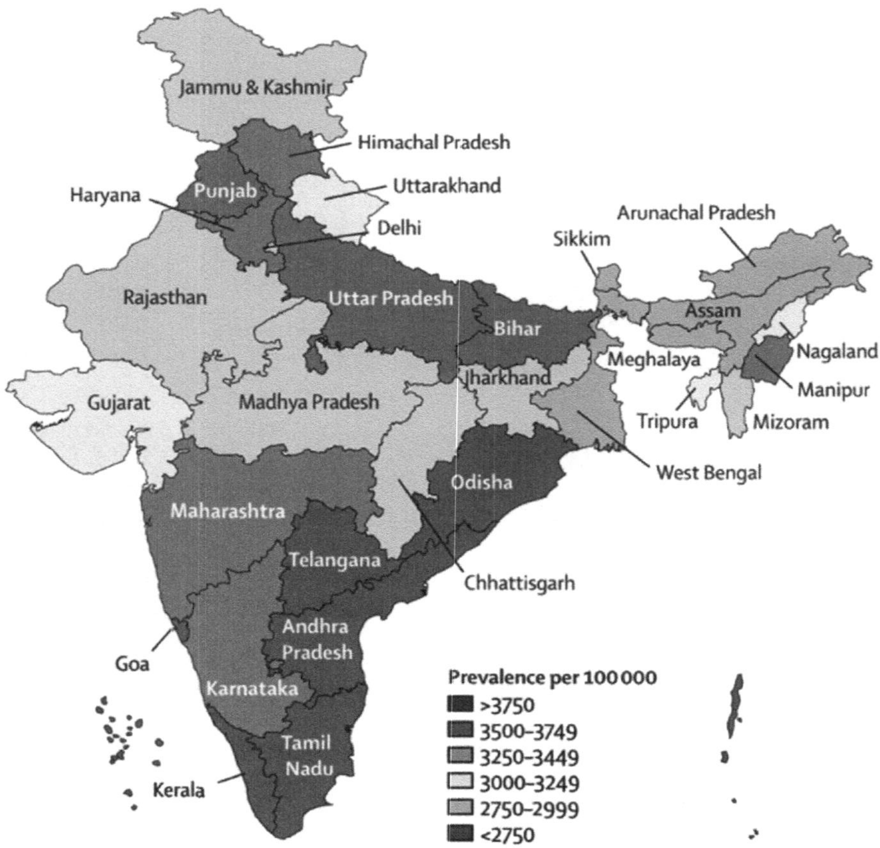

FIGURE 7.1 State wise prevalence of depressive disorders across Indian states.

Source: India State-Level Disease Burden Initiative Mental Disorders Collaborators, 2020. Copyright © 2020 World Health Organization. http://creativecommons.org/licenses/by/3.0/.

The prevalence of current DD varies by state, ranging from 1.2% in Gujarat to 4.7% in Jharkhand. One or more mental health conditions require active attention in over 15% of Indian persons over the age of 18.

It is crucial to raise the curtain of stigma and burden and bring to light the staggering load depression has added on individuals' psychological and physical state in terms of non-communicable diseases, social, and emotional responses.

7.4 AN EXCITING AVENUE TO COMBAT DEPRESSION: GUT-BRAIN AXIS

The hype surrounding the human microbiome and its impact on mental health stems from a long-held belief that manipulating the host's gut microbiota confers positive health benefits. There are almost 50 phyla of bacteria dwelling in the dynamic ecosystem of the human intestine. Trillions of bacteria present in the gut form the link of cross-talk between the gut and brain. This complex community of gut flora interacts with each other and local tissues maintaining homeostasis. Gut flora holds the power to positively or negatively regulate the responses of the central nervous system by acting as a catalyst amid brain-gut communication loop.

Equanimity of the gut is crucial for the proper functionality of the human system. Dysbiosis of the gut microflora leads to an overgrowth of pathogenic bacteria like *Salmonella, Shigella, Clostridia, Staphylococcus aureus, Candida albicans, Campylobacter jejuni, Escherichia coli, Veillonella*, and *Klebsiella,* which secrete proinflammatory cytokines, amyloids, and lipopolysaccharides associated with the pathogenesis of various infections and metabolic diseases of the liver, respiratory tract, gastrointestinal tract, and neurodegenerative diseases. This gut-brain axis is regulated via various direct and indirect routes involving neural, endocrine, immune, and metabolic pathways, making our intestine far beyond the organ for just food processing. As our autonomous enteric nervous system is actually capable of defining one's state of mind, it can undoubtedly be referred to as the second brain.

7.5 MEDIUM OF BIOCHEMICAL SIGNALING BETWEEN THE ENTERIC NERVOUS SYSTEM AND CENTRAL NERVOUS SYSTEM

a) **Vagus nerve** is one of the longest fundamental nerves of the parasympathetic nervous system. This neural circuit connects the brain and gastrointestinal tract by sending crucial messages from inner body organs to the brain within seconds through nerve signals, which travel much faster than hormones. The presence of this direct loop and the speed with which information gets transmitted through vagus nerve signals has been confirmed through research findings. Rabies virus travels and spreads to the brain of a mouse via neurons. A fluorescently-tagged rabies virus was inserted into the colon of mice. As the virus enrooted from enteroendocrine cells of the intestine through the vagus nerve all the way to the brainstem, the traced fluorescent signals indicated the vagus nerve as a direct line of communication between the gut and brain (Kaelberer et al., 2018).

The activity of the vagus nerve is dependent on the gut microbiota metabolites. Patients with IBS generally have low vagal tone favoring peripheral inflammation. A comparative study by Pellissier, in 2014 showed reduced vagal tone and cortisol levels in patients suffering from IBS compared to healthy subjects. The vagus nerve is responsible for anxiogenic and anxiolytic effects, acting as one of the strong linkages between the gut microbiome and the brain. Infection by pathogens, such as *Campylobacter jejuni* and *Citrobacter amalonaticus,* increased anxiety-like behavior. Supplementation with prebiotics improved the mental condition in mice. However, these effects were eliminated on vagotomy (Quigley, 2017).

b) **Neuroendocrine,** the hypothalamic-pituitary-adrenal (HPA) axis is the core of the neuroendocrine system that directs the body's metabolism, development, and stress reactions. The gut microbial ecosystem plays a crucial role in the smooth functioning and maturation of this axis. Recent research has shown that reconstitution with beneficial microbes, such as Bifidobacterium infantis, can reverse the exaggerated HPA response in germ-free (GF) mice (Sudo, 2014). Enteroendocrine cells (EECs) that line the entire gastrointestinal (GI) mucosa release specific neuropeptides, including peptide YY, neuropeptide Y (NPY), cholecystokinin, glucagon-like peptide-1 and -2, and substance P. This secretion directly affects the brain and regulates the microbiota and their bioactive metabolites, which act as messengers of the microbiota-gut-brain axis.

NPY signaling not only tunes neuroprotection, neurogenesis, and neuroinflammation but also contributes to the composition of gut microbiota influencing the brain. Evidence points toward increased colonic synthesis of NPY with depression and anxiety. When gut microbiota was depleted from mice, NPY expression within the amygdala and hypothalamus was disturbed, followed by cognitive deficits (Fröhlich et al., 2016). Proinflammatory cytokines are decreased by the GLP-1 receptor, aiding in neuroprotection. Various pieces of evidence support its ability to reverse depression and cognitive deficits. The ability of this receptor in reversing LPS-induced, diabetes-induced depression, and anxiety-like behavior has been proved. A similar four-week study emphasized the ability of GLP-1 treatment in reversing mood disorders (Sharma et al., 2015).

c) **Neurotransmitters** enable signal transmission between neurons, forming a chemical synapse in the brain. Dopamine and serotonin, also known as the body's "feel-good" hormones, affect overall mental health, including mood, memory, sleep, libido, and appetite. Interestingly, research has shown that gut bacteria have the potential to directly produce a range of major neurotransmitters that affect psychological status, highlighting the connection between the brain and gut. The gut is where the body produces more than 90% of its serotonin and 50% of its dopamine. Germ-free mice show a 50% reduction in serotonin levels (Wikoff et al., 2009). Introduction of *Turicibacter sanguinis* and *Clostridia* signals the gut cells to produce serotonin (Fung et al., 2019). Gamma-aminobutyric acid (GABA) maintains the functionality of both the central and enteric nervous systems. Altered levels of GABA have been shown to interfere with behavior patterns, circadian rhythm, intestinal motility, gastric emptying, nociception, and acid secretion (Hyland and Cryan, 2010), indicating its association with both the gut and brain. Evidence has revealed that the biosynthesis of GABA is dependent on gut bacteria. Luminal and serum levels of GABA were substantially reduced in germ-free mice (Matsumoto et al., 2013).

d) **Short Chain Fatty Acids,** the stomach produces about 500–600 mmol of SCFAs every day, which directly affect the gut–brain axis via many mechanisms. SCFA, particularly acetate, propionate, and butyrate in an approximate molar rate of 60:20:20, respectively, are the main metabolites produced by the microbiota in the large intestine which serve a vast array of roles in protecting gut-brain integrity.

SCFA interact with receptors present on enteroendocrine cells which promote indirect signaling to the brain via the systemic circulation or vagal pathways by inducing the secretion of gut hormones such as glycogen-like peptide 1, peptide YY, γ-aminobutyric acid, and serotonin.

Peripherally, SCFAs stimulate systemic inflammation by modulating interleukin secretion. Monocarboxylate transporters are located on endothelial cells, which aid SCFAs in crossing the blood-brain barrier, influencing its integrity. SCFAs potentially affect the pathophysiology of mental disorders by modulating the levels of neurotrophic factors, increasing and improving neurogenesis, neural homeostasis, and contributing to the biosynthesis of serotonin (Erny et al., 2015). A 2018 study by Skonieczna-Żydecka et al. showcased decreased levels of propionic acid and acetic acid concentration in depressed women when compared with healthy individuals. Sodium butyrate reversed behavioral hyperactivity and acted as an antidepressant in rats (Moretti et al., 2011).

Consistent with these findings, injecting butyrate induced histone hyperacetylation in the hippocampus and frontal cortex of mice (Schroeder et al., 2007). Intraventricular propionic acid injection induced irreversible behavioral changes (Shultz and MacFabe, 2014).

7.6 PREBIOTICS AS A MEDIUM OF COMMUNICATOR BETWEEN GUT MICROBIOME–BRAIN–AXIS

Remodulation of the gut microbiota to achieve a positive balance is key to mental health. The influence of the gut microbiota on brain chemistry has been convincingly demonstrated in rodents and in a few human trials. Nurturing with prebiotics has proven to be an effective strategy that improves psychological healing while providing additional health benefits. Various animal studies and a few human trials have assessed the efficacy of prebiotics in promoting the colonization of beneficial gut bacteria to enhance mental well-being. The administration of prebiotics 3′Sialyllactose and 6′Sialyllactose reduced stress and anxiety responses in mice subjected to an immune challenge stressor (Savignac et al., 2016). The quality of sleep in rats was also enhanced by reducing the impact of stress on diurnal rhythms, which was associated with changes in gut microbiota (Thompson et al., 2017). Prebiotics increased hippocampal BDNF levels relative to controls in rats supplemented with oligosaccharides (Savignac et al., 2013). A 12-week Xylooligosaccharide intervention improved the metabolic parameters and brain pathology of rats. Gut inflammation decreased, hippocampal plasticity was restored, and cognitive function improved (Chunchai et al., 2018). A study in 2021 evaluated the impact of a synbiotic containing prebiotics (fructooligosaccharides) fed to broiler chickens exposed to heat stress.

Compared to control chickens, those fed the synbiotic showed a reduction in stress and fear responses (Mohammed et al., 2021). Administration of the prebiotic Bimuno® galactooligosaccharide (B-GOS) mixture (15 g/L) for 3 weeks to rats ameliorated postoperative cognitive dysfunction (POCD) and neuroinflammation through the microbiota-brain-axis. Proliferation of *Bifidobacterium* and other potentially anti-inflammatory microbes was also reported (Yang et al., 2018). FOS and GOS combination exhibited antidepressant effect, they enhanced the composition of fecal SCFA acetate and propionate, and decreased the levels of elevated corticosterone and proinflammatory cytokines in mice. Increased fecal SCFA acetate and propionate compositions and reduced elevated corticosterone and proinflammatory cytokine levels in mice (Burokas et al., 2017). Supplementing dietary fiber polydextrose to 18 healthy female participants resulted in a modest improvement in cognitive performance (Berding et al., 2021). Ingestion of prebiotic for eight weeks in adequate amount resulted in a significant improvement of brain-derived neurotrophic factor (BDNF) and imparted anxiolytic effects in 25 patients with low to moderate depression (Heidarzadeh et al., 2020). Prebiotic compounds, fructo-oligosaccharides (FOS) and galacto-oligosaccharides (GOS) increased intrinsic enteric microbiota of rats, which increased hippocampal BDNF and NR1 subunit expression relative to controls. Elevation in plasma peptide YY (PYY) after GOS intake was also observed (Savignac et al., 2013). Feeding galactooligosaccharides to anxious mice for 3 weeks relaxed them as compared to controls. This data suggested the potential of GOS in treating neuropsychiatric defects (Savignac et al., 2016). A significant increase in mean log counts of beneficial gut microbiota with improvement in SAGE scores of the Mild to moderate AD subjects was seen by FOS supplementation.

Phil Burnet with Kristin Schmidt at Oxford in England carried out the first and only research on humans regarding the effect of prebiotics on the stress hormone cortisol. A three-week prebiotic (GOS) administration showed a significant reduction in cortisol awakening response and decreased attentional vigilance to negative versus positive information in a dot-probe task (Schmidt et al., 2015). GOS consumption increased fecal *Bifidobacteria* and attenuated stress-induced gastrointestinal distress in the adult human population (Schmidt et al., 2015).

Dwivedi and Sheth (2022) studied the efficacy of fermented beverages and prebiotic supplementation in managing depression by modulating fecal microbial counts (*Lactobacillus*, *Bifidobacteria*, and *E. coli*) as indicated in Figure 7.2(a-h). Using systematic random sampling, subjects were

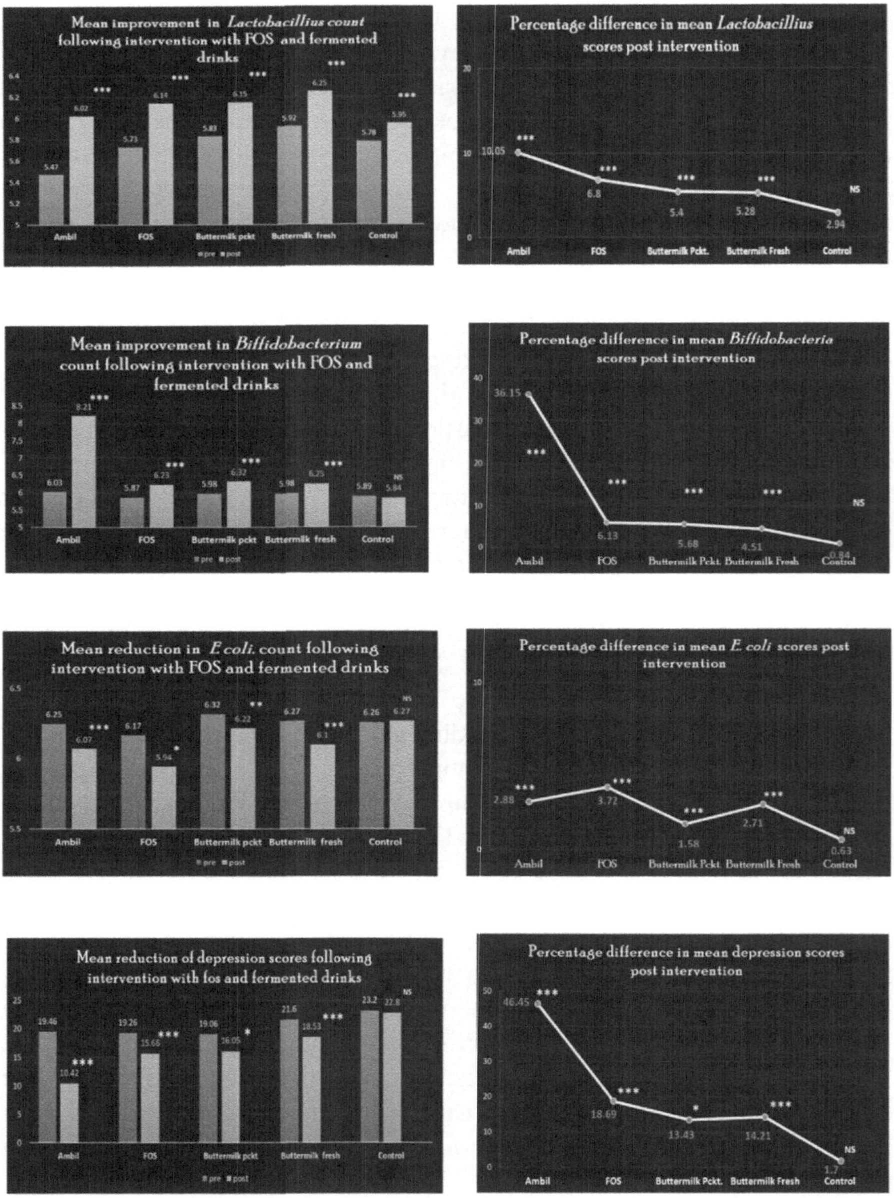

FIGURE 7.2 Pre- and post-assessment of depression, blood serum cortisol, and log count of fecal microbiota with respect to *Lactobacillus*, *Bifidobacteria*, and *E. coli* in mild to moderately depressed subjects post-intervention with fermented beverages and FOS.

segregated into 4 groups of 30 individuals each. They were supplemented for 45 days either with fructooligosaccharide (FOS, prebiotic), ambil (a prebiotic fermented beverage), fresh buttermilk, and packed buttermilk respectively. Gut microflora count, depression status, and blood cortisol levels were determined pre and post-intervention trials. Ambil proved to be the most effective intervention in increasing the colonization of good gut bacteria; *Lactic acid bacteria,* and *Bifidobacterium* by (36.15% and 10.05%) followed by FOS (6.8% and 6.13%), tetra-packed buttermilk (5.4% and 5.68%), and fresh buttermilk (5.28% and 4.51%) respectively. The highest reduction in the count of pathogenic *E. coli* was reported on FOS supplementation (3.72%) followed by ambil (2.88%), fresh buttermilk (2.71%), and tetra-packed buttermilk (1.58%). A downturn in depression rate was observed highest on supplementation with ambil (46.45%) followed by FOS (18.69%), fresh buttermilk (14.21%), and tetra-packed buttermilk (13.43%).

Literature proves the existence of a bidirectional mechanism and a strong positive relation between gut flora and depression. Getting the right gut bacteria is one of the favorable options and the core connection for the gut-brain axis. Research in the area highlights the benefits of psych biotics in strengthening the 'gut-brain axis.' This emphasizes the importance of gut ecology in providing exciting avenues for psychological therapies for conditions like anxiety and depression.

KEYWORDS

- **alcohol use**
- **autonomic nervous system**
- **depression (mental disorders)**
- **disabled people**
- **eating disorders**
- **health**
- **mental health**
- **mental well-being**
- **mice**
- **viruses**

REFERENCES

Berding, K., Long-Smith, C. M., Carbia, C., Bastiaanssen, T. F. S., van de Wouw, M., Wiley, N., Strain, C. R., Fouhy, F., Stanton, C., Cryan, J. F., & Dinan, T. G. (2021). A specific dietary fiber supplementation improves cognitive performance—An exploratory randomized, placebo-controlled, crossover study. *Psychopharmacology, 238*(1), 149–163. https://doi.org/10.1007/s00213-020-05665-y.

Burokas, A., Arboleya, S., Moloney, R. D., Peterson, V. L., Murphy, K., Clarke, G., Stanton, C., Dinan, T. G., & Cryan, J. F. (2017). Targeting the microbiota-gut-brain axis: Prebiotics have anxiolytic and antidepressant-like effects and reverse the impact of chronic stress in mice. *Biological Psychiatry, 82*(7), 472–487. https://doi.org/10.1016/j.biopsych.2016.12.031.

Chunchai, T., Thunapong, W., Yasom, S., Wanchai, K., Eaimworawuthikul, S., Metzler, G., Lungkaphin, A., Pongchaidecha, A., Sirilun, S., Chaiyasut, C., Pratchayasakul, W., Thiennimitr, P., Chattipakorn, N., & Chattipakorn, S. C. (2018). Decreased microglial activation through gut-brain axis by prebiotics, probiotics, or symbiotic effectively restored cognitive function in obese-insulin resistant rats. *Journal of Neuroinflammation, 15*(1). https://doi.org/10.1186/s12974-018-1055-2.

Dwivedi and Sheth (2022). Role of fructooligosaccharide buttermilk and biogenic metabolites released from fermented beverage ambil as a communicator between gut and brain [Doctoral dissertation, Department of Foods and Nutrition, Maharaja Sayajirao University of Baroda]. http://hdl.handle.net/10603/478232

Fröhlich, E. E., Farzi, A., Mayerhofer, R., et al. (2016). Cognitive impairment by antibiotic-induced gut dysbiosis: Analysis of gut microbiota-brain communication. *Brain, Behavior, and Immunity, 56*, 140–155. https://doi.org/10.1016/j.bbi.2016.02.020.

Fung, T. C., Vuong, H. E., Luna, C. D. G., Pronovost, G. N., Aleksandrova, A. A., Riley, N. G., Vavilina, A., McGinn, J., Rendon, T., Forrest, L. R., & Hsiao, E. Y. (2019). Intestinal serotonin and fluoxetine exposure modulate bacterial colonization in the gut. *Nature Microbiology, 4*(12), 2064–2073. https://doi.org/10.1038/s41564-019-0540-4.

Heidarzadeh-Rad, N., Gökmen-Özel, H., Kazemi, A., Almasi, N., & Djafarian, K. (2020). Effects of a psychobiotic supplement on serum brain-derived neurotrophic factor levels in depressive patients: A post hoc analysis of a randomized clinical trial. *Journal of Neurogastroenterology and Motility, 26*(4), 486–495. https://doi.org/10.5056/jnm20079.

Hyland, N. P., & Cryan, J. F. (2010). A gut feeling about GABA: Focus on GABAB receptors. *Frontiers in Pharmacology, OCT*. https://doi.org/10.3389/fphar.2010.00124.

India State-Level Disease Burden Initiative Mental Disorders Collaborators (2020). The burden of mental disorders across the states of India: the Global Burden of Disease Study 1990-2017. Lancet Psychiatry. 7(2):148-161. doi: 10.1016/S2215-0366(19)30475-4. Epub 2019 Dec 23. PMID: 31879245; PMCID: PMC7029418.

Kaelberer, M. M., Buchanan, K. L., Klein, M. E., Barth, B. B., Montoya, M. M., Shen, X., & Bohórquez, D. V. (2018). A gut-brain neural circuit for nutrient sensory transduction. *Science, 361*(6408). https://doi.org/10.1126/science.aat5236.

Keightley, P. C., Koloski, N. A., & Talley, N. J. (2015). Pathways in gut-brain communication: Evidence for distinct gut-to-brain and brain-to-gut syndromes. *Australian & New Zealand Journal of Psychiatry, 49*(3), 207–214. https://doi.org/10.1177/0004867415569801.

Liang, S., Wang, T., Hu, X., Li, W., Jin, F., & Wang, L. (2012). Microorganism and behavior and psychiatric disorders (Article in Chinese). *Advances in Psychological Science, 20*, 75–97.

Mathers, C. D., & Loncar, D. (2006). Projections of global mortality and burden of disease from 2002 to 2030. *PLoS Medicine, 3*(11), e442. https://doi.org/10.1371/journal.pmed.0030442.

Matsumoto, M., Kibe, R., Ooga, T., Aiba, Y., Sawaki, E., Koga, Y., & Benno, Y. (2013). Cerebral low-molecular metabolites influenced by intestinal microbiota: A pilot study. *Frontiers in Systems Neuroscience, 7*. https://doi.org/10.3389/fnsys.2013.00009.

Mohammed, A., Mahmoud, M., Murugesan, R., & Cheng, H. (2021). Effect of a synbiotic supplement on fear response and memory assessment of broiler chickens subjected to heat stress. *Animals, 11*(2), 427. https://doi.org/10.3390/ani11020427.

Moretti, M., Valvassori, S. S., Varela, R. B., Ferreira, C. L., Rochi, N., Benedet, J., & Kapczinski, F. (2011). Behavioral and neurochemical effects of sodium butyrate in an animal model of mania. *Behavioral Pharmacology, 22*(8), 766–772. https://doi.org/10.1097/FBP.0b013e32834d0f1b.

Pellissier, S., Dantzer, C., Mondillon, L., Trocme, C., Gauchez, A. S., Ducros, V., Mathieu, N., Toussaint, B., Fournier, A., Canini, F., & Bonaz, B. (2014). Relationship between vagal tone, cortisol, TNF-alpha, epinephrine and negative effects in Crohn's disease and irritable bowel syndrome. *PLoS ONE, 9*(9), e105328. https://doi.org/10.1371/journal.pone.0105328.

Petschow, B., Doré, J., Hibberd, P., Dinan, T., Reid, G., Blaser, M., Cani, P. D., Degnan, F. H., Foster, J., Gibson, G., Hutton, J., Klaenhammer, T. R., Ley, R., Nieuwdorp, M., Pot, B., Relman, D., Serazin, A., & Sanders, M. E. (2013). Probiotics, prebiotics, and the host microbiome: The science of translation. *Annals of the New York Academy of Sciences, 1306*(1), 1–17. https://doi.org/10.1111/nyas.12303.

Quigley, E. M. M. (2017). Microbiota-brain-gut axis and neurodegenerative diseases. *Current Neurology and Neuroscience Reports, 17*(12), 94. https://doi.org/10.1007/s11910-017-0802-6.

Sagar, R., Dandona, R., Gururaj, G., Dhaliwal, R. S., Singh, A., Ferrari, A., Dua, T., Ganguli, A., Varghese, M., Chakma, J. K., Kumar, G. A., Shaji, K. S., Ambekar, A., Rangaswamy, T., Vijayakumar, L., Agarwal, V., Krishnankutty, R. P., Bhatia, R., Charlson, F., & Dandona, L. (2020). The burden of mental disorders across the states of India: The Global Burden of Disease Study 1990–2017. *The Lancet Psychiatry, 7*(2), 148–161. https://doi.org/10.1016/S2215-0366(19)30475-4.

Savignac, H. M., Corona, G., Mills, H., Chen, L., Spencer, J. P. E., Tzortzis, G., & Burnet, P. W. J. (2013). Prebiotic feeding elevates central brain-derived neurotrophic factor, N-methyl-D-aspartate receptor subunits, and D-serine. *Neurochemistry International, 63*(8), 756–764. https://doi.org/10.1016/j.neuint.2013.10.006.

Savignac, H. M., Couch, Y., Stratford, M., Bannerman, D. M., Tzortzis, G., Anthony, D. C., & Burnet, P. W. J. (2016). Prebiotic administration normalizes lipopolysaccharide (LPS)-induced anxiety and cortical 5-HT2A receptor and IL1-β levels in male mice. *Brain, Behavior, and Immunity, 52*, 120–131. https://doi.org/10.1016/j.bbi.2015.10.007.

Schmidt, K., Cowen, P. J., Harmer, C. J., Tzortzis, G., Errington, S., & Burnet, P. W. J. (2015). Prebiotic intake reduces the waking cortisol response and alters emotional bias in healthy volunteers. *Psychopharmacology, 232*(10), 1793–1801. https://doi.org/10.1007/s00213-014-3810-0.

Schroeder, F. A., Lin, C. L., Crusio, W. E., & Akbarian, S. (2007). Antidepressant-like effects of the histone deacetylase inhibitor, sodium butyrate, in the mouse. *Biological Psychiatry, 62*(1), 55–64. https://doi.org/10.1016/j.biopsych.2006.06.036.

Sharma, A. N., Pise, A., Sharma, J. N., & Shukla, P. (2015). Glucagon-like peptide-1 (GLP-1) receptor agonist prevents development of tolerance to anti-anxiety effect of ethanol and withdrawal-induced anxiety in rats. *Metabolic Brain Disease, 30*(3), 719–730. https://doi.org/10.1007/s11011-014-9627-z.

Shultz, S. R., & MacFabe, D. F. (2014). Propionic acid animal model of autism. In Comprehensive guide to autism (pp. 1755–1778). Springer New York. https://doi.org/10.1007/978-1-4614-4788-7_106.

Skonieczna-Żydecka, K., Łoniewski, I., Misera, A., Stachowska, E., Maciejewska, D., Marlicz, W., & Galling, B. (2019). Second-generation antipsychotics and metabolism alterations: A systematic review of the role of the gut microbiome. *Psychopharmacology, 236*(5), 1491–1512. https://doi.org/10.1007/s00213-018-5102-6.

Smith, K. (2014). Mental health: A world of depression. *Nature, 515*(7526), 181. https://doi.org/10.1038/515180a.

Sudo, N. (2014). Microbiome, HPA axis and production of endocrine hormones in the gut. *Advances in Experimental Medicine and Biology, 817*, 177–194. https://doi.org/10.1007/978-1-4939-0897-4_8.

Thompson, R. S., Roller, R., Mika, A., Greenwood, B. N., Knight, R., Chichlowski, M., Berg, B. M., & Fleshner, M. (2017). Dietary prebiotics and bioactive milk fractions improve NREM sleep, enhance REM sleep rebound and attenuate the stress-induced decrease in diurnal temperature and gut microbial alpha diversity. *Frontiers in Behavioral Neuroscience, 10*. https://doi.org/10.3389/fnbeh.2016.00240.

Wikoff, W. R., Anfora, A. T., Liu, J., Schultz, P. G., Lesley, S. A., Peters, E. C., & Siuzdak, G. (2009). Metabolomics analysis reveals large effects of gut microflora on mammalian blood metabolites. *Proceedings of the National Academy of Sciences of the United States of America, 106*(10), 3698–3703. https://doi.org/10.1073/pnas.0812874106.

Yang, X. D., Wang, L. K., Wu, H. Y., & Jiao, L. (2018). Effects of prebiotic galacto-oligosaccharide on postoperative cognitive dysfunction and neuroinflammation through targeting of the gut-brain axis. *BMC Anesthesiology, 18*(1). https://doi.org/10.1186/s12871-018-0642-1.

CHAPTER 8

Psychobiotics for Manipulating Gut–Brain Axis in Alzheimer's Disease

SHRUTI DWIVEDI AND MINI K. SHETH

Department of Foods and Nutrition, The Maharaja Sayajirao University of Baroda, Gujarat, India

Alzheimer's disease (AD) is the leading cause of dementia and is estimated to affect 10% of the individuals older than 65 years and nearly 50% of the individuals older than 85 years (WHO, 2019 factsheet). Gut-brain communication suggests the promising role of microbial-based therapies in the treatment of neurodegenerative disease.

8.1 INTRODUCTION

One of the most prevalent types of senile dementia is Alzheimer's disease (AD). The World Health Organization reports that AD accounts for 60–80% of cases worldwide and approximately 46.8 million people around the globe have dementia. It is a syndrome characterized by loss of memory and cognitive skills that affects a person's ability to perform daily activities. Recent knowledge on gut-brain communication suggests the promising role of microbial-based therapies for the treatment of neurodegenerative diseases. This bi-directional communication enables signals from the brain that can modify the motor, sensory, and secretory modalities of the gastrointestinal

Prebiotic Foods for a Healthy Gut: Scientific Studies to Manage Metabolic Disorders.
Mini K. Sheth and Shruti Dwivedi (Eds.)

tract; conversely, visceral messages from the gut to the central nervous system (CNS) can modulate and affect brain functions. Prebiotics, probiotics, and synbiotics can alter gut flora and improve host cognition possibly by the production of SCFA such as butyrate, acetate, and propionate, which reduce inflammation and stimulate the secretion of neurotransmitters. Psychobiotics seem to be a safer mode for the management of AD than conventional pharmacological medication while adding extra health benefits. More research needs to focus on the use of specific microbial strains in the treatment of AD and other related neurodegenerative disorders.

One of the leading causes of dementia is Alzheimer's disease (AD) and is estimated to affect 10% of individuals older than 65 years and nearly 50% of individuals older than 85 years (WHO, 2019 factsheet). It is clinically characterized by cognitive impairment that affects the functionality of the brain and behavior, marked by a progressive decline in carrying out daily activities significantly (Imbimbo et al., 2015). Every 3 seconds, a new case of dementia is diagnosed in the world. In 2015, 46.8 million people worldwide were living with dementia, which increased to 50 million by 2018. It is predicted that the numbers will double every 20 years, reaching 74.7 million by 2030 and surpassing 152 million by 2050 (The World Alzheimer's Report, 2018).

Dr. Alois Alzheimer, a neuroanatomist, reported extensive atrophy of the cerebral cortex in a 50-year-old woman in 1906. She had a follow-up for memory disturbance, aggression, paranoia, progressive sleep, and confusion until her death. On examination, distinctive plaques and neurofibrillary tangles around nerve cells were observed (Hippius and Neundörfer, 2003). Later, this disease was termed Alzheimer's disease. Brain degeneration begins more than 20 years before the symptoms appear; this is characterized as atypical Alzheimer's (Alzheimer's Association, 2018). The disease begins by affecting the medial temporal lobe and gradually spreads to the lateral temporal and parietal lobes, marked by loss of cognition and memory. As it progresses to the middle stage, the frontal lobe of the brain gets affected, hampering speech, reasoning, sensory processing, and consciousness. Some people may exhibit impulsive behavior due to hallucinations, delusions, and paranoia. During the late stage or severe Alzheimer's disease, patients become completely dependent on others for their care as the occipital lobe gets affected, affecting vision and communication (Lancet, 2011). Alzheimer's disease is not a normal part of aging, but age is the strongest known risk factor. About 50% of people have the apoE gene's e4 allele, which triples the risk of developing the disease as people age (Ballard et al., 2011). Sharp &

Gatz (2011) concluded that lower education increases the risk of dementia after reviewing 71 articles from January 1985 to July 2010. A cohort study of 6434 subjects carried out for 5 years concluded that moderate and high levels of physical activity were associated with a significantly lower risk for Alzheimer's and dementia (Laurin et al., 2001).

Various diseases have been linked with an increased risk of developing AD. Hypertension disrupts cerebral blood vessels, which damages the white matter region critical for cognitive function, promoting Alzheimer's pathology (Elias et al., 2003; Iadecola et al., 2016). Diabetes directly affects the accumulation of beta-amyloid in the brain (Toro et al., 2009). Kopf and Frölich (2009) reviewed fourteen prospective studies and concluded that patients associated with diabetes mellitus are more prone to develop AD. Another systematic review of 18 prospective studies with 14,331 participants found an association between total serum cholesterol, dementia, and cognitive decline (Anstey et al., 2008). The depressed Alzheimer's disease patients were more cognitively impaired and disabled than the non-depressed patients (Rovner et al., 1991). There have been compelling shreds of evidence for microbiome-based therapeutic strategies in treating neurodegenerative diseases, including Alzheimer's.

8.2 PATHOPHYSIOLOGY OF BRAIN IN ALZHEIMER'S DISEASE

Two major histopathological hallmarks often cited in the progression of AD are plaques and tangles. Amyloid Precursor Protein (APP), present in the cell membrane of the brain's neuron, helps with its repair and growth. APP breaks down and dissolves over time in the presence of α and γ secretase. If β secretase teams up with γ secretase, APP breaks down to its monomer amyloid beta. It clumps together outside the neuron, leading to the formation of beta amyloid plaques. These plaques interrupt neuron-to-neuron signaling, causing inflammation which hampers the brain's function and memory.

The cytoskeleton, made up of microtubules, keeps the neurons connected and ships nutrients and molecules along the length of the cell. A special protein called tau ensures to keep microtubules intact. Extracellular β-amyloid (Aβ) present between the neurons leads to activation of enzyme kinase that transfers a phosphate group to the tau protein, which reshapes and clumps tau protein, and neurofibrillary tangles are formed. This clumping leads to atrophy of the brain, gyri get narrower, sulci widen, and ventricles enlarge, which is responsible for deficits in cognition and synaptic plasticity (Shoji et al., 1992; Qiu et al., 1995; Chow, 2010; Reitz et al., 2011; Selkoe, 2016).

8.3 COMPOSITION OF ALZHEIMER'S GUT MICROBIOTA

A healthy gut is abundant in *Actinobacteria, Bacteroidetes, Firmicutes, Proteobacteria,* and *Verrucomicrobia.* This profile differs in the gut of patients with neurodegenerative disorders (Brandscheid et al., 2017; Parashar and Udayabanu, 2017; Wang et al., 2017).

There is impaired diversity of gut microbiota in Alzheimer's. A mice model with severe amyloid pathology has shown an increment in the phyla of *Firmicutes* and *Tenericutes* with a notable decline in Bacteroidetes and Actinobacteria (Harach et al., 2015; Brandscheid et al., 2017). Another study in support of the notion that AD decreases the abundance of anti-inflammatory bacterial species and increases the abundance of pro-inflammatory species by Cattaneo et al. (2017) revealed a reduction in *Eubacterium rectale* and *Bacteroides fragilis* in patients with amyloidosis while an increment in the number of *Escherichia/Shigella* was seen.

The amyloid transgenic drosophila model has also revealed a change in the diversity of microbial flora (Wu et al., 2017). Gut profile of humans suffering from AD was analyzed by Vogt et al. in 2017 using 16S ribosomal RNA gene sequencing. Data showed altered microflora composition where *Firmicutes* (78%) and *Bacteroidetes* (15%) were the most dominant phyla. *Lachnospiraceae* (39.1%), *Ruminococcaceae* (29.6%), and *Bacteroidaceae* (9.8%) were the most dominant families. In germ-free transgenic mice, colonization with microbiota from conventional APP transgenic mice increased cerebral A pathology, whereas control colonization with microbiota from wild-type mice had less impact on cerebral A levels (Harach et al., 2017). Pathogenic bacteria populating the gut microbiota can secrete large amounts of amyloids and lipopolysaccharides, which modulate signaling pathways and produce proinflammatory cytokines associated with the pathogenesis of AD (de De-Paula et al., 2018). This impairment has been linked with aggregated Aβ and hyperphosphorylated tau causing degenerated brain capillaries and vessels which hampers the brain blood supply accounting for cognitive decline (Kim et al., 2018).

8.4 MANAGEMENT OF ALZHEIMER'S DISEASE

Cognitive rehabilitation therapy, music therapy, caregiver training programs, social stimulation, and diet have demonstrated a positive effect on behavior and cognition in AD (Fang et al., 2017; Leggieri et al., 2019; Choi et al., 2013).

Five medications have been approved by the U.S. Food and Drug Administration (FDA), namely donepezil, galantamine, memantine, rivastigmine, and valium, which regulate the activity of chemical messengers involved in memory and learning (Atri et al., 2008). Crosstalk between the brain and gut has emerged as another popular and effective approach in improving the condition of AD.

8.5 MODULATING THE GUT MICROBIOTA TO IMPROVE AD

Our brain contains billions of neurons, but we less often talk about the fact that trillions of good bacteria are alive in our intestinal tracts. There are about a hundred trillion viable bacteria in the adult human gastrointestinal tract. These live bacteria account for about 90% of total cells in the human body, constituting two pounds of a body's weight (Frank and Pace, 2008). Gut microbiota and brain are involved in bidirectional communication with each other, linking central and enteric nervous systems with peripheral intestinal functions through the 'Gut-Brain-gut-axis' (Carabotti et al., 2015). Regulating gut commensal flora has proved to be a promising therapy for various neurodegenerative diseases like Parkinson's, depression, anxiety, autism, etc. (Bienenstock et al., 2015; Ambrosini et al., 2019; Dwivedi and Sheth, 2019).

The vagus nerve originating from the foregut and colon serves as a major pathway for the transmission of signals responsible for primary parasympathetic control, which is known to affect the brain. Altering the bacterial profile of the gut stimulates vagal afferent fibers, triggering the secretion of neurotransmitters – serotonin, cholecystokinin, glucagon-like peptide-1, peptide YY, ghrelin, and orexin into the intercellular space (Goehler et al., 2005). Lactobacillus johnsonii increases the activity of the gastric vagus nerve (Tanida et al., 2005).

The increased permeability of the gut and blood-brain barrier (BBB) induced by microbiota dysbiosis may mediate or affect AD pathogenesis (Hu et al., 2016; Jiang et al., 2017). The intestinal mucosal barrier is formed of epithelial cell lines in the gastrointestinal tract. These epithelial cells have tight junctions between them which protect the host by preventing health-hampering chemicals, bacteria, and other organisms from entering the blood circulation (Turner, 2009). Disruption of intestinal permeability causes toxins to leak out in the gut, accompanied by an increase in inflammatory levels (Kelly et al., 2015). This induced inflammation eventually threatens the integrity of the BBB, affecting the brain's health and leading to a leaky brain. Braniste et al. (2014) altered the gut microbiota of the GF mice model with good bacteria that decreased BBB permeability and up-regulated the expression of tight junction proteins.

The primary element of gram-negative bacteria's cell wall is lipopolysaccharide (LPS), an endotoxin. LPS is capable of inducing a serious body inflammatory response and it finds its way into the bloodstream through increased permeability of tight junctions caused by leaky gut. LPS has been associated with amyloid plaque formation, a protein linked with Alzheimer's disease. Mice injected with intraperitoneal injections of LPS altered the BBB transport of protein by increasing blood-to-brain influx, decreasing brain-to-blood efflux, and increasing neuronal Aβ production associated with cognitive decline and memory impairment (Lee et al., 2008; Kahn et al., 2012). Studies reported that patients with AD have three times the levels of LPS in plasma compared to the controls (Zhang et al., 2009; Hu and Jin, 2016).

Severe cases of AD have shown to reduce levels of GABA and other neurotransmitters in the frontal, temporal, and parietal cortex of the brain in postpartum studies (Solas et al., 2015). GABA is the major inhibitory neurotransmitter and its dysfunction has shown a direct contribution to cognitive decline (Lanctot et al., 2004). Modulation of gut microbiota through beneficial microbiota intervention may delay the progression of AD. Probiotics, prebiotics, and synbiotics (psychobiotics) alter the gut flora and provide an effective therapeutic strategy for neurodegenerative disorders.

8.6 PROBIOTICS IN GUT BRAIN AXIS

Probiotics to reduce depression can be as effective as pharmacological treatment. *Lactobacillus rhamnosus,* used to ferment milk into yogurt, contains GABA, which reduces anxiety. Valium and other anti-anxiety drugs also work by focusing on GABA receptors in the brain (Griffin, 2013).

Kollath introduced the term 'probiotics' in 1953 after Elie Metchnikoff postulated in 1907 that lactic acid bacteria could normalize bowel health and increase longevity by replacing or reducing the quantity of "putrefactive" bacteria in the gut. The FAO/WHO joint report 2001 defines Probiotics as "Living microorganisms which when administered in adequate amount confer health benefits on the host".

Researchers have claimed that probiotics restore gut microflora balance, which attributes towards a healthy intestine, strong immunity, balanced cholesterol, and confers health benefits in various psychiatric illnesses including AD. Strains of *Lactobacillus* and *Bifidobacterium* can improve symptoms associated with anxiety in patients with chronic fatigue syndrome, reduce depression, inflammation, and epithelial barrier function, and potentially ameliorate behavioral symptoms among children with autism and overcome Parkinson's (Critchfield et al., 2011; Messaoudi et al., 2011; Misra and Mohanty, 2019). A face-matching attention task and functional magnetic resonance imaging (fMRI), which scans and records brain activity, revealed lower levels of activity in several brain regions in the group supplemented with probiotics. Particularly noticeable changes in the periaqueductal gray region of the midbrain, a region of the midbrain that is involved in pain regulation, were seen (Tillisch et al., 2013).

8.7 ROLE OF PROBIOTICS IN AD

The mechanism by which the gut affects brain physiology and vice versa is partially modulated by short-chain fatty acid (SCFA) production. Probiotics are microbiome modulators, which restore normal microbial composition, increasing native SCFA production (Vieira et al., 2016). SCFA provides benefits in AD by offering alternative energy substrates for brain energy metabolism (Zilberter and Zilberter, 2017), attenuating inflammation by modulating the activation of microglia (Sánchez et al., 2017), and normalizing histone acetylation (Den Besten et al., 2015). Valeric acid, isovaleric acid, isobutyric acid, butyric acid, propionic acid, acetic acid, and formic acid have been shown to reduce inflammation and aggregations of Aβ and tau by inhibiting histone deacetylase in microglia, affecting the pathogenesis of AD (Cummings et al., 1987; Macfarlane and Macfarlane, 2012; Martins and Binosha, 2014).

Valeric acid, butyric acid, and propionic acid interfere with the assembly of Aβ1–40 and Aβ1–42 peptides into neurotoxic Aβ aggregates, attenuating Aβ-mediated pathologic processes (Ho et al., 2018). SCFA also affects the homeostasis of the CNS through vagal afferent nerve fibers.

Probiotics increase the production of neurotransmitters. Severe cases of AD have shown a reduction in levels of gamma-aminobutyric acid (GABA) in the frontal, temporal, and parietal cortex of the brain in postpartum studies (Lanctot et al., 2004 and Solas et al., 2015). GABA is the major inhibitory neurotransmitter, and its dysfunction has shown a direct contribution to cognitive decline (Lanctot et al., 2004). Gram-positive gut microbes such as *Bifidobacterium* and *Lactobacillus* metabolize glutamate to produce GABA. There is a direct correlation between the increase in GABA production in the GI tract with the amount of GABA in the CNS and vice versa. Microbiota also influences the production of other neurotransmitters such as serotonin, acetylcholine, dopamine, histamine, and noradrenaline, all of which have been hypothesized to regulate mental ability and cognition (Matsumoto et al., 2013; Yano et al., 2015). Selective serotonin reuptake inhibitors (SSRIs) have shown to reduce Aβ protein production in the brain, indicating that the increased extracellular serotonin 5-hydroxytryptamine (5-HT) levels can effectively reduce Aβ plaque formation, decreasing the risk of AD (Cirrito et al., 2011). More than 95% of serotonin is secreted in the human gut. Consuming *Lactobacillus rhamnosus* ameliorated brain inflammation with increased expression of GABA in the cingulate cortex, providing therapeutic benefit for cognitive deficit AD subjects (Bravo et al., 2011).

Probiotics also protect the structural and functional integrity of neurons by regulating the ionic concentration gradient through their antioxidant property. An intervention trial with *L. plantarum* MTCC1325 on rats for 60 days showed significant improvement in brain regions associated with the membrane transport ATPase system in AD (Mallikarjuna et al., 2016). This strain can also produce acetylcholine via both external and internal pathways (Girvin et al., 1954). Studies have shown that strong antioxidants decrease acetylcholinesterase levels and increase the Na^+, K^+-ATPase activities, which ultimately improve cognition by enhancing cholinergic transmission (Anand et al., 2011).

A study to evaluate the effect of supplementing probiotics on learning and memory was carried out on the rat model induced with amyloid beta (1–42) using the Morris water maze test. The group received 500 mg of probiotics (15×10^9 CFU) daily for 4 weeks before and 2 weeks after injection. Long-term synaptic plasticity and pre-synaptic mechanisms were evaluated using long-term potentiation (LTP) and paired-pulse facilitation (PPF) ratios. Probiotic supplementation positively increased learning and PPF scores. The serum lipid profile also showed improvement. This marked improvement was most likely via the release of neurotransmitters through presynaptic mechanisms (Leblhuber et al., 2017). Mice who were treated with probiotic formulation (SLAB51) showed alteration in the composition of gut microbiota affecting the partial restoration of two neural pathways, namely the ubiquitin-proteasome system and autophagy. Accumulation of amyloid-beta reduced and an increase in cognition was seen (Bonfili, 2018). This data was supported by various other mice model studies which have also shown the protective effect against AD by D-Galactose and scopolamine (Jung, 2012; Woo et al., 2014). *L. Plantarum* NDC75017 has also improved the learning and memory ability of AD rats (Peng et al., 2014). Probiotic supplementation to mice coupled with exercise showed improvement in learning and performance skills along with a significant decrease in the size of beta amyloid plaque area in the hippocampus (Abraham et al., 2016). The efficacy of probiotic supplementation in the improvement of memory and learning aspects of human AD patients was looked at by Abkari et al., in 2016. Intervention with 200 ml/day of milk enriched with *Lactobacillus acidophilus, Lactobacillus casei, Bifidobacterium bifidum,* and *Lactobacillus fermentum* for 12 weeks revealed significant improvement in Mini-Mental State Examination in both components by 27.90% and 8.07%, respectively, compared to the control.

Prebiotics are dietary fibers that allow specific changes, both in the composition and activity of the gastrointestinal microflora through their fermentation, and confer health benefits in many diseases including AD. Patients supplemented with prebiotic fructooligosaccharide (FOS) have shown better establishment of *Bifidobacteria* (131%) and *Lactobacillus* (32.6%) (Sheth et al., 2015), and they were less attentive to negative stimuli (Tuohy et al., 2005). Consuming biscuits containing 6.6 g FOS per day reported a significant increase in *Bifidobacterium* levels (Schmidt et al., 2015). Symbiotic supplementation, a combination of 200 ml probiotic buttermilk and 10 ml of liquid FOS, intervened with AD patients for 45 days, increasing total mean SAGE scores by 38%, with high improvements in memory (462%) and executive problem-solving (215%) components. The colonization of *Bifidobacterium* and Lactic acid bacteria improved significantly by 3.32% and 1.52%, along with a reduction in *E. coli* by 2.99% in the gut after the intervention trial (Gandhi and Sheth, 2019). Desharwala and Sheth (2018) also showed that supplementation of FOS (10 ml) followed for 45 days increased the total score of the Self-Administered Gero cognitive Exam (SAGE) by 27% (Tables 8.1 and 8.2).

The incidence of Alzheimer's increases exponentially with age. The onset can be prevented or delayed by focusing on strategies to check cognition and dementia by taking precautions beforehand, through lifestyle and dietary modifications. Evidence suggests the potential for psychobiotics-based therapeutic strategies in the augmentation of beneficial human microbiome helps in the management of various neurodegenerative diseases through the gut-brain axis. Probiotic supplementation beneficially affects microbes, which in turn increase the secretion of neurotransmitters, suppress inflammation, ameliorate oxidative stress, and decrease levels of beta-amyloids, thereby attenuating brain functional loss marked in AD.

TABLE 8.1 Fecal Log Counts of Mild to Moderate AD Subjects Before and After FOS Supplementation

Parameters		Mean log10 CFU/g
Lactic acid bacteria	Pre	8.01
	Post	8.14
	Paired 't' test	4.43***
	Difference	↑0.13
	% difference	↑1.67
Bifidobacteria	Pre	7.94
	Post	8.13
	Paired 't' test	6.23***
	Difference	↑0.19
	%difference	↑2.40
E coli	Pre	8.29
	Post	8.10
	Paired 't' test	8.85***
	Difference	↓0.19
	% difference	↓2.24

NOTE: Dissimilar superscripts within the rows indicate significant difference among the values; NS is non-significant; * Significant at 0.05; ** Significant at 0.01; *** Significant at 0.001.

TABLE 8.2 Changes in SAGE Score of Subjects with Mild – Moderate Alzheimer's Disease Before and After Supplementation of Prebiotic

Components of SAGE Scale	Maximum Score		Value	Paired T – Test T Critical Two-Tail: 2.009
Orientation	4	Pretest	3.27	1.20917^{NS}
		Post test	3.62	
		Difference	↑0.35	
		% difference	↑10.59	
Naming	2	Pretest	1.46	1.73916^{NS}
		Post test	1.73	
		Difference	↑0.27	
		% difference	↑18.42	
Similarities	2	Pretest	1.42	2.28695*
		Post test	1.81	
		Difference	↑0.39	
		% difference	↑27.1	

TABLE 8.2 *(Continued)*

Components of SAGE Scale	Maximum Score		Value	Paired T – Test T Critical Two-Tail: 2.009
Calculation A	1	Pretest	0.81	0.7581[NS]
		Post test	0.88	
		Difference	↑0.08	
		% difference	↑9.52	
Calculation B	1	Pretest	0.96	1[NS]
		Post test	1.00	
		Difference	↑0.04	
		% difference	↑4.00	
Construction:3-D figure	2	Pretest	0.92	3.033780*
		Post test	1.58	
		Difference	↑0.11	
		% difference	↑71.74	
Construction: Clock	2	Pretest	1.69	0.61546[NS]
		Post test	1.77	
		Difference	↑0.08	
		% difference	↑4.55	
Verbal fluency	2	Pretest	1.73	0.247234[NS]
		Post test	1.77	
		Difference	↑0.04	
		% difference	↑2.22	
Executive: Modified trials	2	Pretest	0.88	3.62933*
		Post test	1.69	
		Difference	↑0.81	
		% difference	↑91.04	
Executive: Problem solving	2	Pretest	0.58	1.18187[NS]
		Post test	0.85	
		Difference	↑0.27	
		% difference	↑46.67	
Memory	2	Pretest	0.38	3.378623*
		Post test	1.23	
		Difference	↑0.85	
		% difference	↑220.00	
Total SAGE Score	22	Pretest	14.12	6.53754*
		Post test	17.92	
		Difference	↑3.80	
		% difference	↑26.95	

KEYWORDS

- **Alzheimer's disease**
- **brain**
- **dementia**
- **diseases**
- **dissimilarity**
- **inflammation**
- **motor functions**
- **neurodegenerative diseases**
- **prebiotics**

REFERENCES

Abraham, D., Radak, Z., & Feher, J. (2016). Exercise and probiotics attenuate the development of Alzheimer disease in mice. *Free Radical Biology and Medicine, 100*, S165. https://doi.org/10.1016/j.freeradbiomed.2016.10.433.

Akbari, E., Asemi, Z., Kakhaki, R. D., Bahmani, F., Kouchaki, E., Tamtaji, O. R., Hamidi, G. A., & Salami, M. (2016). Effect of probiotic supplementation on cognitive function and metabolic status in Alzheimer's disease: A randomized, double-blind and controlled trial. *Frontiers in Aging Neuroscience, 8*(NOV), 256. https://doi.org/10.3389/fnagi.2016.00256.

Ambrosini, Y. M., Borcherding, D., Kanthasamy, A., Kim, H. J., Willette, A. A., Jergens, A., Allenspach, K., & Mochel, J. P. (2019). The gut-brain axis in neurodegenerative diseases and relevance of the canine model: A review. *Frontiers in Aging Neuroscience, 11*(JUN). https://doi.org/10.3389/fnagi.2019.00130.

Anand, P., Singh, B., & Singh, N. (2012). A review on coumarins as acetylcholinesterase inhibitors for Alzheimer's disease. *Bioorganic and Medicinal Chemistry, 20*(3), 1175–1180. https://doi.org/10.1016/j.bmc.2011.12.042.

Anstey, K. J., Lipnicki, D. M., & Low, L. F. (2008). Cholesterol as a risk factor for dementia and cognitive decline: A systematic review of prospective studies with meta-analysis. *American Journal of Geriatric Psychiatry, 16*(5), 343–354. https://doi.org/10.1097/jgp.0b013e31816b72d4.

Atri, A., Shaughnessy, L. W., Locascio, J. J., & Growdon, J. H. (2008). Long-term course and effectiveness of combination therapy in Alzheimer disease. *Alzheimer Disease and Associated Disorders, 22*(3), 209–221. https://doi.org/10.1097/WAD.0b013e31816653bc.

Ballard, C., Gauthier, S., Corbett, A., Brayne, C., Aarsland, D., & Jones, E. (2011). Alzheimer's disease. *The Lancet, 377*(9770), 1019–1031. https://doi.org/10.1016/S0140-6736(10)61349-9.

Bienenstock, J., Kunze, W., & Forsythe, P. (2015). Microbiota and the gut-brain axis. *Nutrition Reviews, 73*, 28–31. https://doi.org/10.1093/nutrit/nuv019.

Bonfili, L., Cecarini, V., Cuccioloni, M., Angeletti, M., Berardi, S., Scarpona, S., Rossi, G., & Eleuteri, A. M. (2018). SLAB51 probiotic formulation activates SIRT1 pathway promoting antioxidant and neuroprotective effects in an AD mouse model. *Molecular Neurobiology, 55*(10), 7987–8000. https://doi.org/10.1007/s12035-018-0973-4.

Brandscheid, C., Schuck, F., Reinhardt, S., Schäfer, K. H., Pietrzik, C. U., Grimm, M., Hartmann, T., Schwiertz, A., & Endres, K. (2017). Altered gut microbiome composition and tryptic activity of the 5xFAD Alzheimer's mouse model. *Journal of Alzheimer's Disease, 56*(2), 775–788. https://doi.org/10.3233/JAD-160926.

Braniste, V., Al-Asmakh, M., Kowal, C., Anuar, F., Abbaspour, A., Tóth, M., Korecka, A., Bakocevic, N., Guan, N. L., Kundu, P., Gulyás, B., Halldin, C., Hultenby, K., Nilsson, H., Hebert, H., Volpe, B. T., Diamond, B., & Pettersson, S. (2014). The gut microbiota influences blood-brain barrier permeability in mice. *Science Translational Medicine, 6*(263). https://doi.org/10.1126/scitranslmed.3009759.

Bravo, J. A., Forsythe, P., Chew, M. V., Escaravage, E., Savignac, H. M., Dinan, T. G., Bienenstock, J., & Cryan, J. F. (2011). Ingestion of Lactobacillus strain regulates emotional behaviour and central GABA receptor expression in a mouse via the vagus nerve. *Proceedings of the National Academy of Sciences of the United States of America, 108*(38), 16050–16055. https://doi.org/10.1073/pnas.1102999108.

Carabotti, M., Scirocco, A., Maselli, M. A., & Severi, C. (2015). The gut-brain axis: Interactions between enteric microbiota, central and enteric nervous systems. *Annals of Gastroenterology, 28*(2), 203–209.

Cattaneo, A., Cattane, N., Galluzzi, S., Provasi, S., Lopizzo, N., Festari, C., Ferrari, C., Guerra, U. P., Paghera, B., Muscio, C., Bianchetti, A., Volta, G. D., Turla, M., Cotelli, M. S., Gennuso, M., Prelle, A., Zanetti, O., Lussignoli, G., Mirabile, D., & Frisoni, G. B. (2017). Association of brain amyloidosis with pro-inflammatory gut bacterial taxa and peripheral inflammation markers in cognitively impaired elderly. *Neurobiology of Aging, 49*, 60–68. https://doi.org/10.1016/j.neurobiolaging.2016.08.019.

Choi, J., & Twamley, E. W. (2013). Cognitive rehabilitation therapies for Alzheimer's disease: A review of methods to improve treatment engagement and self-efficacy. *Neuropsychology Review, 23*(1), 48–62. https://doi.org/10.1007/s11065-013-9227-4.

Chow, V. W., Mattson, M. P., Wong, P. C., & Gleichmann, M. (2010). An overview of APP processing enzymes and products. *Neuromolecular Medicine, 12*(1), 1–12. https://doi.org/10.1007/s12017-009-8104-z.

Cirrito, J. R., Disabato, B. M., Restivo, J. L., Verges, D. K., Goebel, W. D., Sathyan, A., Hayreh, D., D'Angelo, G., Benzinger, T., Yoon, H., Kim, J., Morris, J. C., Mintun, M. A., & Sheline, Y. I. (2011). Serotonin signaling is associated with lower amyloid-β levels and plaques in transgenic mice and humans. *Proceedings of the National Academy of Sciences of the United States of America, 108*(36), 14968–14973. https://doi.org/10.1073/pnas.1107411108.

Critchfield, J. W., Van Hemert, S., Ash, M., Mulder, L., & Ashwood, P. (2011). The potential role of probiotics in the management of childhood autism spectrum disorders. *Gastroenterology Research and Practice, 2011*, 161358. https://doi.org/10.1155/2011/161358.

Cummings, J. H., Pomare, E. W., Branch, H. W. J., Naylor, C. P. E., & MacFarlane, G. T. (1987). Short chain fatty acids in human large intestine, portal, hepatic and venous blood. *Gut, 28*(10), 1221–1227. https://doi.org/10.1136/gut.28.10.1221.

De-Paula, V. D. R., Forlenza, A. S., & Forlenza, O. V. (2018). Relevance of gut microbiota in cognition, behavior and Alzheimer's disease. *Pharmacological Research, 136*, 29–34. https://doi.org/10.1016/j.phrs.2018.07.007.

Den Besten, G., Bleeker, A., Gerding, A., Van Eunen, K., Havinga, R., Van Dijk, T. H., Oosterveer, M. H., Jonker, J. W., Groen, A. K., Reijngoud, D. J., & Bakker, B. M. (2015). Short-chain fatty acids protect against high-fat diet-induced obesity via a PPARγ-dependent switch from lipogenesis to fat oxidation. *Diabetes, 64*(7), 2398–2408. https://doi.org/10.2337/db14-1213.

Dwivedi, S., & Sheth, M. (2019). Fermented milk derived biogenic metabolites and their probable impact on cortisol levels and depression scores. *International Journal of Research and Analytical Review, 6*(1).

Elias, M. F., Elias, P. K., Sullivan, L. M., Wolf, P. A., & D'Agostino, R. B. (2003). Lower cognitive function in the presence of obesity and hypertension: The Framingham Heart Study. *International Journal of Obesity, 27*(2), 260–268. https://doi.org/10.1038/sj.ijo.802225.

Fang, R., Ye, S., Huangfu, J., & Calimag, D. P. (2017). Music therapy is a potential intervention for cognition of Alzheimer's Disease: A mini-review. *Translational Neurodegeneration, 6*(1). https://doi.org/10.1186/s40035-017-0073-9.

Frank, D. N., & Pace, N. R. (2008). Gastrointestinal microbiology enters the metagenomics era. *Current Opinion in Gastroenterology, 24*(1), 4–10. https://doi.org/10.1097/MOG.0b013e3282f2b0e8.

Girvin, G. T., & Stevenson, J. W. (1954). Cell free choline acetylase from *Lactobacillus plantarum*. *Canadian Journal of Biochemistry and Physiology, 32*, 131–146. https://doi.org/10.1139/y54-015.

Goehler, L. E., Gaykema, R. P. A., Opitz, N., Reddaway, R., Badr, N., & Lyte, M. (2005). Activation in vagal afferents and central autonomic pathways: Early responses to intestinal infection with *Campylobacter jejuni*. *Brain, Behavior, and Immunity, 19*(4), 334–344. https://doi.org/10.1016/j.bbi.2004.09.002.

Griffin, C. E., Kaye, A. M., Rivera Bueno, F., & Kaye, A. D. (2013). Benzodiazepine pharmacology and central nervous system-mediated effects. *Ochsner Journal, 13*(2), 214–223.

Hanish Singh, J. C., Alagarsamy, V., Diwan, P. V., Sathesh Kumar, S., Nisha, J. C., & Narsimha Reddy, Y. (2011). Neuroprotective effect of *Alpinia galanga* (L.) fractions on Aβ (25–35) induced amnesia in mice. *Journal of Ethnopharmacology, 138*(1), 85–91. https://doi.org/10.1016/j.jep.2011.08.048.

Harach, T., Marungruang, N., Duthilleul, N., Cheatham, V., McCoy, K. D., Frisoni, G., Neher, J. J., Fåk, F., Jucker, M., Lasser, T., & Bolmont, T. (2017). Reduction of Abeta amyloid pathology in APPPS1 transgenic mice in the absence of gut microbiota. *Scientific Reports, 7*. https://doi.org/10.1038/srep41802.

Hippius, H., & Neundörfer, G. (2003). The discovery of Alzheimer's disease. *Dialogues in Clinical Neuroscience, 5*(1), 101–108.

Ho, L., Ono, K., Tsuji, M., Mazzola, P., Singh, R., & Pasinetti, G. M. (2018). Protective roles of intestinal microbiota derived short chain fatty acids in Alzheimer's disease-type beta-amyloid neuropathological mechanisms. *Expert Review of Neurotherapeutics, 18*(1), 83–90. https://doi.org/10.1080/14737175.2018.1400909.

Hu, X., Wang, T., & Jin, F. (2016). Alzheimer's disease and gut microbiota. *Science China Life Sciences, 59*(10), 1006–1023. https://doi.org/10.1007/s11427-016-5083-9.

Iadecola, C., Yaffe, K., Biller, J., Bratzke, L. C., Faraci, F. M., Gorelick, P. B., Gulati, M., Kamel, H., Knopman, D. S., Launer, L. J., Saczynski, J. S., Seshadri, S., Zeki Al Hazzouri, A., & American Heart Association Council on Hypertension; Council on Clinical Cardiology; Council on Cardiovascular Disease in the Young; Council on Cardiovascular and Stroke Nursing; Council on Quality of Care and Outcomes Research; and Stroke Council. (2016). Impact of hypertension on cognitive function: A scientific statement from the American Heart Association. *Hypertension, 68*(6), e67–e94. https://doi.org/10.1161/HYP.0000000000000053.

Imbimbo, B. P., Lombard, J., & Pomara, N. (2005). Pathophysiology of Alzheimer's disease. *Neuroimaging Clinics of North America, 15*(4), 727–753. https://doi.org/10.1016/j.nic.2005.09.009.

Jiang, C., Li, G., Huang, P., Liu, Z., & Zhao, B. (2017). The gut microbiota and Alzheimer's disease. *Journal of Alzheimer's Disease, 58*(1), 1–15. https://doi.org/10.3233/JAD-161141.

Kahn, M. S., Kranjac, D., Alonzo, C. A., Haase, J. H., Cedillos, R. O., McLinden, K. A., Boehm, G. W., & Chumley, M. J. (2012). Prolonged elevation in hippocampal Aβ and cognitive deficits following repeated endotoxin exposure in the mouse. *Behavioral Brain Research, 229*(1), 176–184. https://doi.org/10.1016/j.bbr.2012.01.010.

Kelly, J. R., Kennedy, P. J., Cryan, J. F., Dinan, T. G., Clarke, G., & Hyland, N. P. (2015). Breaking down the barriers: The gut microbiome, intestinal permeability and stress-related psychiatric disorders. *Frontiers in Cellular Neuroscience, 9*. https://doi.org/10.3389/fncel.2015.00392.

Kim, J., & Lee, B. (2018). Identification of Alzheimer's disease and mild cognitive impairment using multimodal sparse hierarchical extreme learning machine. *Human Brain Mapping, 39*(9), 3728–3741. https://doi.org/10.1002/hbm.24207.

Kopf, D., & Frölich, L. (2009). Risk of incident Alzheimer's disease in diabetic patients: A systematic review of prospective trials. *Journal of Alzheimer's Disease, 16*(4), 677–685. https://doi.org/10.3233/JAD-2009-1011.

Lancet, T. (2011). The three stages of Alzheimer's disease. *The Lancet, 377*(9776), 1465. https://doi.org/10.1016/S0140-6736(11)60582-5.

Lanctôt, K. L., Herrmann, N., Mazzotta, P., Khan, L. R., & Ingber, N. (2004). GABAergic function in Alzheimer's disease: Evidence for dysfunction and potential as a therapeutic target for the treatment of behavioral and psychological symptoms of dementia. *Canadian Journal of Psychiatry, 49*(7), 439–453. https://doi.org/10.1177/070674370404900705.

Laurin, D., Verreault, R., Lindsay, J., MacPherson, K., & Rockwood, K. (2001). Physical activity and risk of cognitive impairment and dementia in elderly persons. *Archives of Neurology, 58*(3), 498–504. https://doi.org/10.1001/archneur.58.3.498.

Leblhuber, F., Egger, M., Schuetz, B., & Fuchs, D. (2018). Commentary: Effect of probiotic supplementation on cognitive function and metabolic status in Alzheimer's disease: A randomized, double-blind and controlled trial. *Frontiers in Aging Neuroscience, 10*, 54. https://doi.org/10.3389/fnagi.2018.00054.

Lee, Y., Kim, J., Han, E. S., Chae, S., Ryu, M., Ahn, K. H., & Park, E. J. (2015). Changes in physical activity and cognitive decline in older adults living in the community. *Age, 37*(2). https://doi.org/10.1007/s11357-015-9759-z.

Leggieri, M., Thaut, M. H., Fornazzari, L., Schweizer, T. A., Barfett, J., Munoz, D. G., & Fischer, C. E. (2019). Music intervention approaches for Alzheimer's disease: A review of the literature. *Frontiers in Neuroscience, 13*, 132. https://doi.org/10.3389/fnins.2019.00132.

Macfarlane, G. T., & Macfarlane, S. (2012). Bacteria, colonic fermentation, and gastrointestinal health. *Journal of AOAC International, 95*(1), 50–60. https://doi.org/10.5740/jaoacint.SGE_Macfarlane.

Mallikarjuna, N., Kukkarasapalli, P., & Kuna, Y. (2016). Role of *Lactobacillus plantarum* MTCC1325 in membrane-bound transport ATPases system in Alzheimer's disease-induced rat brain. *BioImpacts, 6*(4), 203–209. https://doi.org/10.15171/bi.2016.27.

Martins, I. J., & Fernando, W. M. A. D. B. (2014). High fibre diets and Alzheimer's disease. *Food and Nutrition Sciences, 5*(4), 410–424. https://doi.org/10.4236/fns.2014.54049.

Matsumoto, M., Kibe, R., Ooga, T., Aiba, Y., Sawaki, E., Koga, Y., & Benno, Y. (2013). Cerebral low-molecular metabolites influenced by intestinal microbiota: A pilot study. *Frontiers in Systems Neuroscience, 7*. https://doi.org/10.3389/fnsys.2013.00009.

Messaoudi, M., Lalonde, R., Violle, N., Javelot, H., Desor, D., Nejdi, A., Bisson, J. F., Rougeot, C., Pichelin, M., Cazaubiel, M., & Cazaubiel, J. M. (2011). Assessment of psychotropic-like properties of a probiotic formulation (*Lactobacillus helveticus* R0052 and *Bifidobacterium longum* R0175) in rats and human subjects. *British Journal of Nutrition, 105*(5), 755–764. https://doi.org/10.1017/S0007114510004319.

Misra, S., & Mohanty, D. (2019). Psychobiotics: A new approach for treating mental illness? *Critical Reviews in Food Science and Nutrition, 59*(8), 1230–1236. https://doi.org/10.1080/10408398.2017.1399860.

Nimgampalle, M., & Kuna, Y. (2017). Anti-Alzheimer properties of probiotic, *Lactobacillus plantarum* MTCC 1325 in Alzheimer's disease induced albino rats. *Journal of Clinical and Diagnostic Research, 11*(8), 01–05.

Parashar, A., & Udayabanu, M. (2017). Gut microbiota: Implications in Parkinson's disease. *Parkinsonism and Related Disorders, 38*, 1–7. https://doi.org/10.1016/j.parkreldis.2017.02.002.

Peng, X., Meng, J., Chi, T., Liu, P., Man, C., Liu, S., Guo, Y., & Jiang, Y. (2014). *Lactobacillus plantarum* NDC 75017 alleviates the learning and memory ability in aging rats by reducing mitochondrial dysfunction. *Experimental and Therapeutic Medicine, 8*(6), 1841–1846. https://doi.org/10.3892/etm.2014.2000.

Qiao, W. Q., Ferreira, A., Miller, C., Koo, E. H., & Selkoe, D. J. (1995). Cell-surface β-amyloid precursor protein stimulates neurite outgrowth of hippocampal neurons in an isoform-dependent manner. *Journal of Neuroscience, 15*(3 II), 2157–2167. https://doi.org/10.1523/jneurosci.15-03-02157.1995.

Reitz, C., Brayne, C., & Mayeux, R. (2011). Epidemiology of Alzheimer disease. *Nature Reviews Neurology, 7*(3), 137–152. https://doi.org/10.1038/nrneurol.2011.2.

Rovner, B. W., German, P. S., Brant, L. J., Clark, R., Burton, L., & Folstein, M. F. (1991). Depression and all-cause mortality in nursing homes. *JAMA, 265*(8), 993–996. https://doi.org/10.1001/jama.265.8.993.

Sánchez, B., Delgado, S., Blanco-Míguez, A., Lourenço, A., Gueimonde, M., & Margolles, A. (2017). Probiotics, gut microbiota, and their influence on host health and disease. *Molecular Nutrition and Food Research, 61*(1). https://doi.org/10.1002/mnfr.201600240.

Schmidt, K., Cowen, P. J., Harmer, C. J., Tzortzis, G., Errington, S., & Burnet, P. W. J. (2015). Prebiotic intake reduces the waking cortisol response and alters emotional bias in healthy volunteers. *Psychopharmacology, 232*(10), 1793–1801. https://doi.org/10.1007/s00213-014-3810-0.

Selkoe, D. J., & Hardy, J. (2016). The amyloid hypothesis of Alzheimer's disease at 25 years. *EMBO Molecular Medicine, 8*(6), 595–608. https://doi.org/10.15252/emmm.201606210.

Sharp, E. S., & Gatz, M. (2011). Relationship between education and dementia: An updated systematic review. *Alzheimer Disease and Associated Disorders, 25*(4), 289–304. https://doi.org/10.1097/WAD.0b013e318211c83c.

Sheth, M., Chand, V., & Thakuria, A. (2015). Inflated levels of SCFA, bifidobacteria and lactobacillus improves the status of pre hypertension and type 2 diabetes mellitus in subjects residing in North East India – A randomized control trial with symbiotic supplementation. *International Journal of Current Pharmaceutical Research.* Retrieved from: https://innovareacademics.in/journals/index.php/ijcpr/article/view/7915/10995 (accessed on 19 July 2024).

Shoji, M., Golde, T. E., Ghiso, J., Cheung, T. T., Estus, S., Shaffer, L. M., Cai, X. D., McKay, D. M., Tintner, R., Frangione, B., & Younkin, S. G. (1992). Production of the Alzheimer amyloid β protein by normal proteolytic processing. *Science, 258*(5079), 126–129. https://doi.org/10.1126/science.1439760.

Solas, M., Puerta, E., & Ramirez, M. (2015). Treatment options in Alzheimer's disease: The GABA story. *Current Pharmaceutical Design, 21*(34), 4960–4971. https://doi.org/10.2174/1381612821666150914121149.

Tanida, M., Yamano, T., Maeda, K., Okumura, N., Fukushima, Y., & Nagai, K. (2005). Effects of intraduodenal injection of *Lactobacillus johnsonii* La1 on renal sympathetic nerve activity and blood pressure in urethane-anesthetized rats. *Neuroscience Letters, 389*, 109–114. https://doi.org/10.1016/j.neulet.2005.07.036.

Tillisch, K., Labus, J., Kilpatrick, L., Jiang, Z., Stains, J., Ebrat, B., Guyonnet, D., Legrain-Raspaud, S., Trotin, B., Naliboff, B., & Mayer, E. A. (2013). Consumption of fermented milk product with probiotic modulates brain activity. *Gastroenterology, 144*(7), 1394–1401.e1-4. https://doi.org/10.1053/j.gastro.2013.02.043.

Toro, P., Schönknecht, P., & Schröder, J. (2009). Type II diabetes in mild cognitive impairment and Alzheimer's disease: Results from a prospective population-based study in Germany. *Journal of Alzheimer's Disease, 16*(4), 687–691. https://doi.org/10.3233/JAD-2009-0981.

Tuohy, K., Rouzaud, G., Bruck, W., & Gibson, G. (2005). Modulation of the human gut microflora towards improved health using prebiotics: Assessment of efficacy. *Current Pharmaceutical Design, 11*(1), 75–90. https://doi.org/10.2174/1381612053382331.

Turner, J. R. (2009). Intestinal mucosal barrier function in health and disease. *Nature Reviews Immunology, 9*(11), 799–809. https://doi.org/10.1038/nri2653.

Vieira, A. T., Galvão, I., Macia, L. M., Sernaglia, É. M., Vinolo, M. A. R., Garcia, C. C., Tavares, L. P., Amaral, F. A., Sousa, L. P., Martins, F. S., Mackay, C. R., & Teixeira, M. M. (2017). Dietary fiber and the short-chain fatty acid acetate promote resolution of neutrophilic inflammation in a model of gout in mice. *Journal of Leukocyte Biology, 101*(1), 275–284. https://doi.org/10.1189/jlb.3a1015-453rrr.

Vogt, N. M., Kerby, R. L., Dill-McFarland, K. A., Harding, S. J., Merluzzi, A. P., Johnson, S. C., Carlsson, C. M., Asthana, S., Zetterberg, H., Blennow, K., Bendlin, B. B., & Rey, F. E. (2017). Gut microbiome alterations in Alzheimer's disease. *Scientific Reports, 7*(1). https://doi.org/10.1038/s41598-017-13601-y.

Wang, J., Wang, Y., Gao, W., Wang, B., Zhao, H., Zeng, Y., Ji, Y., & Hao, D. (2017). Diversity analysis of gut microbiota in osteoporosis and osteopenia patients. *PeerJ, 2017*(6). https://doi.org/10.7717/peerj.3450.

Woo, J. Y., Gu, W., Kim, K. A., Jang, S. E., Han, M. J., & Kim, D. H. (2014). *Lactobacillus pentosus* var. *plantarum* C29 ameliorates memory impairment and inflammaging in a d-galactose-induced accelerated aging mouse model. *Anaerobe, 27*, 22–26. https://doi.org/10.1016/j.anaerobe.2014.03.003.

Wu, X., Li, M., Li, Z., Lv, L., Zhang, Y., & Li, C. (2017). Amyloid-graphene oxide as immobilization platform of Au nanocatalysts and enzymes for improved glucose-sensing activity. *Journal of Colloid and Interface Science, 490*, 336–342. https://doi.org/10.1016/j.jcis.2016.11.058.

Yano, J. M., Yu, K., Donaldson, G. P., Shastri, G. G., Ann, P., Ma, L., Nagler, C. R., Ismagilov, R. F., Mazmanian, S. K., & Hsiao, E. Y. (2015). Indigenous bacteria from the gut microbiota regulate host serotonin biosynthesis. *Cell, 161*(2), 264–276. https://doi.org/10.1016/j.cell.2015.02.047.

Zhang, R., Miller, R. G., Gascon, R., Champion, S., Katz, J., Lancero, M., Narvaez, A., Honrada, R., Ruvalcaba, D., & McGrath, M. S. (2009). Circulating endotoxin and systemic immune activation in sporadic amyotrophic lateral sclerosis (sALS). *Journal of Neuroimmunology, 206*(1–2), 121–124. https://doi.org/10.1016/j.jneuroim.2008.09.017.

Zilberter, Y., & Zilberter, M. (2017). The vicious circle of hypometabolism in neurodegenerative diseases: Ways and mechanisms of metabolic correction. *Journal of Neuroscience Research, 95*, 2217–2235.

CHAPTER 9

Prebiotics in Management of Diabetes

RUCHI VAIDYA

Ramaiah International Center for Public Health Innovations, Bangalore, Karnataka, India

Diabetes is an important public health problem, one of four priority non- communicable diseases (NCDs) targeted for action by world leaders. There is strong evidence observed for the potential of prebiotics to impact on metabolic effects relevant to Type 2 Diabetes.

9.1 INTRODUCTION TO DIABETES MELLITUS

Diabetes Mellitus is a chronic disease that occurs when the pancreas fails to produce enough insulin (a hormone that regulates blood sugar, or glucose) or when the body is unable to effectively use the insulin that is produced. Although several other categorizations exist, diabetes is typically divided into four main groups: type 1 diabetes, type 2 diabetes, gestational diabetes, and "other types of diabetes" (IDF, 2019). Autoimmune destruction of pancreatic beta cells causes type 1 diabetes while insulin resistance causes type 2 diabetes. Gestational diabetes occurs when glucose intolerance is identified during pregnancy. Diabetes is thought to also be caused by various other etiologies, including genetic defects, diseases of the exocrine pancreas, endocrinopathies, infections, and drugs. The genetic defects include mature-onset diabetes of the young (MODY), which exhibits impaired insulin secretion with little or no insulin resistance. In addition, diseases such as pancreatic cancer or pancreatic trauma can also lead to diabetes. In the presence of pre-existing beta-cell

Prebiotic Foods for a Healthy Gut: Scientific Studies to Manage Metabolic Disorders.
Mini K. Sheth and Shruti Dwivedi (Eds.)

failure, other endocrinopathies, such as cortisol, growth hormone, glucagon, and epinephrine, can exacerbate insulin resistance. Infectious diseases, such as congenital rubella, can predispose to the development of diabetes. Additionally, various drugs can induce insulin resistance and impair beta-cell function (Khanna and Tosh, 2014). Recent studies have also shown that alterations in gut microbiota, predominantly due to poor lifestyle, can also lead to insulin resistance and the manifestation of diabetes in individuals. Diabetes can cause complications in many parts of the body and raise the overall risk of death. Heart attack, stroke, kidney failure, leg amputation, vision loss, and nerve damage are all possible complications. Poorly controlled diabetes increases the risk of fetal death and other complications during pregnancy.

9.2 PREVALENCE OF DIABETES

Over the last 20 years, the global diabetes population has more than doubled. According to the most recent IDF estimates, 463 million adults worldwide have diabetes. With 77 million diabetics and 43.9 million undiagnosed diabetics, India ranks second in the world after China. Approximately 90% of people have type 2 diabetes (IDF, 2019). The emergence of type 2 diabetes in children, adolescents, and young adults is one of the most concerning aspects of this rapid rise. Although traditional risk factors for type 2 diabetes (such as genetic, lifestyle, and behavioral risk factors) have been studied, recent research has focused on identifying the contributions of epigenetic mechanisms and the effect of the intrauterine environment. The so-called "south Asian" or "Asian-Indian" phenotype makes this ethnic group more susceptible to type 2 diabetes and premature coronary artery disease than white Caucasians (Unnikrishnan et al., 2014). This phenotype is distinguished by increased insulin resistance, high diabetes rates, and dyslipidemia, as evidenced by elevated serum triglycerides and low levels of high-density lipoprotein (Mohan et al., 1986). While genetic factors may play a role in the South Asian phenotype, the current diabetes epidemic is primarily driven by lifestyle, specifically an unhealthy diet and physical inactivity (Mohan et al., 2016). There is an urgent need to explore alternative mechanisms that entail the risks, prevention, and management of diabetes.

9.3 GUT MICROBIOTA AND DIABETES

Recent evidence suggests a link between human metabolic health and bacterial populations in the gut, and it is primarily controlled by the nutritional quality of the diet. The gut microbiome is host-specific, evolves throughout an individual's lifetime, and is susceptible to both exogenous (e.g., antibiotic use) and endogenous (e.g., systemic disease) factors. Gut bacteria have an important relationship, whether symbiotic or pathogenic, to the host and play a vital role in an individual's GI and systemic health. *Firmicutes, Bacteroidetes, Actinobacteria*, and *Proteobacteria* comprise the four major bacterial phyla in the human gut. When there is a disruption in the microbiota homeostasis, caused by either an excess or depletion of specific bacterial species, systemic complications and disease may follow. One important condition that has been shown to be associated with an out-of-balance microbiome is diabetes mellitus.

The two variations of diabetes mellitus, type 1 and type 2, while differing in their respective etiologies and disease manifestations, have both been linked to alterations in gut microbiota. The predominant effector mechanism (although not necessarily the cause) of type 1 diabetes is autoimmunity; therefore, numerous studies investigating the etiology of type 1 diabetes have focused specifically on the immune system function in the gut. The GI tract contains Paneth cells, which secrete antimicrobial peptides in response to foreign pathogens. These proteins disrupt the luminal microbiology, which, in turn, can affect the gut microbiome. Recently, there has been interest in the association of the gram-negative bacteria *B. dorei* with normal intestinal function in type 1 diabetes, as studies have shown significant differences in *B. dorei* levels in individuals with and without type 1 diabetes. These observations have led researchers to hypothesize that elevated levels of *B. dorei* could be a risk factor for the onset of type 1 diabetes (Davis-Richaardson et al., 2014).

In contrast, in the case of type 2 diabetes, metabolic endotoxemia, nonalcoholic fatty liver disease (NAFLD), and obesity have been shown to alter the gut microbiome, and each can result in the development of type 2 diabetes. It has been shown that a high-fat diet increases the number of lipopolysaccharides (LPS) in the body and, as a result, affects the metabolism and percentage of body fat. The consumption of excess LPS and bacterial lipopeptides is strongly associated with metabolic endotoxemia. Metabolic endotoxemia is a state in which there is an elevation in the normal amounts of endotoxins present in the gut epithelium. This is a strong contributing factor to diabetes, obesity, and NAFLD (Figure 9.1). It is reported that in gut dysbiosis, there is an increase in LPS production, pro-inflammatory cytokines, and a series of nonspecific inflammatory responses occur in the body, which interferes with insulin signal transduction, in turn, results in insulin resistance and leads to hyperinsulinemia, as well as triggers chronic low-level inflammatory responses of islet cells and metabolic endotoxemia, and ultimately, leads to destruction and apoptosis of islet β cells.

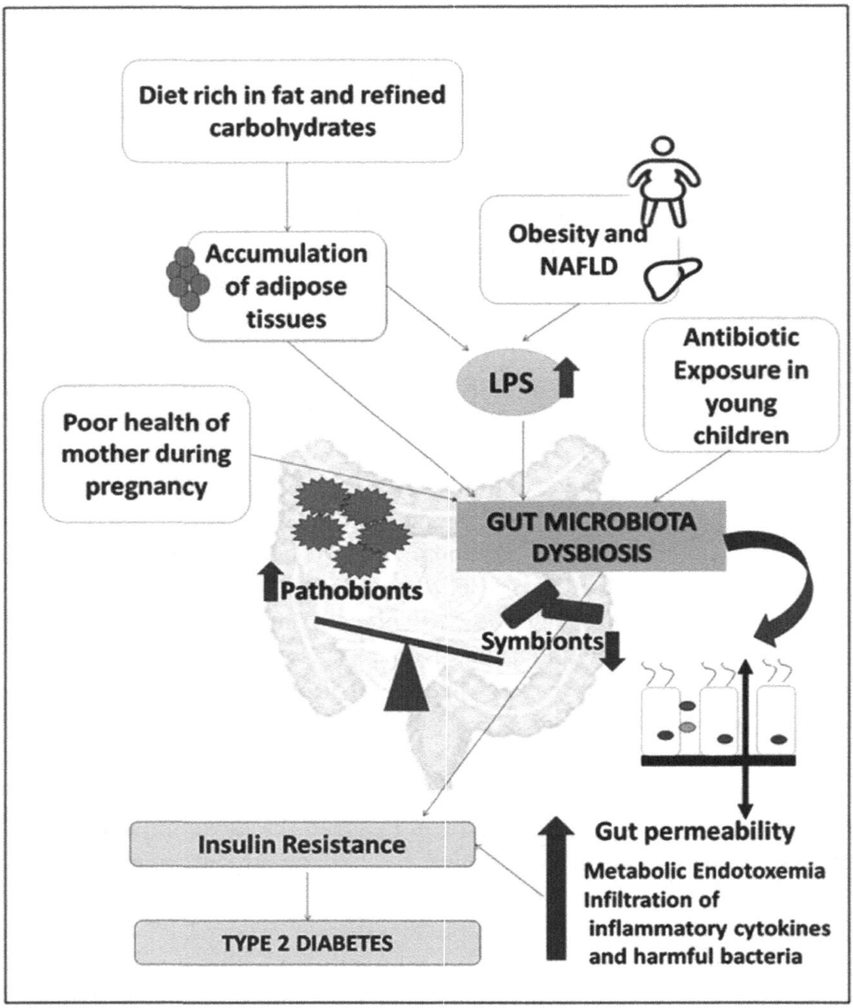

FIGURE 9.1 Role of gut microbiota in development of type 2 diabetes.

Generally speaking, obese individuals have a higher *Firmicutes-to-Bacteroidetes* ratio in the gut. As people begin to lose weight, this ratio begins to normalize to a more equal distribution (Million et al., 2013). This has caused researchers to look more closely at the *Firmicutes* bacteria because thousands of *Lactobacilli* are within this family, and *Lactobacilli* are a significant portion of the microbiota found in the digestive, urinary, and reproductive systems. The higher levels of *Bacteroidetes* in the body are an indicator of a healthier gut microbiome and play a potential role in preventing the development of obesity, insulin resistance, and ultimately diabetes.

Short-chain fatty acids (SCFAs), for example, are microbial products that may influence metabolism by regulating appetite, lipogenesis, gluconeogenesis, inflammation, and producing GLP-1 to regulate glucose metabolism. In contrast, intestinal dysbiosis can affect how well the intestinal barrier and the gut-associated lymphoid tissues (GALT) function by allowing the passage of bacterial structural components like lipopolysaccharides (LPS), which trigger inflammatory pathways that may aid in the development of CVD, insulin resistance, and change the production of gastrointestinal peptides that are involved in satiety, food intake, and body fat metabolism.

9.4 PREBIOTICS IN MANAGEMENT OF DIABETES AND ITS COMPLICATIONS

Gibson & Roberfroid first introduced the concept of "prebiotic" in 1995 as a non-digestible food component that positively affects the host by selectively stimulating the growth and/or activity of gut bacteria. Initially, the microbial population was confined to *Lactobacillus* and/or *Bifidobacterium* spp., but now includes a wide range of *Firmicutes* including Roseburia, *Eubacterium*, or *Faecalibacterium* spp. The concept of prebiotics has evolved with the updated consensus from 2017 (Gibson et al., 2017) as a "substrate that is selectively utilized by host-microorganisms conferring a health benefit." Initially, all prebiotics were carbohydrates, typically also fulfilling the CODEX Alimentarius definition of dietary fiber. The prebiotics are broadly classified into the following categories:

1. Non-starch polysaccharides including cellulose, hemicelluloses, pectins, and other hydrocolloids (i.e., β-glucans and mucilages), inulin, and fructans.
2. Resistant oligosaccharides including fructo-oligosaccharides (FOS) and galactooligosaccharides (GOS).
3. Resistant starch and dextrins.
4. Others: There is a growing body of literature reporting the prebiotic potential of polyphenols (found in foods such as tea, cocoa, wine, and coffee) and polyunsaturated fatty acids (PUFA), which may now fit into this expanded definition.

There is strong evidence on the prebiotic benefits of fructo-oligosaccharides (FOS), galacto-oligosaccharides (GOS), inulin, beta-glucan, pectin, and resistant starch (RS) (Davani-Davari, D et al., 2019; Vaidya and Sheth, 2011), while other types of prebiotics need to be studied further for their beneficial effects.

9.5 EVIDENCE OF THE EFFICACY OF PREBIOTICS IN MANAGEMENT OF METABOLIC REGULATION IN TYPE 2 DIABETES

There is strong evidence observed in animal models for the potential of prebiotics to impact on metabolic effects relevant to Type 2 diabetes (Le-Bourgot et al., 2018). In addition to the gut microbiota dysbiosis linked to Type 2 diabetes, other significant confounding factors and antibiotic-like effects of medications that change metabolic processes complicate the relationship between prebiotics and the microbiota (Maniar et al., 2017). The chapter discusses the meta-analysis published recently to understand health benefits of prebiotics in type 2 diabetes and has illustrated a few individual RCT studies performed to study the efficacy of different types of prebiotics in type 2 diabetes. Probiotics and prebiotics work on almost similar mechanisms in the management of metabolic endotoxemia and metabolic regulations. However, the dosage, the mode of delivery of the prebiotics and probiotics, the characteristics of the host, and the severity of the conditions of the host are strongly associated with studying the beneficial effects of the prebiotics and probiotics. According to a meta-analysis of 28 RCT trials measuring impact on glycemic control in people with prediabetes and type 2 diabetes, prebiotics appear to have a greater impact on glycemic outcomes in people with poorly controlled diabetes (fasting blood glucose (FBG) >130 mg/dL) (Rittiphairoj et al., 2021). Patients with type 2 diabetes who need additional insulin therapy may have impaired beta-cell function. Probiotics may have glycemic effects by increasing insulin sensitivity; as a result, they are less likely to significantly lower blood glucose in diabetics who are taking insulin.

Meta-analysis by Wang et al., in 2019 included 34 RCTs and 1346 participants from the healthy, obese, and T2D groups. They found that prebiotic supplementation, in particular FOS, reduced HbA1c (95% CI –0.83, –0.32%) by 0.58%. An increasing number of individual studies have made it possible for a sub-group analysis to elucidate specifics regarding dosage and duration of supplementation (Table 9.1). This proved that the most significant effects on glucose control required a daily dose of 10 g/day of FOS supplementation.

The supplements were more successful at enhancing glucose control when administered in liquids rather than solids, which raises critical questions about how these prebiotics should be consumed in everyday life and the role that the food matrix plays.

The advantages of resistant starch and dextrins on metabolic controls are well documented in the literature. This form of prebiotic has characteristics that are similar to those of oligo-fructans. Strong evidence that resistant starch is a beneficial addition to a daily diet because of its health-promoting properties was presented in the meta-analysis on the beneficial effect of resistant starch (RS) and dextrins by Halajzadeh et al. in 2020. Additionally, gut bacteria produce SCFAs, methane, and hydrogen from the resistant starch. SCFAs influence the balance of the microflora colonizing the large intestine, lowering luminal pH, promoting the growth of beneficial bacterial phyla, and limiting the growth of pathogenic microorganisms. The meta-analysis showed that RS has positive effects on levels of TNF-a, total cholesterol, LDL cholesterol, fasting glucose, fasting insulin, and HbA1c.

9.6 CONCLUSION

Type 2 diabetes is one of the leading non-communicable diseases globally, and the projections will double in the next decade. There is a growing body of literature that supports the beneficial effects of prebiotics on glycemic and insulinemic regulations among adults with prediabetes and type 2 diabetes. The studies and meta-analysis have demonstrated that a minimum dosage of 10 g/day for at least 6 weeks is required for effective glycemic and insulinemic regulations among prediabetes and type 2 diabetes. Prebiotics are naturally present in grains, vegetables, and many fruits, but they are not in enough concentration to meet the requirements for clinical management. Therefore, it is advisable to supplement the prebiotics in the form of powder or liquid for effective results. All these important findings provide practical information and indicate that prebiotics can be used as an adjuvant therapy for glycemic control, especially for patients with prediabetes or T2DM in clinical practice. However, there is a lot of individual variation in how people react to prebiotics, both microbially and physiologically, so more research over a longer period of time is needed to understand this complexity and figure out how to use it as a dietary strategy to improve glycemic control.

TABLE 9.1 Evidence of the Efficacy of Prebiotics in Metabolic Parameters in Type 2 Diabetes Adults From Individual Studies

Study	Type of prebiotic	Dosage	Study participants	Results
Sheth et al., 2016	Liquid FOS	10 ml/day (45 days)	Adults with Type 2 diabetes with pre-hypertension	TG, LDL, VLDL – 8% decrease (p<0.001) HDL – 25% increase (p<0.001)
Sheth et al., 2015	Liquid FOS	10 ml/day (45 days)	Adults with Type 2 diabetes with pre-hypertension	FBS – 6% decrease; TG, TC, LDL – 9–14% decrease (p<0.001)
Dewulf et al. 2013	Inulin	10 g/day (56 days)	Adult females with Type 2 diabetes	FBS (5% decrease), FIN,HOMA-IR (2.4% decrease), HBA1C (1.2% decrease) (P<0.001)
Sheth et al., 2019	FOS added barley films	10 ml/day (45 days)	Adults with type 2 diabetes	Significant reduction in weight, BMI, SBP, FBS, ABG, HbA1c and fructosamine by 1.1%,0.7%,3.5%,14%,4.3%,3.8% and 3% respectively. hs-CRP and homocysteine levels reduced non-significantly by 8% and 9% respectively.
Dehgan et al,2016	Inulin, oligofructose	10 g / day (60 days)	Adults with type 2 diabetes	FBS – 6% decrease (p<0.001) No significant difference in HOMA IR and fasting insulin
Asemi et al., 2016	Inulin and *Lactobacillus sporogenes*	8.4 g/days (56 days)	Adults with type 2 diabetes	FBS – no significant change Fasting Insulin – 3.4% decrease (p<0.01)
Vulevic et al., 2013	GOS	5.5 g/days (84 days)	Overweight adults	FBS – no significant change Fasting Insulin – 4.5% decrease (p<0.01)
Kwak et al., 2012	Resistant starch	6.5 g /day (28 days)	Adults with newly diagnosed type 2 diabetes	IFG and IGT reduces significantly (p<0.001)
Sheth et al., 2012	Resistant starch	10 g/day (90 days)	Middle aged adults with type 2 diabetes	FBS, PPBS – 8–10% decrease (p<0.001)
Alfa et al., 2018	Resistant starch	30 g/day (84 days)	Young adults with prediabetes	Significant reduction in HOMA IR (p<0.001) and TNF – alpha (p<0,001)

KEYWORDS

- **blood**
- **blood sugar**
- **diabetes**
- **fatty acids**
- **glucose**
- **hormonal effects**
- **hormones**
- **insulin**
- **insulin resistance**
- **overweight**

REFERENCES

Alfa, M. J., Strang, D., Tappia, P. S., Olson, N., DeGagne, P., Bray, D., & Hiebert, B. (2018). A randomized placebo controlled clinical trial to determine the impact of digestion resistant starch MSPrebiotic® on glucose, insulin, and insulin resistance in elderly and mid-age adults. *Frontiers in Medicine, 4*, 260.

Asemi, Z., Alizadeh, S. A., Ahmad, K., Goli, M., & Esmaillzadeh, A. (2016). Effects of beta-carotene fortified synbiotic food on metabolic control of patients with type 2 diabetes mellitus: A double-blind randomized cross-over controlled clinical trial. *Clinical Nutrition, 35*(4), 819–825.

Davani-Davari, D., Negahdaripour, M., Karimzadeh, I., Seifan, M., Mohkam, M., Masoumi, S. J., Berenjian, A., & Ghasemi, Y. (2019). Prebiotics: Definition, types, sources, mechanisms, and clinical applications. *Foods, 8*(3), 92. https://doi.org/10.3390/foods8030092.

Davis-Richardson, A. G., Ardissone, A. N., & Dias, R. (2014). Bacteroides dorei dominates gut microbiome prior to autoimmunity in Finnish children at high risk for type 1 diabetes. *Frontiers in Microbiology, 5*, 678.

Dehghan, P., Farhangi, M. A., Tavakoli, F., Aliasgarzadeh, A., & Akbari, A. M. (2016). Impact of prebiotic supplementation on T-cell subsets and their related cytokines, anthropometric features and blood pressure in patients with type 2 diabetes mellitus: A randomized placebo-controlled trial. *Complementary Therapies in Medicine, 24*, 96–102.

Dewulf, E. M., Cani, P. D., Claus, S. P., Fuentes, S., Puylaert, P. G., Neyrinck, A. M., et al. (2013). Insight into the prebiotic concept: Lessons from an exploratory, double blind intervention study with inulin-type fructans in obese women. *Gut, 62*(8), 1112–1121.

Gibson, G. R., Hutkins, R., Sanders, M. E., Prescott, S. L., Reimer, R. A., Salminen, S. J., et al. (2017). Expert consensus document: The International Scientific Association for

Probiotics and Prebiotics (ISAPP) consensus statement on the definition and scope of prebiotics. *Nature Reviews Gastroenterology & Hepatology, 14*(8), 491–502.

Halajzadeh, J., Milajerdi, A., Reiner, Ž., Amirani, E., Kolahdooz, F., Barekat, M., & Asemi, Z. (2020). Effects of resistant starch on glycemic control, serum lipoproteins and systemic inflammation in patients with metabolic syndrome and related disorders: A systematic review and meta-analysis of randomized controlled clinical trials. *Critical Reviews in Food Science and Nutrition, 60*(18), 3172–3184.

International Diabetes Federation. (2019). *IDF Diabetes Atlas* (9th ed.).

Khanna, S., & Tosh, P. K. (2014). A clinician's primer on the role of the microbiome in human health and disease. *Mayo Clinic Proceedings, 89*(1), 107–114.

Kwak, J. H., Paik, J. K., Kim, H. I., Kim, O. Y., Shin, D. Y., Kim, H. J., Lee, J. H., & Lee, J. H. (2012). Dietary treatment with rice containing resistant starch improves markers of endothelial function with reduction of postprandial blood glucose and oxidative stress in patients with prediabetes or newly diagnosed type 2 diabetes. *Atherosclerosis, 224*(2), 457–464. https://doi.org/10.1016/j.atherosclerosis.2012.08.003.

Le Bourgot, C., Apper, E., Blat, S., & Respondek, F. (2018). Fructo-oligosaccharides and glucose homeostasis: A systematic review and meta-analysis in animal models. *Nutrition & Metabolism, 15*, 9.

Maniar, K., Moideen, A., Mittal, A., Patil, A., Chakrabarti, A., & Banerjee, D. (2017). A story of metformin-butyrate synergism to control various pathological conditions as a consequence of gut microbiome modification: Genesis of a wonder drug? *Pharmacological Research, 117*, 103–128.

Million, M., Lagier, J. C., Yahav, D., & Paul, M. (2013). Gut bacterial microbiota and obesity. *Clinical Microbiology and Infection, 19*(4), 305–313.

Mohan, V., Ruchi, V., Gayathri, R., Bai, M. R., Sudha, V., Anjana, R. M., et al. (2016). Slowing the diabetes epidemic in the World Health Organization South-East Asia Region: The role of diet and physical activity. *WHO South-East Asia Journal of Public Health, 5*(1), 5–16.

Mohan, V., Sharp, P. S., Cloke, H. R., Burrin, J. M., Schumer, B., & Kohner, E. M. (1986). Serum immunoreactive insulin responses to a glucose load in Asian Indian and European type 2 (non-insulin-dependent) diabetic patients and control subjects. *Diabetology, 29*(4), 235–237.

Rittiphairoj, T., Pongpirul, K., Janchot, K., Mueller, N. T., & Li, T. (2021). Probiotics contribute to glycemic control in patients with type 2 diabetes mellitus: A systematic review and meta-analysis. *Advances in Nutrition, 12*(3), 722–734.

Sheth, M., & Patel, M. (2019). Beta glucan and FOS added barley slims improves glycemia, gut health and inflammatory status of type 2 diabetes mellitus individuals residing in urban Baroda. *Indian Journal of Research–Paripex, 8*, 1–9.

Sheth, M., & Vaidya, R. (2012). Effect of resistant starch supplementation on glycemia and gut microflora in diabetic adults. *CTRI/2012/01/002359*. Retrieved from: http://www.ctri.nic.in.

Sheth, M., Khan, M. R., Dehingia, M., Thakuria, A., & Chand, V. (2015). A comparative study of synbiotic and prebiotic supplementation on gut health, SCFA, hs-CRP and lipid profile of type 2 diabetic subjects with pre hypertension. *European Journal of Biomedical and Pharmaceutical Sciences, 3*, 1–12.

Sheth, M., Parnami, S., Prakash, P., Quraishi, S., Kumar, A., Jain, M., Gote, M., & Dubey, A. (2017). Daily intake of fructoligosaccharide reduces hsCRP and other inflammatory markers in grade 1 sedentary obese individuals residing in urban Vadodara (Gujarat, India)–A double blind placebo control trial. *European Journal of Biomedical and Pharmaceutical Sciences, 4*, 1–3.

Sheth, M., Thakuria, A., Chand, V., & Nath, M. P. (2015). Fructo-oligosaccharide (FOS)-a smart strategy to modulate inflammatory marker and lipid profile in non-insulin dependent diabetes mellitus (NIDDM) subjects residing in Assam, India-a randomized control trial. *World Journal of Pharmaceutical Research, 4*(5), 2673–2678.

Unnikrishnan, R., Anjana, R. M., & Mohan, V. (2014). Diabetes in South Asians: Is the phenotype different? *Diabetes, 63*(1), 53–55. https://doi.org/10.2337/db13-1592.

Vaidya, R. H., & Sheth, M. K. (2011). Processing and storage of Indian cereal and cereal products alters its resistant starch content. *Journal of Food Science and Technology, 48*(5), 622–627.

Vulevic, J., Juric, A., Tzortzis, G., & Gibson, G. R. (2013). A mixture of trans-galactooligosaccharides reduces markers of metabolic syndrome and modulates the fecal microbiota and immune function of overweight adults. *Journal of Nutrition, 143*(3), 324–331.

Wang, L., Yang, H., Huang, H., Zhang, C., Zuo, H. X., Xu, P., et al. (2019). Inulin-type fructans supplementation improves glycemic control for the prediabetes and type 2 diabetes populations: Results from a GRADE-assessed systematic review and dose-response meta-analysis of 33 randomized controlled trials. *Journal of Translational Medicine, 17*(1), 410.

CHAPTER 10

Prebiotics in Obesity Management

GEETA DHARMATTI

Geeta Nutri Heal Consultancy, Pune, Maharashtra, India

Obesity prevalence has more than doubled in the last three decades, with over 1900 million (39%) people over the age of 18 being overweight and 600 million (13%) being obese (WHO, 2016).

10.1 INTRODUCTION

Obesity is an epidemiological syndrome characterized by increased fat mass, especially as visceral fat or central obesity. Obesity is currently recognized as a global problem by the WHO. An improper or excessive deposit of fat that poses a risk to one's health is referred to as obesity. Due to the increased carbohydrates from cereals during the Green Revolution, obesity increased in frequency after the industrial revolution of the eighteenth century. By 2035, 39% of today's population will be obese, according to the World Health Organization (WHO). Despite the fact that its etiology has been connected to an energy imbalance between calories taken and calories expended, primarily as a result of a poor diet and insufficient physical activity, it is actually the result of a medley of factors involving interactions between heredity, hormones, and the environment.

10.2 PATHOPHYSIOLOGY OF OBESITY

Obesity pathophysiology is multifactorial, involving poor lifestyle choices, genetic factors, physiological changes, and epigenetics which result from

Prebiotic Foods for a Healthy Gut: Scientific Studies to Manage Metabolic Disorders.
Mini K. Sheth and Shruti Dwivedi (Eds.)

energy imbalance in the human body. Growing evidence suggests that energy intake and output are regulated and interconnected by the central nervous system, pituitary, hypothalamus, and limbic system to regulate food intake and metabolism. Signals that determine food intake and non-activity-associated energy expenditure are metabolic and physiological energy demands and the maintenance of adequate energy stores. These regulatory processes appear to be controlled by a number of hormones and neuronal circuits, including leptin, which provides feedback from fat, ghrelin, various intestinal peptides, and several appetite-regulating neuropeptides (Maruvada et al., 2017).

Nutrient digestion and absorption take place primarily in the stomach and proximal small intestine in humans. Carbohydrates are one of the primary sources of energy for humans. Due to the complex structure of saccharides, humans have limited ability to digest and utilize these starches. Gut microbiota, known as saccharolytic microorganisms, help in the digestion of these complex saccharides (glycans, fiber, etc.), providing SCFA (Short-Chain Fatty Acids) to humans. Before entering the large intestine, 66–95% of proteins, 85% of carbohydrates, and 95% of fats are absorbed in healthy people (Clercq et al., 2016).

The human intestinal microbiota is a complex ecosystem composed of microorganisms such as bacteria, viruses, fungi, protists, nematodes, and phages that are deeply involved in various host metabolism functions. The gut flora is primarily involved in the fermentation of indigestible starch (prebiotics) into SCFA, which have an impact on energy homeostasis and good gut health. Prominent SCFAs are acetate, butyrate, and propionate, which are substrates for sugar metabolism; these SCFAs account for more than 95% of the total SCFA. Many animal and human studies have found higher SCFA fecal concentrations, particularly propionate, in obese individuals, implying that higher SCFA fecal concentrations are associated with obesity.

Their microbial inhabitants play an important role in metabolic processes, such as energy extraction from food, and are now widely regarded as the most important environmental factor linking dietary factors to the host. The link between the microbiota and metabolic inflammation, as seen in obesity, NAFLD (Non-Alcoholic Fatty Liver Disease), Cardiovascular disease, and Type 2 diabetes mellitus, is becoming clearer and more clinically supported. Indigestible, fermentable starch when adequately consumed, selectively modulates the growth and activity of specific microbiota members. Certain prebiotics, such as inulin or trans-galacto-oligosaccharides, that are resistant to gastric digestion and hydrolysis, can have a positive effect on the gut flora.

Obesity is linked to both insulin resistance and chronic low-grade inflammation. White adipose tissue, which is significant in contributing to the inflammatory state and expresses pro-inflammatory cytokines like TNF- and interleukins [IL]-1, IL-6, IL-10, and IL-12, is the most significant adipose tissue (Clercq et al., 2016).

White adipose tissue in obese individuals causes increased cytokine production as well as macrophage infiltration, which increases pro-inflammatory cytokines and, as a result, insulin resistance. The composition of gut flora changes with the onset of chronic low-grade inflammation. These changes cause increased gut permeability, also known as gut barrier imbalance. In this regard, the intestinal microbiota has recently received increased attention, particularly as a metabolic gateway between the outside environment and the host in terms of chronic low-grade inflammation, energy balance, and homeostasis.

10.3 GUT MICROBIOTA AND OBESITY

Short-term dietary changes can alter the gut microbiota. Dietary intervention with low calorie intake and modulating the gut microbiota to aid weight loss could be a good strategy for treating obesity. Obesity causes increased cytokine production in white adipose tissue as well as macrophage infiltration, which increases pro-inflammatory cytokines and, as a result, insulin resistance. High Fat Diet (increased n6-n3 ratio) alters the composition of gut flora, which can be a contributing factor to the onset of low-grade inflammation. These changes cause increased gut permeability, also known as gut barrier dysfunction. Gut dysbiosis and leaky gut are factors that further contribute to insulin resistance and oxidative stress (Virgin et al., 2018).

The intestinal microbiota has recently received increased attention, particularly as a metabolic gateway between the outside environment and the host in terms of inflammation, energy metabolism, and homeostasis. Surprisingly, the microbiome has properties in common with both the environment (it is, after all, a part of the human environment) and genes. It has been proposed by some researchers that the microbial genetic material we carry with us functions as an extension of our genome (a "meta-genome"). The gut microbiome (total genomic content of microorganisms) in humans contains 40 trillion microorganisms. The dominant bacterial phyla in humans, accounting for 90% of the gut microbiota, are *Firmicutes* and *Bacteroidetes*. There are currently >274 genera within the *Firmicutes* phylum, including

Bacillus, Lactobacillus, Mycoplasma, and *Clostridium.* The most abundant genus in the human gut is Bacteroidetes, which has 20 genera (Kobyliak et al., 2016).

The gut microbiota is crucial for the digestion, absorption, storage, and utilization of energy from dietary intake. Animal studies have also shown that the gut microbiome regulates food intake by influencing hormones and neurotransmitters that affect metabolic function and eating behavior (Khan et al., 2016).

This "gut microbiota-brain axis" is a bidirectional signaling-behavior axis that balances appetite, storage, and energy expenditure to regulate body weight. Changes in microbial diversity and richness are believed to be linked to changes in SCFA composition, calorie homeostasis, and inflammation. The relationship between gut microbiome composition and calorie homeostasis is complex, with variables such as genetic constitution, age, and food intake having a significant impact on gut microbiota function. Both metabolic disorders and stress-related neuropsychiatric disorders, such as depression and anxiety, can cause changes in the human gut microbiota composition; these pathologies are characterized by changes in eating behavior. The gut-brain axis has a significant physiological impact on mood, behavior, and stress responsiveness. Both short-term and long-term stress exposure can alter the type and amount of calorie intake, and these changes in energy homeostasis can affect how we feel emotionally.

An optimally functioning metabolism and homeostasis depend on a healthy gut microbiota, which also significantly benefits the host in return for living and growing in the intestinal environment. The structure of the intestinal mucosa has been protected and supported by commensal gut bacteria throughout the evolution of the human body. As a result, gut bacteria are increasingly acknowledged as important controllers of host physiology and pathophysiology and unquestionably play a part in both health and disease.

10.4 MECHANISMS PUT FORTH TO ACCOUNT FOR HOW THE GUT MICROBIOTA AFFECTS OBESITY

1. Energy regulation and the microorganism's capacity to ferment dietary polysaccharides that are difficult to digest by humans.
2. SCFAs are produced as a result of the fermentation of dietary fiber. Once absorbed, SCFAs work through molecular pathways to

promote lipogenesis and raise triglyceride stores. The sterol regulatory element-binding transcription factor 1 and the carbohydrate-responsive element-binding protein, both involved in lipogenesis, have been shown to be activated by SCFAs. SCFAs also inhibit the fasting-induced adipocyte factor, which blocks lipoprotein lipase and causes triglyceride accumulation in host adipocytes (Torres et al., 2015).

3. Conjugated fatty acids have peripheral effects and modulate the brain via direct or indirect mechanisms, altering host metabolism and the central control of hunger and food intake. Along with SCFAs, there are other bioactive metabolites made by the intestinal microbiota in a diet-dependent manner (Al-Assal et al., 2018).
4. Another mechanism to decode the link between the gut microbiota and obesity is its ability to suppress adenosine monophosphate kinase, which reduces liver fatty acid oxidation (AMPK). AMP Kinase is found in the liver and muscle fibers and functions as a cellular energy indicator. This results in decreased fatty acid oxidation and, hence, increased fat accumulation because of AMPK inhibition (Pineiro et al., 2018).

Dietary habits alter the composition of the human gut microbiota, and hence are the primary contributors to its diversity. Short-term dietary changes can alter the gut microbiota. Dietetic intervention, by reducing calorie intake and potentially modulating the gut microbiota to aid weight loss, can be considered for treating obesity.

10.5 PREBIOTICS AND PROBIOTICS IN OBESITY MANAGEMENT

Obesity has been identified as a major public health problem in all age groups, prompting significant efforts to implement obesity prevention programs and identify new therapeutic and intervention targets. Behavior modification, dietary changes, physical activity, and medical management are traditional treatments for obesity. The bariatric surgery approach also requires discipline and lifestyle management. It appears that lifestyle intervention is always the primary treatment option for obesity management.

Because the gut microbiota differs markedly between obese and lean individuals, specific modulation of gut microbiota has emerged as a potential therapy with pre- and probiotics for weight gain control in both obese and susceptible-to-obesity subjects. Prebiotics are non-digestible, non-hydrolyzable carbohydrate forms that can reach the last part of the human GI tract and be used as nutrients by intestinal host bacteria. Prebiotics' anti-obesogenic effects include improved lipid and glucose metabolism, in addition to modulating gut microbiota composition. These effects were linked to changes in gene expression involved in lipid metabolism, such as the upregulation of peroxisome proliferator-activated receptor (PPAR) and PPAR, as well as the downregulation of sterol regulatory element-binding protein-1c (SREBP-1c) and fatty acid synthase, which can explain part of the anti-obesogenic effect of cyclodextrins.

Prebiotics, which are commonly found in fruits and vegetables, may provide many health benefits. Among the benefits of prebiotics are the reduction of blood low-density lipoprotein levels, the stimulation of the immune system, the improved bioavailability of calcium, the maintenance of a correct intestinal pH value, and the low energy value. Recent research suggests that the mechanisms in which prebiotics support benefits to the host are mediated by microbial metabolic products, specifically SCFA, which promote ion and trace element absorption such as calcium, iron, and magnesium, and immune system regulation, including an increase in IgA production and modulating cytokine production (Figure 10.1).

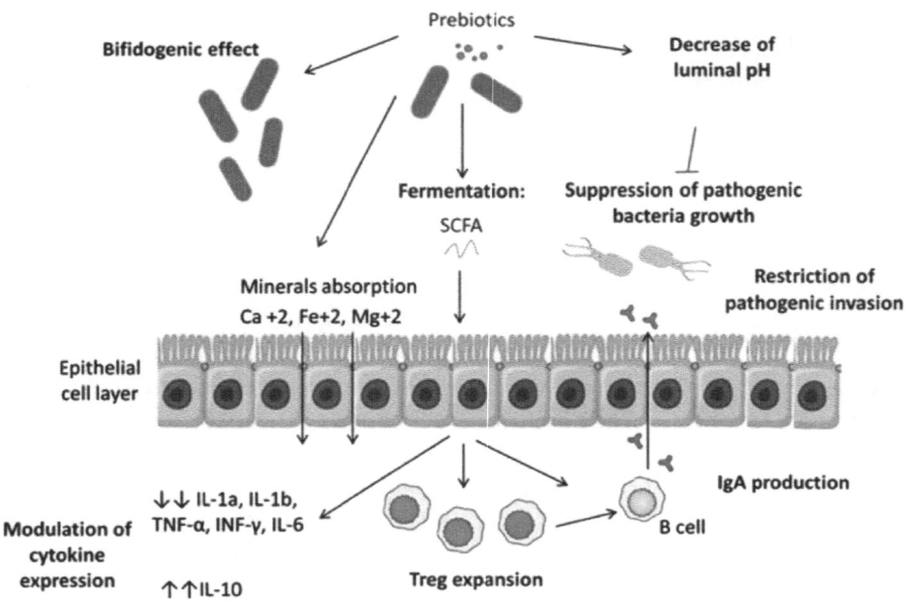

FIGURE 10.1 Mechanism of prebiotic action.

Prebiotics have a bifidogenic effect, meaning they provide a fermentable food source that promotes specific beneficial microbial populations like *Lactobacillus* and *Bifidobacteria* growth. This can be seen when there is a comparison of different types of infant feeding, where those fed formula supplemented with GOS and FOS had higher levels of *Bifidobacterium* than those fed formulae without prebiotics.

Prebiotics, on the other hand, have the ability to improve ion and trace element absorption. Connie et al. investigated the potential effect of GOS on bone mineral properties in mice. Dietary GOS reduced cecal pH while increasing total magnesium absorption, calcium and magnesium retention, and femur and tibia strength. Probiotic bacteria use prebiotic fibers as a source to generate large amounts of SCFAs (lactate, pyruvate, and acetate). These are in turn used as starting units for propionate and butyrate production by other colon bacteria. Butyrate is one of the most important colon metabolites because of its anti-inflammatory properties. This includes the promotion of regulatory T-cell expansion via the inhibition of histone deacetylation.

Despite a large number of human and animal studies aimed at elucidating the potential mechanisms of prebiotics, their long-term effects on host health remain unknown, particularly in early life (fetal and neonatal periods), infants, and young children. Thus, more research is needed to prove their mechanisms and effects on health. Prebiotics can be used as an alternative to or in addition to probiotics. The development of biotherapeutic formulas containing microbial strains and synergistic prebiotics can lead to an increase in probiotic effects in the small intestine and colon. Such "enhanced" probiotic products may be even more effective, with protective and stimulatory effects that outperform their constituents when taken separately. Further research into the interactions of probiotics and prebiotics, along with the development of synbiotics, may help to explain the mechanisms of action of these components, which may benefit human health.

As previously stated, the composition of gut microbiota improved with prebiotics. There was an improvement in the activity of entero-endocrine cells, which release hormones for appetite regulation, energy homeostasis, and body weight. Prebiotic-induced changes in gut microbiota composition improved the activity of entero-endocrine cells, which release hormones that regulate food intake, energy homeostasis, and body weight. As a result, the anti-obesity properties of prebiotics appear to be strongly related to the regulation of satiety hormones.

Prebiotics aid in the absorption of minerals like calcium, magnesium, zinc, and iron as they have the ability to bind to minerals. This mechanism prevents mineral absorption in the small intestine, and upon reaching the colon, they are released and absorbed. In the distal part of the intestine, due to fermentation caused by gut microbiota, SCFA (Short Chain Fatty Acids) are produced, which lowers luminal pH. This increases calcium bioavailability and passive absorption in the colon. Gut microbiota also produces a compound known as "calcium phytate," which hydrolyzes to improve calcium bioavailability. The osmotic effect of probiotics causes an increase in the volume of water in the colon, improving calcium absorption. This increase in calcium absorption benefits bone health by increasing bone mass and postponing the onset of osteoporosis.

This is significant when considering growth periods, which are when postmenopausal women and the elderly reach their peak bone mass. Iron absorption increases in the colon due to an increase in the soluble fraction in the cells with the stimulation of prebiotics. Zinc is required for skeleton development and maturation, and prebiotics increase zinc bioavailability. As a result, the diet becomes more adequate with the bioavailability of minerals. Also, prebiotics are linked to a different set of properties related to specific systemic disorders. Prebiotic starches (GOS, FOS, and inulin) lower blood pressure, glucose, cholesterol, triglycerides, and phospholipid levels, as well as fat synthesis in the liver, lowering the risk of diabetes, obesity, and atherosclerosis.

The composition of the gut microbiota is different in obese and lean individuals. Animal and human studies have shown a link between dysbiosis and obesity-related metabolic problems. However, it is still unknown which gut microbiota components cause weight gain and deranged glucose and fat metabolism and which are protective against metabolic changes. Several studies have shown that probiotics and/or prebiotics may have therapeutic effects on obesity, central obesity, adipogenesis, and inflammation which can lead to strategies for obesity management and related metabolic disorders.

10.6 INFLUENCE OF PREBIOTICS ON GUT HORMONES

Feeding is controlled by the central nervous system, but it is heavily influenced by a variety of peripheral physiological signals that either promote or limit energy intake (Hameed and Dhillo, 2009). In general, these gut hormones communicate information about changing energy status to the brain's periphery via a neuroendocrine mechanism. Few of these peptides are made by the gastrointestinal tract. The majority of these gastrointestinal-derived signals, such as cholecystokinin, glucagon-like peptide-1, and peptide YY, promote meal termination; in contrast, the hunger hormone ghrelin promotes food intake when readily available energy is low (Mietlicki-Baase and Hayes, 2015).

Glucagon-like peptide-1 (GLP-1) is a neuropeptide and incretin derived from the proglucagon gene transcription product. The intestinal L cell, which secretes GLP-1 as a gut hormone, is the primary source of GLP-1 in the periphery. Probiotic bacteria colonization increases GLP-1 synthesis in the proximal colon via a mechanism related to the differentiation of precursor cells into entero-endocrine L-cells. GLP-1 is known to affect the satiety center and is likely to affect obesity in the long run, resulting in lower food intake, body weight, fat mass development, restored beta cell mass, and glucose-induced insulin secretion (Chaudhri et al., 2008).

Lipopolysaccharide (LPS) plays a role in metabolic endotoxemia because it is a major molecular component of Gram-negative bacteria's outer membrane and acts as a physical barrier protecting the bacteria from its surroundings. LPS is continuously produced within the gut by the death of Gram-negative bacteria and is physiologically carried into intestinal capillaries via a TLR4-dependent mechanism (Neal et al., 2006). LPS is transported from the intestine to target tissues via a mechanism aided by lipoproteins, specifically chylomicrons synthesized from epithelial intestinal cells in response to fats.

10.7 GUT MICROFLORA, LOW INFLAMMATION, AND OBESITY: A VICIOUS CYCLE

In humans, the gut microflora has been linked to chronic diseases such as obesity (Cani et al., 2007). Obese and lean subjects have different gut microflora composition profiles, according to research. Obese people have a lower proportion of Bacteroidetes than lean people (Angelakis et al., 2012). Weight imbalance and impaired metabolism are exacerbated by differences in microbiota composition. Animal model evidence suggests that the microbiota of obese subjects may have a greater capacity to derive energy from the food by providing substrates that can activate adipogenesis. Furthermore, microorganisms can influence lipoprotein lipase activity, interfering with triglyceride accumulation in adipose tissue. The interaction of gut microflora with the endocannabinoid system provides a pathway for altering intestinal permeability. Increased intestinal permeability allows endotoxins into the bloodstream. This has been linked to insulin resistance and inflammation in mice (Zao, 2013).

LPS are unique glycolipids found in the cell walls of gram-negative/pathogenic bacteria. LPS molecules, also known as bacterial endotoxins, can cause acute and chronic inflammation by activating immune cells and releasing cytokines. One of the most important factors in the elimination of LPS molecules from circulation is HDL cholesterol. Primarily, LPS is bound to HDL in healthy subjects, but it is redistributed to LDL and VLDL lipoproteins in patients with sepsis (TranDinh et al., 2013). High LPS activity in conjunction with low HDL levels raises the risk of type II diabetes and cardiovascular disease (Jayashree et al., 2014). In mammals, LPS infusion causes the appearance of factors known to be associated with MetS: elevated levels of pro-inflammatory markers, dyslipidemia, fasting hyperglycemia, insulin resistance, and obesity (Jayashree et al., 2014; Parekh et al., 2014) (Figure 10.2).

(1) Chronic bacterial translocation as a result of increased intestinal permeability causes systemic inflammation that results in an influx of macrophages in visceral tissue, hepatic Kupffer cell activation, and insulin resistance.
(2) By reducing free fatty acid production by visceral tissue, short-chain fatty acids normalize intestinal permeability with an impact on gluco- and lipogenesis.

Prebiotic treatment boosts *Bifidobacterium* spp., lowers plasma LPS levels, and improves insulin sensitivity, steatosis, and low-grade inflammation.

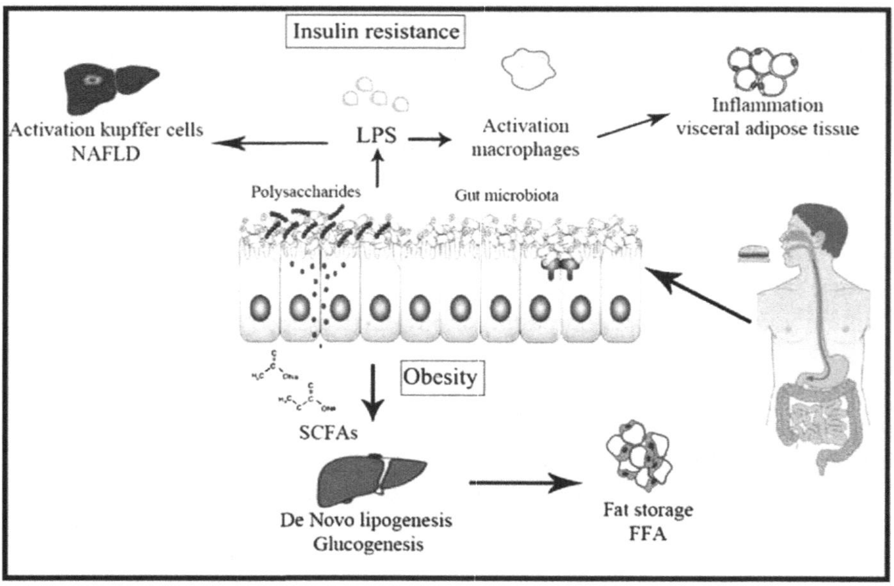

FIGURE 10.2 The mechanism by which intestinal microbiota alter human metabolism, resulting in obesity and insulin resistance.

Source: Reprinted from Moreno-Indias (2014). https://creativecommons.org/licenses/by/3.0/

10.8 EFFECT OF FRUCTOOLIGOSACCHARIDE ON OBESITY AND LOW-GRADE INFLAMMATION

Oligofructose administration to genetically obese mice increased the levels of *Lactobacillus*, *Bifidobacterium*, *C. coccoides*, and *E. rectale*, which resulted in a reduced intestinal permeability and an improvement in tight junction integrity and inflammatory markers like lipopolysaccharides and cytokines (Cani et al., 2007; Million et al., 2013).

Parnell and Reimer (2009) investigated the effect of oligofructose supplement on individual body weight, ghrelin, and PYY concentrations as a marker of satiety in obese adults. FOS-induced carbohydrate fermentation produces SCFAs (Tolhurst et al., 2012), which in turn regulate gut hormones such as glucagon-like peptide (GLP-1) and peptide YY (PYY) (Bomhof et al., 2014). These gut hormones control satiety by regulating the secretion of digestive enzymes (Baggio and Drucker, 2014). For 12 weeks, 48 healthy adults (BMI > 25 kg/m^2) were randomly assigned to receive oligofructose (21 g daily) or a placebo (maltodextrin). They observed a reduction of 1.03 ± 0.43 kg in oligofructose patients compared to an increase of 0.4 ± 0.31 kg ($p = 0.01$). This led to the conclusion that oligofructose supplementation may benefit both weight loss and glucose metabolism in obese adults (Parekh et al., 2014).

FOS has attracted special attention because of its prebiotic properties (Yun, 1996). It acts as a functional food ingredient with specific physiological effects, such as stimulating the growth of beneficial *Bifidobacteria* in the digestive tract, lowering total cholesterol and lipid levels in serum, relieving constipation, and generally improving human health (Tomomatsu, 1994).

To determine the role of metabolic endotoxemia as a triggering factor in the development of metabolic disorders associated with obesity, scientists created a mouse model chronically infused with a very low dose of LPS to achieve the same plasma LPS levels as the high-fat diet-fed mice. Fasting hyperglycemia, obesity, steatosis, adipose tissue macrophage infiltration, hepatic insulin resistance, and hyperinsulinemia were all mimicked by a four-week chronic low-dose LPS infusion, which mimicked the phenotype of mice fed with a high-fat diet (Cani et al., 2007). Kok et al. discovered that feeding rats a prebiotic fiber called oligofructose (OFS) increased total caecal GLP-1 and jejunum GIP concentrations.

Inulin-type polysaccharide fructans have been well studied and shown to be effective in human and animal models in stimulating the growth of health-promoting *Bifidobacteria* and *Lactobacilli* and modulating the gut hormone GLP-1 (Macfarlane et al., 2006). Many studies have found that prebiotic supplementation increases gut hormone signaling (GLP-1, PYY). In one study, rats fed prebiotics (resistant starch) had higher levels of GLP-1 and PYY (Cani et al., 2009).

Studies confirmed that mice fed a high-fat diet have higher endotoxemia, which is completely eliminated by dietary supplementation with prebiotic dietary fibers. *Bifidobacteria* significantly and positively correlated with improved glucose homeostasis and normal low-grade inflammation in prebiotic-treated mice (decreased endotoxemia, plasma, and adipose tissue proinflammatory cytokines). The researchers also discovered a negative correlation between metabolic endotoxemia and *Bifidobacterium* spp. (Tuohy and Gibson, 2005). As shown in Table 10.2, the gut microbe profile of normal weight employees had 4.27% more *Bifidobacterium* colonization and 8.17% more *Bacteroides* colonization. However, when compared to non-obese employees, colonization of the gut in overweight employees was highlighted by higher counts of pathogenic bacteria (*Clostridium*) (Assudani and Sheth, 2018) (Table 10.2).

FOS supplementation reduced body weight, WC, WHR, and body fat by 1.06%, 1.66%, 2.13%, and 4.23% in obese subjects, respectively. GLP-1 levels increased by 17% on average, while LPS levels decreased by 2.81% (p 0.05). The fecal log counts changed significantly after supplementation, with an increase in lactic acid bacteria (13.84%) and *Bifidobacteria* (10.30%) and a decrease in enteric pathogens (20.24%). A daily intake of 12 g of FOS for more than 50 days reduced body weight, hunger, plasma LPS, and plasma GLP-1 levels, as well as improved colonization of beneficial gut microorganisms such as LAB and *Bifidobacteria*. As shown in Figures 10.3–10.5, significant reductions in weight, waist circumference, and waist-hip ratio ($p < 0.001$) were also observed, indicating the potential of FOS for overall metabolic control in obese individuals (Gupta and Sheth, 2016).

TABLE 10.2 Mean Values of Gut-Microbial Profiles and Differences in Their Gut Colonization Patterns of Non-Obese and Obese Bank Employees

Parameter	Non-obese [N=115] Log 10values [CFU/g] Mean ± SD	obese [N=115] Log 10values [CFU/g] Mean ± SD	't' test Value	Percent Difference
Bifidobacteria	12.63 ± 1.68	12.09 ± 1.12	2.85**	− 4.27
Lactobacillus	*11*.84 ± 1.54	11.99 ± 1.61	0.98 NS	+ 1.27
Clostridium	11.82 ± 1.54	12.33 ± 1.15	2.70**	+ 4.32
Bacteroides	12.85 ± 1.44	11.80 ± 1.55	5.19***	+ 8.17

NOTE: NS non- significant , $p < 0.05$.* , $p < 0.01$.**, $p < 0.001$.***

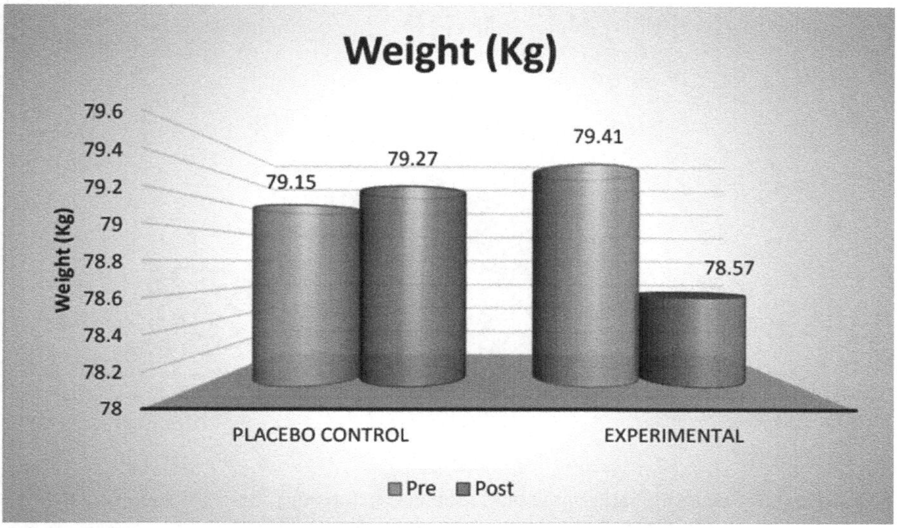

FIGURE 10.3 Weight (kg) of the obese subjects before and after FOS supplementation.

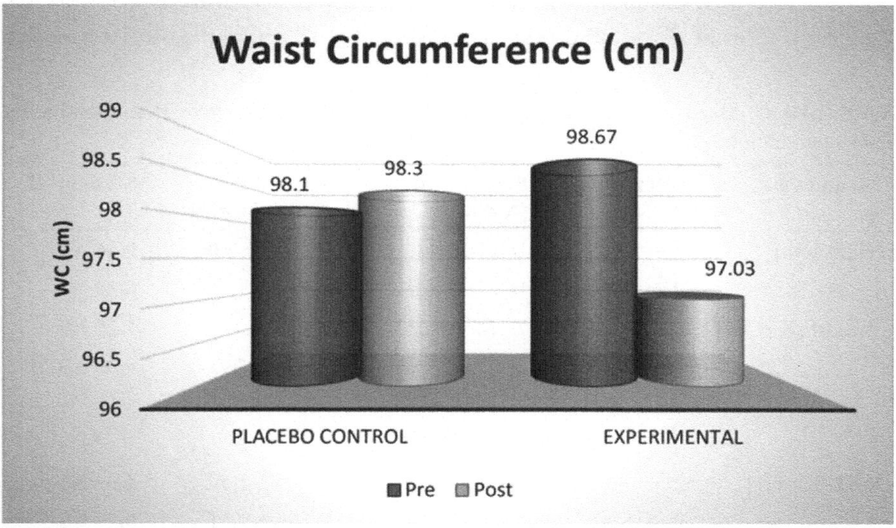

FIGURE 10.4 Waist circumference (cm) of the obese subjects before and after FOS supplementation.

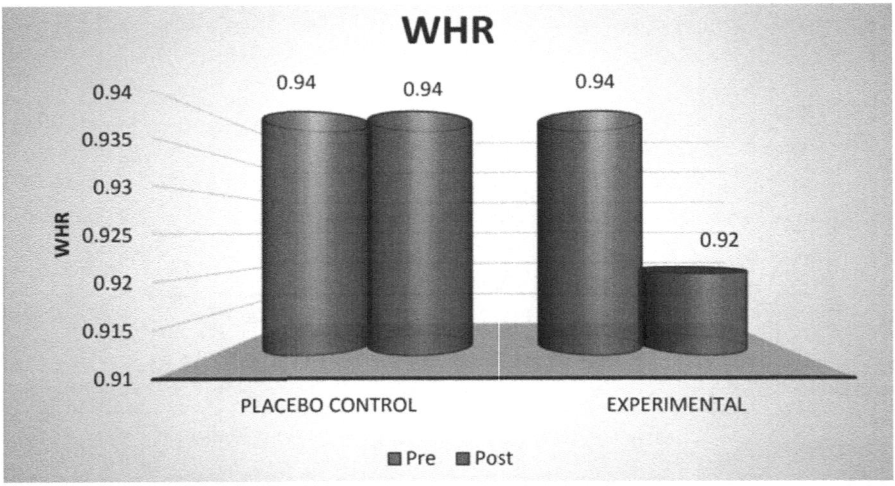

FIGURE 10.5 Waist hip ratio of the obese subjects before and after FOS supplementation.

A similar study by Assudani and Sheth (2018) found that FOS intervention for 90 days was an effective supplement for weight loss. Weight loss was 3.25%, with reductions in BMI (3.25%), waist circumference (2.31%), waist-hip ratio (1.07%), and body fat percentage (3.39%) (Table 10.3).

TABLE 10.3 Mean Anthropometric Values of Obese Bank Employees Before and After FOS Intervention

Parameter	Intervention Phase	Place Group [N=70]	Experiment Group [n=72]	Student 't' Test
Hight [cm]	pre	168.78 ± 11.32	169.49 ± 7.01	**0.44 NS**
	Post	168.78 ± 11.32	169.49 ± 7.01	**0.44 NS**
Weight [Kg]	Pre	77.74 ± 10.82	80.58 ±	**1.75 NS**
	Post	77.74 ± 10.82	77.96 ± 7.93	**0.36 NS**
	Paired 't'	**1.24 NS**	**27.61*****	
	% ↑/↓	**0.24 % ↓**	**3.25 % ↓**	
BMI [Kg/m² |	Pre	27.18 ± 1.38	27.95 ± 1.50	**1.63 NS**
	Post	27.12 ± 1.44	27.09 ± 1.41	**0.09 NS**
	Paired 't'	**1.10 NS**	**29.14*****	
	% ↑/↓	**0.22 % ↓**	**3.25 % ↓**	
WC [cm]	Pre	94.31 ± 7.43	96.74 ± 7.57	**1.92 NS**
	Post	93.94 ± 6.29	94.51 ± 7.67	**0.49 NS**

TABLE 10.3 *(Continued)*

Parameter	Intervention Phase	Place Group [N=70]	Experiment Group [n=72]	Student 't' Test
	Paired 't'	**1.16 NS**	**7.92*** **	
	% ↑/↓	**0.39 % ↓**	**2.31 % ↓**	
HC [cm]	Pre	102.06 ± 7.49	103.35 ± 5.69	**1.15 NS**
	Post	102.03 ± 7.55	103.03 ±5.48	**1.75 NS**
	Paired 't'	**0.62 NS**	**1.83 NS**	
	% ↑/↓	**0.03 % ↓**	**0.31% ↓**	
WHR	Pre	0.92 ± 0.04	0.93 ± 0.05	**1.48 NS**
	Post	0.92 ± 0.05	0.92 ± 0.05	**1.68 NS**
	Paired 't'	**0.66 NS**	**5.43*** **	
	% ↑/↓	**0 % ↓**	**1.07 % ↓**	

NOTE: NS -non-significant, p < 0.05:* .p < 0.01:**. p < 0.001:***

It can be concluded that FOS is an attractive therapy for the management of obesity. It also helps reduce hunger pangs and achieve early satiety during lunch and dinner.

KEYWORDS

- **bacteria**
- **cardiovascular diseases**
- **fats**
- **global problems**
- **hormones**
- **inflammation**
- **metabolic disorders**
- **metabolism**
- **obesity**
- **prebiotics**

REFERENCES

Al-Assal, K., Martinez, A. C., Torrinhas, R. S., Cardinelli, C., & Waitzberg, D. (2018). Gut microbiota and obesity. *Clinical Nutrition Experimental, 20*, 60–64. https://doi.org/10.1016/j.yclnex.03.001.

Angelakis, E., Armougom, F., Million, M., & Raoult, D. (2012). The relationship between gut microbiota and weight gain in humans. *Future Microbiology, 7*(1), 91–109. https://doi.org/10.2217/fmb.11.142.

Assudani, A., & Sheth, M. (2018). Retrieved from: http://dspace.hmlibrary.ac.in:8080/jspui/bitstream/123456789/1409/18/18_Appendices.pdf (accessed on 19 July 2024).

Baggio, L. L., & Drucker, D. J. (2014). Glucagon-like peptide-1 receptors in the brain: Controlling food intake and body weight. *Journal of Clinical Investigation, 124*(10), 4223–4226. https://doi.org/10.1172/JCI78371.

Bomhof, M. R., Saha, D. C., Reid, D. T., Paul, H. A., & Reimer, R. A. (2014). The combined effects of oligofructose and Bifidobacterium animalis on gut microbiota and glycemia in obese rats. *Obesity (Silver Spring), 22*(3), 763–771. https://doi.org/10.1002/oby.20632.

Bomhof, M. R., Saha, D. C., Reid, D. T., Paul, H. A., & Reimer, R. A. (2014). Combined effects of oligofructose and Bifidobacterium animalis on gut microbiota and glycemia in obese rats. *Obesity (Silver Spring), 22*(3), 763–771. https://doi.org/10.1002/oby.20632.

Bray, G. A., & Ryan, D. H. (2000). Clinical evaluation of the overweight patient. *Endocrine, 13*(2), 167–186.

Bray, G. A., Kim, K. K., & Wilding, J. P. H. (2017). Obesity: A chronic, relapsing, progressive disease process. A position statement of the World Obesity Federation. *Obesity Reviews, 18*(7), 715–723.

Cani, P. D., Amar, J., Iglesias, M. A., Poggi, M., Bastelica, D., Neyrinck, A. M., Fava, F., Tuohy, K. M., Chabo, C., Waget, A., Delmée, E., Cousin, B., Sulpice, T., Chamontin, B., Ferrières, J., Tanti, J. F., Gibson, G. R., Casteilla, L., Delzenne, N. M., & Burcelin, R. (2007). Metabolic endotoxemia initiates obesity and insulin resistance. *Diabetes, 56*(7), 1761–1772.

Cardona, F., Tinahones, F. J., & Isabel, M. (2014). Impact of the gut microbiota on the development of obesity and type 2 diabetes mellitus. *Frontiers in Microbiology, 5*. https://doi.org/10.3389/fmicb.2014.00190.

Chaudhri, O. B., Parkinson, J. R., Kuo, Y., Druce, M. R., Herlihy, A. H., Bell, J. D., Dhillo, W. S., Stanley, S. A., Ghatei, M. A., & Bloom, S. R. (2006). Differential hypothalamic neuronal activation following peripheral injection of GLP-1 and oxyntomodulin in mice detected by manganese-enhanced magnetic resonance imaging. *Biochemical and Biophysical Research Communications, 350*(2), 298–306. https://doi.org/10.1016/j.bbrc.2006.09.033.

Corzo, N., Alonso, J. L., Azpiroz, F., Calvo, M. A., Cirici, M., & Leis, R. (2015). Prebióticos: Concepto, propiedades, y beneficial effects. *Nutrición Hospitalaria, 31*, 99–118.

de Clercq, N. C., Groen, A. K., Romijn, J. A., & Nieuwdorp, M. (2016). Gut microbiota in obesity and undernutrition. *Advances in Nutrition, 6*, 1080–1089.

Duca, F. A., & Lam, T. K. (2014). Gut microbiota, nutrient sensing, and energy balance. *Diabetes, Obesity and Metabolism, 16*(Suppl 1), 68–76.

Gibson, G. R., Hutkins, R., Sanders, M. E., Prescott, S. L., Reimer, R. A., Salminen, S. J., Scott, K., Stanton, C., Swanson, K. S., Cani, P. D., Verbeke, K., & Reid, G. (2017). The International Scientific Association for Probiotics and Prebiotics (ISAPP) consensus statement on the definition and scope of prebiotics. *Nature Reviews Gastroenterology & Hepatology, 14*(8), 491–502.

Hildebrandt, M. A., Hoffmann, C., Sherrill-Mix, S. A., Keilbaugh, S. A., Hamady, M., Chen, Y. Y., Knight, R., Ahima, R. S., Bushman, F., & Wu, G. D. (2009). High-fat diet determines the composition of the murine gut microbiome independently of obesity. *Gastroenterology, 137*(5), 1716–1724.e1.

Jayashree, S., Arindam, M., & Vijay, K. V. (2015). Genetic epidemiology of coronary artery disease: An Asian Indian perspective. *Journal of Genetics, 94*, 539–549.

Khan, M. J., Gerasimidis, K., Edwards, C. A., & Shaikh, M. G. (2016). Role of gut microbiota in the aetiology of obesity. *Journal of Obesity, 2016*, 7353642.

Kobyliak, N., Conte, C., Cammarota, G., Haley, A. P., Styriak, I., Gaspar, L., Fusek, J., Rodrigo, L., & Kruzliak, P. (2016). Probiotics in prevention and treatment of obesity: A critical view. *Nutrition & Metabolism (London), 13*, 14.

Krajmalnik-Brown, R., Ilhan, Z. E., Kang, D. W., & DiBaise, J. K. (2012). Effects of gut microbes on nutrient absorption and energy regulation. *Nutrition in Clinical Practice, 27*(2), 201–214.

Macfarlane, S., Macfarlane, G. T., & Cummings, J. H. (2006). Review article: Prebiotics in the gastrointestinal tract. *Alimentary Pharmacology & Therapeutics, 24*(5), 701–714. https://doi.org/10.1111/j.1365-2036.2006.03042.x.

Maruvada, P., Leone, V., Kaplan, L. M., & Chang, E. B. (2017). The human microbiome and obesity: Moving beyond associations. *Cell Host & Microbe, 22*(5), 589–599.

Mietlicki-Baase, E. G., Reiner, D. J., Cone, J. J., Olivos, D. R., McGrath, L. E., Zimmer, D. J., Roitman, M. F., & Hayes, M. R. (2015). Amylin modulates the mesolimbic dopamine system to control energy balance. *Neuropsychopharmacology, 40*(2), 372–385. https://doi.org/10.1038/npp.2014.180.

Million, M., Lagier, J. C., Yahav, D., & Paul, M. (2013). Gut bacterial microbiota and obesity. *Clinical Microbiology and Infection, 19*(4), 305–313. https://doi.org/10.1111/1469-0691.12172.

Neal, M. D., Leaphart, C., Levy, R., Prince, J., Billiar, T. R., Watkins, S., Li, J., Cetin, S., Ford, H., Schreiber, A., & et al. (2006). Enterocyte TLR4 mediates phagocytosis and translocation of bacteria across the intestinal barrier. *Journal of Immunology, 176*, 3070–3079.

Oozeer, R., van Limpt, K., Ludwig, T., Ben Amor, K., Martin, R., Wind, R. D., Boehm, G., & Knol, J. (2013). Intestinal microbiology in early life: Specific prebiotics can have similar functionalities as human-milk oligosaccharides. *American Journal of Clinical Nutrition, 98*(2), 561S–571S.

Parekh, P. J., Arusi, E., Vinik, A. I., & Johnson, D. A. (2014). The role and influence of gut microbiota in pathogenesis and management of obesity and metabolic syndrome. *Frontiers in Endocrinology, 5*, 47. https://doi.org/10.3389/fendo.2014.00047.

Pineiro, M., Asp, N. G., Reid, G., Macfarlane, S., Morelli, L., Brunser, O., & Tuohy, K. (2008). FAO technical meeting on prebiotics. *Journal of Clinical Gastroenterology, 42*(Suppl 3), S156–S159.

Tomomatsu, H. (1994). Health effects of oligosaccharides. *Food Technology, 48*(10), 61–65.

Torres-Fuentes, C., Schellekens, H., Dinan, T. G., & Cryan, J. F. (2017). The microbiota-gut-brain axis in obesity. *The Lancet Gastroenterology & Hepatology, 2*(10), 747–756.

Tran-Dinh, A., Diallo, D., Delbosc, S., Varela-Perez, L. M., Dang, Q. B., Lapergue, B., Burillo, E., Michel, J. B., Levoye, A., Martin-Ventura, J. L., & Meilhac, O. (2013). HDL and endothelial protection. *British Journal of Pharmacology, 169*(3), 493–511. https://doi.org/10.1111/bph.12174.

Tuohy, K. M., Rouzaud, G. C., Brück, W. M., & Gibson, G. R. (2005). Modulation of the human gut microflora towards improved health using prebiotics: Assessment of efficacy. *Current Pharmaceutical Design, 11*(1), 75–90. https://doi.org/10.2174/1381612053382331.

Virgin, H. W., & Todd, J. A. (2011). Metagenomics and personalized medicine. *Cell, 147*(1), 44–56.

Weaver, C. M., Martin, B. R., Nakatsu, C. H., Armstrong, A. P., Clavijo, A., McCabe, L. D., McCabe, G. P., Duignan, S., Schoterman, M. H., & van den Heuvel, E. G. (2011). Galactooligosaccharides improve mineral absorption and bone properties in growing rats through gut fermentation. *Journal of Agricultural and Food Chemistry, 59*(12), 6501–6510.

World Health Organization. (2000). Obesity: Preventing and managing the global epidemic. World Health Organization.

Younis, K., Ahmad, S., & Jahan, K. (2015). Health benefits and application of prebiotics in foods. *Journal of Food Processing & Technology, 6*, 1–5.

Yun, J. W. (1996). Fructooligosaccharides: Occurrence, preparation, and application. *Enzyme and Microbial Technology, 19*, 107–117.

CHAPTER 11

Probiotics During Pregnancy

MEGHANA PATEL

S.B.K.S. Medical Institute and Research Center, Gujarat, India

> ***Pregnancy complications can significantly impact the health of both the mother and child. Probiotics offer a promising way to support maternal health during pregnancy, potentially reducing risks and improving outcomes for both.***

11.1 INTRODUCTION

Several immunological, hormonal, and metabolic changes brought on by pregnancy are necessary for the fetus's optimal development, prompt commencement of labor, and safe delivery (Koren et al., 2012). These can exert substantial changes in women. Changes in the levels of circulating hormones, elevated intravascular volume, and compression due to the expanding uterus underlie the complex physiological adaptations that are important for the growth and development of the fetus. Frequent alterations in the endocrine system are observed for the developing fetus. Many endocrine glands' functions are altered by steroid hormones and proteins secreted by the placenta. Every so often, the changes in certain hormone levels have effects on their target organ which lead to gestational diabetes mellitus and gestational hypertension. Furthermore, higher body fat during the early stages of pregnancy leads to decreased insulin sensitivity in later pregnancy. This decrease in sensitivity has been associated with alterations

Prebiotic Foods for a Healthy Gut: Scientific Studies to Manage Metabolic Disorders.
Mini K. Sheth and Shruti Dwivedi (Eds.)

in immune status, including heightened levels of circulating cytokines, which are thought to contribute to metabolic inflammation related to obesity (Kohlhepp et al., 2018).

11.2 COMPLICATIONS DURING PREGNANCY

Vascular dysfunction of the placental tissue can arise when conditions like maternal obesity, decreased insulin sensitivity, or a leaky gut trigger the maternal inflammatory state to shift from a physiologic to an excessive level, which can have harmful effects like restricted fetal development, preeclampsia, GDM, and other conditions (Kashtanova et al., 2016). Recently, researchers have shown that these pregnancy-specific complications tend to affect maternal health post-pregnancy as well. According to the studies, women who have experienced adverse pregnancy outcomes are at a higher risk of developing cardiovascular and metabolic diseases later in life. Additionally, though pregnancy is an exciting time for pregnant mothers, depression and anxiety often coexist with it. Lately, research is emerging in the area of gut microbiome in pregnancy, and it has been identified as a key factor contributing to pregnancy complications.

11.3 GUT MICROFLORA LANDSCAPE DURING PREGNANCY

Studies have revealed that the microorganisms inhabiting the gut play various crucial roles in the metabolism of the host, defense system, and overall health. The 100 trillion bacteria reside in the human intestinal tract, which is referred to as gut microbiota that are indispensable for human body metabolism and immunity (Power et al., 2014). The mother's microbiome, the delivery method, mother's diet, and environmental exposures have an impact on how the human gut microbiome develops before birth and over the course of infancy (Brown et al., 2013). The gut microbiome has been identified as an important factor for maintaining health before and during pregnancy and could contribute to complications during pregnancy too. Metabolic alterations during pregnancy, including changes in energy homeostasis, storage of fat, and hormonal profiles, lead to increased glycemic levels, insulin resistance, glucose intolerance, and weight gain. Variations in metabolism, such as those seen in obesity, metabolic syndrome, and diabetes, have an impact on gut microbiota as well. Thus, these changes occurring in pregnancy are expected to influence the composition of microbiota. Without a strong intestinal wall, bacteria enter through the gut into the general circulation during pregnancy, causing metabolic and immunological changes that worsen systemic inflammation. The "leaky gut" theory of intestinal permeability is the method through which this movement takes place wherein protein-based junctions connecting adjacent intestinal cells weaken and stop functioning as a vital barrier to advancing pathogens. Obesity and Type 2 DM and its accompanying illnesses are examples of low-grade inflammatory disorders that are linked to increased intestinal permeability (Mokkala et al., 2016). Bacterial components, such as the lipopolysaccharides (LPS) found in the cell walls of gram-negative bacteria, can leak through increased permeability and/or cause a harmful systemic inflammatory syndrome known as metabolic endotoxemia (Cani et al., 2012). Numerous clinical and physiological disorders, including obesity and the metabolic syndrome, have been linked to microbial alterations known as dysbiosis. Therefore, it is not surprising that the microbiome undergoes major changes during pregnancy.

Observations indicate that during the transition from the first to third trimester of pregnancy, there is a gradual and significant transformation in the gut microbiota (Crusell et al., 2018). These alterations may be brought on by changes in food, use of antibiotics, gestational diabetes, or pre-pregnancy body mass index. During the first trimester, the gut microflora is similar to that of healthy non-pregnant women in many aspects (Walters et al., 2014). Metabolic hormonal shifts and alterations in the gastrointestinal mucosa can lead to a mild pro-inflammatory condition during typical pregnancies. This state may facilitate enhanced glucose diffusion from the gut epithelium into the lumen, potentially contributing to weight gain and influencing changes in the gut microbiota. Additionally, studies have indicated that the gut microbiota during pregnancy plays a crucial role in shaping the health of the child and could potentially influence the child's susceptibility to atopy and autoimmune traits (Nyangahu & Jaspan, 2019).

Gut microbiota-targeted indicators may be used as possible predictors of Gestational Diabetes Mellitus since a peculiar gut microflora interaction is linked to the condition before it manifests (Ma et al., 2020). Early in pregnancy, there is an increase in insulin sensitivity, which encourages the uptake of glucose into adipose storage for the energy needs in the third trimester. Due to the antagonistic effects of insulin-like compounds, pregnant women's insulin sensitivity decreases in the later stages of pregnancy. The demand for insulin increases consequently to maintain normal blood glucose levels (Ma et al., 2020). However, the normal insulin levels during gestation do not effectively occur in all pregnancies, except in those with pre-existing insulin resistance or women who cannot increase insulin secretion, resulting in Gestational Diabetes Mellitus (GDM). Research has shown that GDM detected in the third trimester is associated with interrupted gut microflora composition compared with normoglycemic pregnant women and also during pregnancy, some species of bacteria showed divergent associations with hs-CRP (Crusell et al., 2018). Short-chain fatty acids (SCFAs) such as butyrate, acetate, or propionate may have an impact on both the mother and the newborn by down-regulating pro-inflammatory responses at specific sites because the gut microbiota stimulates immune function and can therefore modulate the response. As we all know, the gut microbiome is shaped by diet and genetics. The outcome of food and nutrients on intestinal microbiota is rapid, and alteration in the gut microbiota can occur within days after modifying the diet.

11.4 PROBIOTICS AND GUT: THE PERFECT DUO

High-fat, salt, and sugar diets encourage weight gain and dysbiosis, which is associated with adverse maternal and child health outcomes. Conversely, short-chain fatty acids (SCFA) like acetate, propionate, and butyrate, produced through the fermentation of dietary fiber (prebiotics), act as an energy supply for beneficial intestinal bacteria residing within epithelial cells. Moreover, they promote stronger connections among these cells (David et al., 2014). Elevated levels of SCFA in the bloodstream have a beneficial impact on the metabolic alterations observed during pregnancy, such as weight gain during pregnancy, glucose metabolism, and levels of various metabolic hormones (Koren et al., 2012). Probiotics, the good bacteria, provide a health benefit to the host when given in adequate doses. The most commonly used probiotic bacteria *Lactobacillus* and *Bifidobacterium* have been established as potential probiotics in recent years.

Gestational Diabetes Mellitus has been found to be a major complication during pregnancy due to significant risks that it poses for pregnancy outcomes and the long-term health of both the mother and child. The potential role of the gut microflora in controlling glycemic status has come to light, and the use of probiotics during pregnancy for the improvement in maternal metabolism through the modification of gut microbiota has been recently reviewed (Lindsay et al., 2014). A meta-analysis showed that probiotic supplementation significantly lowered serum glucose, insulin levels, and HOMA-IR index in pregnant women with GDM, improving glucose metabolism (Nordqvist et al., 2018). Similarly, a randomized double-blind placebo-controlled trial revealed a significant reduction in FBS, TG levels, and increased insulin sensitivity compared to the placebo group upon supplementation with probiotics (Łagowska et al., 2020).

Numerous studies have demonstrated that probiotics can indeed lessen the symptoms of depression and anxiety as well. Recently, it has been studied that probiotic supplementation showed a positive impact on the pre-existing symptoms of anxiety or depression in pregnant women (Slykerman et al., 2017).

The level of mild systemic inflammation during pregnancy has been associated with dietary factors and conditions such as preeclampsia, obesity, and gestational diabetes, all of which are recognized as fetal stressors capable of inducing developmental adaptations (Ahmadi et al., 2016).

Particularly in cases of severe and early-onset preeclampsia, an increased inflammatory response appears to be more important. Pre-eclampsia patients were found to have significantly different levels of several harmful microorganisms compared to healthy pregnant women. Pre-eclampsia development was somewhat correlated with increased levels of harmful bacteria such as *Clostridium perfringens, Bulleidia moorei*, and decreased probiotic *Coprococcus catus* (Badehnoosh et al., 2018). One study revealed that consuming probiotic milk early in pregnancy reduced the risk of preterm delivery, and consumption later in pregnancy demonstrated a decreased risk of preeclampsia (Nordqvist et al., 2018).

Probiotics have been shown to offer significant benefits during pregnancy, promoting both maternal and fetal health. These good bacteria aid in maintaining a balanced gut microbiome, which positively influences the immune system and digestion. By reducing the risk of gestational diabetes, preeclampsia, and infections, probiotics can contribute to a healthier pregnancy. They also alleviate common gastrointestinal issues such as constipation and bloating.

11.5 CONCLUSION

Pregnancy complications are associated with adverse short-term and long-term outcomes for both the mother and child. The gut microbiome has been identified as a vital aspect for maintaining maternal health during pregnancy and could contribute to alleviating pregnancy complications. Probiotics have the potential to be a safe treatment intervention for both preventing and lowering the risk of pregnancy issues and adverse outcomes related to maternal metabolism.

KEYWORDS

- **bacteria**
- **diabetes**
- **fetus**
- **hormones**
- **immunology**
- **maternity**
- **metabolism**
- **obesity**
- **pregnancy**
- **women**

REFERENCES

Ahmadi, S., Jamilian, M., Tajabadi-Ebrahimi, M., Jafari, P., & Asemi, Z. (2016). The effects of synbiotic supplementation on markers of insulin metabolism and lipid profiles in gestational diabetes: A randomized, double-blind, placebo-controlled trial–Corrigendum. *The British Journal of Nutrition, 116*(11), 1998. https://doi.org/10.1017/S0007114516004323.

Badehnoosh, B., Karamali, M., Zarrati, M., Jamilian, M., Bahmani, F., Tajabadi-Ebrahimi, M., Jafari, P., Rahmani, E., & Asemi, Z. (2018). The effects of probiotic supplementation on biomarkers of inflammation, oxidative stress, and pregnancy outcomes in gestational diabetes. *Journal of Maternal-Fetal & Neonatal Medicine, 31*(9), 1128–1136. https://doi.org/10.1080/14767058.2017.1310193.

Brown, J., de Vos, W. M., DiStefano, P. S., Dore, J., Huttenhower, C., Knight, R., & Turnbaugh, P. (2013). Translating the human microbiome. *Nature Biotechnology, 31*(4), 304–308. https://doi.org/10.1038/nbt.2543.

Cani, P. D., Osto, M., Geurts, L., & Everard, A. (2012). Involvement of gut microbiota in the development of low-grade inflammation and type 2 diabetes associated with obesity. *Gut Microbes, 3*(4), 279–288. https://doi.org/10.4161/gmic.19625.

Crusell, M., Hansen, T. H., Nielsen, T., Allin, K. H., Rühlemann, M. C., Damm, P., Vestergaard, H., Rørbye, C., Jørgensen, N. R., Christiansen, O. B., & Pedersen, O. (2018). Gestational diabetes is associated with change in the gut microbiota composition in third trimester of pregnancy and postpartum. *Microbiome, 6*, 89. https://doi.org/10.1186/s40168-018-0472-x.

David, L. A., Maurice, C. F., Carmody, R. N., Gootenberg, D. B., Button, J. E., Wolfe, B. E., & Turnbaugh, P. J. (2014). Diet rapidly and reproducibly alters the human gut microbiome. *Nature, 505*(7484), 559–563. https://doi.org/10.1038/nature12820.

Kashtanova, D. A., Popenko, A. S., Tkacheva, O. N., Tyakht, A. B., Alexeev, D. G., & Boytsov, S. A. (2016). Association between the gut microbiota and diet: Fetal life, early childhood, and further life. *Nutrition, 32*(6), 620–627. https://doi.org/10.1016/j.nut.2015.12.037.

Kohlhepp, L. M., Hollerich, G., Vo, L., Hofmann-Kiefer, K., Rehm, M., Louwen, F., & Weber, C. F. (2018). Physiological changes during pregnancy. *Der Anesthetist, 67*(5), 383–396.

Koren, O., Goodrich, J. K., Cullender, T. C., Spor, A., Laitinen, K., Bäckhed, H. K., Gonzalez, A., Werner, J. J., Angenent, L. T., Knight, R., & Bäckhed, F. (2012). Host remodeling of the gut microbiome and metabolic changes during pregnancy. *Cell, 150*(3), 470–480. https://doi.org/10.1016/j.cell.2012.07.008.

Łagowska, K., Malinowska, A. M., & Zawieja, B. (2020). Improvement of glucose metabolism in pregnant women through probiotic supplementation depends on gestational diabetes status: Meta-analysis. *Scientific Reports, 10*, 17796. https://doi.org/10.1038/s41598-020-74860-2.

Lindsay, K. L., Kennelly, M., Culliton, M., Smith, T., Maguire, O. C., Shanahan, F., Brennan, L., & McAuliffe, F. M. (2014). Probiotics in obese pregnancy do not reduce maternal fasting glucose: A double-blind, placebo-controlled, randomized trial (Probiotics in Pregnancy Study). *The American Journal of Clinical Nutrition, 99*(6), 1432–1439. https://doi.org/10.3945/ajcn.113.079723.

Ma, S., You, Y., Huang, L., Long, S., Zhang, J., Guo, C., Zhang, N., Wu, X., Xiao, Y., & Tan, H. (2020). Alterations in gut microbiota of gestational diabetes patients during the first trimester of pregnancy. *Frontiers in Cellular and Infection Microbiology, 10*, 58. https://doi.org/10.3389/fcimb.2020.00058.

Mokkala, K., Roytio, H., Munukka, E., Pietila, S., Ekblad, U., Ronnemaa, T., & Laitinen, K. (2016). Gut microbiota richness and composition and dietary intake of overweight pregnant women are related to serum zonulin concentration, a marker for intestinal permeability. *The Journal of Nutrition, 146*(9), 1694–1700. https://doi.org/10.3945/jn.116.235358.

Nordqvist, M., Jacobsson, B., Brantsæter, A. L., Myhre, R., Nilsson, S., & Sengpiel, V. (2018). Timing of probiotic milk consumption during pregnancy and effects on the incidence of preeclampsia and preterm delivery: A prospective observational cohort study in Norway. *BMJ Open, 8*(1), e018021. https://doi.org/10.1136/bmjopen-2017-018021.

Nyangahu, D. D., & Jaspan, H. B. (2019). Influence of maternal microbiota during pregnancy on infant immunity. *Clinical and Experimental Immunology, 198*(1), 47–56. https://doi.org/10.1111/cei.13351.

Power, S. E., O'Toole, P. W., Stanton, C., Ross, R. P., & Fitzgerald, G. F. (2014). Intestinal microbiota, diet, and health. *The British Journal of Nutrition, 111*(3), 387–402. https://doi.org/10.1017/s0007114513002560.

Slykerman, R. F., Hood, F., Wickens, K., Thompson, J., Barthow, C., Murphy, R., Kang, J., Rowden, J., Stone, P., Crane, J., Stanley, T., Abels, P., Purdie, G., Maude, R., Mitchell, E. A., & Probiotic in Pregnancy Study Group. (2017). Effect of Lactobacillus rhamnosus HN001 in pregnancy on postpartum symptoms of depression and anxiety: A randomized double-blind placebo-controlled trial. *EBioMedicine, 24*, 159–165. https://doi.org/10.1016/j.ebiom.2017.09.013.

Walters, W. A., Xu, Z., & Knight, R. (2014). Meta-analyses of human gut microbes associated with obesity and IBD. *FEBS Letters, 588*(22), 4223–4233. https://doi.org/10.1016/j.febslet.2014.09.039.

CHAPTER 12

Cancer and Prebiotics

SUJOYA SUKUL

Nutrition Consultant, Jaipur, Rajasthan, India

The mammalian gastrointestinal tract is home to a diverse community of microorganisms from the mouth to the anus, but the colon is by far the primary location of this microbial colonization. Over the last 20–25 years, the complexity of this microbiota has been studied extensively.

12.1 UNDERSTANDING CANCER – THE DISEASE THAT TAKES LIVES BY MILLIONS

Cancer is a global health problem that affects millions of lives each year. In the most basic and simple terms, cancer is a disease condition wherein there is rapid and uncontrolled growth of cells that function in an abnormal manner. Cancer is not one disease but rather a collection of diseases that can occur anywhere in the body (National Cancer Institute). Understanding cancer, its metastasis, and the way our lifestyle habits and other causative factors play a role can help us take better control over this malignant disease.

12.2 CANCER PREVALENCE AND WORLDWIDE STATISTICS

The global statistics on cancer reveal an alarmingly high incidence of cancer rates among the global population. The increase in age and growth of the

Prebiotic Foods for a Healthy Gut: Scientific Studies to Manage Metabolic Disorders.
Mini K. Sheth and Shruti Dwivedi (Eds.)

population, coupled with adherence to cancer-causing behaviors, majorly contributes to the huge spike in cancer rates globally (Jemal et al., 2011).

Undoubtedly, cancer is the leading cause of death globally. According to the World Health Organization, in 2020, it accounted for about 10 million deaths worldwide. The following list shows the number of cases based on the type of cancer in the year 2020.

Types of cancer and statistics are:

1. Breast cancer – 2.26 million cases
2. Lung cancer – 2.21 million cases
3. Colon and rectal cancer – 1.93 million cases
4. Prostate cancer – 1.41 million cases
5. Skin cancer – 1.20 million cases
6. Stomach cancer – 1.09 million cases

The above statistics reveal alarming data about cancer and its prevalence. The data clearly reveals that colon and rectal cancer, after breast and lung cancer, also majorly contribute to the cancer rates significantly. Having said this, we need to understand the various ways by which we can prevent the cancer rates.

12.3 PREVENTING CANCER – A POSSIBILITY

Cancers are preventable and can be easily done by adopting a few simple lifestyle changes. Research says that about 50% of all cancer cases and 50% of cancer-related deaths can be prevented if effective steps are taken at the right time. Among the several changes that one needs to adopt are having a healthier lifestyle with exercise, maintaining a healthy body weight, and carefully considering what goes onto our plate each day. Research has also stated that dietary factors contribute to about 30% of cancers in Western countries (World Health Organization, 2020). On the other hand, according to the WHO, cancer rates are lower in India compared to its Western counterparts. However, with the sudden drastic changes in lifestyle factors and dietary habits due to rapid urbanization, the incidence of cancer rates in India is also on the rise.

Of the several preventive factors, diet particularly has a significant role to play in preventing stomach cancer, colorectal cancer, cancer of the pancreas, and liver cancer (Koumbi, 2017; Li et al., 2015; Casari and Falasca, 2015). Faulty dietary habits such as high consumption of red meat, sugar-rich foods, processed foods, and going low on fiber-rich foods such as fruits and vegetables pave the way for the development of cancer.

12.4 CANCER AND GUT MICROBIOTA –AN AMAZING ASSOCIATION

Our gut microbiota refers to the one hundred trillion live bacteria that colonize the human intestine. While our intestine offers a nutrient-rich, protective, and warm environment (Zitvogel et al., 2017).

Stomach infections, pathogenic bacteria, as well as antibiotics can affect the population of good bacteria. The beneficial bacteria residing in our gut offer numerous health benefits. They aid in the digestion of complex carbohydrates and also offer protection against pathogenic microorganisms from entering the human gut. These intestinal microorganisms are an essential entity for the functioning of a healthy human gut. Intestinal disorders, metabolic syndrome, neurodegenerative diseases, and cancer can result from a breakdown in the communication between the intestinal microflora, gut microbiota, and the host immune system (Samanta et al., 2022).

Intestinal microbiota is composed of a pretty large population of microorganisms and is extremely important for maintaining human health and development of diseases. There is much increasing evidence that shows that the dysbiosis of the intestinal microbiota paves the way for the development of colorectal cancers. The world is witnessing a steady rise in the number of cases of colorectal cancers and is estimated to increase to 2.2 million new cases and about 1.1 million deaths worldwide by 2030 (Arnold et al., 2017).

As a general rule, the intestinal mucosal barrier is known to separate the intestinal microbiota from the immune cells. However, it has also been studied that this intestinal mucosal barrier is highly permeable and if the barrier function is disrupted owing to disease conditions such as colitis, it can then make the host more susceptible to develop colorectal cancers. Prolonged infections by pathogenic bacteria due to poor intestinal microbiota are also one of the several factors that can pose as a risk factor for gut cancers.

12.5 GUT MICROBIOTA AND ITS ROLE IN CANCER TREATMENT

As already discussed in the previous section, the gut microbiota plays an important role in enhancing immunity and also in the prevention as well as in the progression of cancers of various origins. However, little did we know about the role of the gut microbiota in cancer treatment. It has been said that an altered gut microbiota does not respond effectively to anticancer drugs. Additionally, an altered gut microbiota shows significant resistance to chemo drugs, making cancer treatment difficult for patients (Cheng et al., 2020). The gut microbiota regulates the host response to chemotherapy treatment through several mechanisms such as xenometabolism, immunomodulation, and the alteration of community structure (Louise et al., 2018). In yet another study carried out on mice in 2013, it was concluded that mice with depleted gut microbiomes had a much weaker immune response to chemotherapy treatment (Viaud et al., 2013).

Having said this, it becomes important to mention here that a healthy gut microbiota is important, not only for the gastrointestinal tract, but also to maintain the healthy functioning of distal organs such as the liver, pancreas, and the brain. Therefore, it becomes extremely important to take necessary steps to maintain a healthy gut microbiota. Making prebiotic-rich foods a regular part of our diet is one of the major steps to build a healthy gut microbiome. The healthier, flourishing beneficial bacteria in our gut, the stronger our gut immunity would be, which would in turn help in strengthening our overall immune system.

12.6 PREBIOTICS AND GUT MICROBIOTA

The gut microbiota of humans contains a pretty diverse microbial community that provides protective functions essential for human health. This gut microbiota is influenced by a number of genetic, physiological, and environmental factors. Interestingly, diet plays an extremely major role in the healthy maintenance of this gut microbiota. Consumption of good amounts of dietary fibers and prebiotics supports the healthy development of the gut microbiota, which then eventually works towards protecting the human body against infections and other chronic diseases (Jumpertz et al., 2011).

Prebiotics are defined as "selectively fermented ingredients that result in specific changes in the composition or activity of the gastrointestinal microbiota, thus conferring benefits upon the host's health" (Scott et al., 2010).

Several kinds of prebiotics are available that are known to enhance our gut immunity and improve health. Of the several prebiotics, inulin, fructooligosaccharide (FOS), and Galactooligosaccharide (GOS) are not only available in high purity, but are also reported to be safe and are universally agreed upon. Other prebiotics include xylooligosaccharide (XOS), glucooligosaccharide, isomaltooligosaccharide (IMOS), pectin oligosaccharides (POS), to name a few (Carlson et al., 2018).

12.7 PREBIOTICS AND CANCER – LIMITING CANCER GROWTH BY STRENGTHENING GUT MICROBIOME

Prebiotics are known to effectively alter the gut microflora and its dynamics by stimulating *Bifidobacteria* and *Lactobacilli*, which bind and eliminate carcinogens. Several studies carried out in animal models have confirmed that feeding prebiotic supplements has a profound and positive impact on the treatment as well as the prevention of colorectal cancers. In research by Bonnet et al. in 2014, it was stated that gut microbiota strongly influences the incidence as well as the progression of colorectal carcinoma. These and many more such studies strongly support the role of a healthy gut microbiome in the prevention as well as the treatment of cancers of various origins.

Prebiotics, being indigestible, have been associated with improved bowel function. Studies carried out on rats (Liong, 2008) showed lowered tumor numbers in rats that were fed with cereal bran, a prebiotic-rich source. Prebiotics such as inulin, GOS, and FOS are known to have a profound effect on bowel function. These prebiotics, upon reaching the large intestine, undergo fermentation, which in turn stimulates accelerated microbial growth, favoring increased bacterial mass, peristalsis, and fecal bulking (Niittynen et al., 2016). Fecal bulking helps in the prevention of cancer as it reduces the transit time of feces and the contact time of mutagens with the intestinal lining.

Animal studies have also demonstrated that inulin-type fructans, extracted from chicory roots, are prebiotic food ingredients and have exceptional anticarcinogenic properties (Pool et al., 2007). Yet another type of prebiotics, galactooligosaccharides, has also shown promising effects by making favorable changes in the colonic environment. Several *in vitro* studies have shown that bacterial fermentation of oligosaccharides stimulates the production of short-chain fatty acids, which in turn significantly lowers the pH of the colon. A lower colonic pH is extremely favorable for the growth of beneficial bacteria such as *Bifidobacteria* and *Lactobacilli* in the gut, and in turn suppresses the growth of unwanted bacteria.

12.8 PREBIOTICS AND ITS ROLE IN PREVENTION OF COLORECTAL CANCER

The third most common cancer in men is colorectal cancer (CRC). According to estimates, there will be more than 2.2 million cases worldwide by the year 2030, with an estimated death rate of 1.1 million. Its incidence rates are rising with each passing year. The past few decades have also witnessed a dramatic rise in the number of cases of early-onset colorectal cancer, which is usually defined as CRC diagnosed in the population aged 50 years or younger. Among the several factors known to play a role, diet is one that is thought to be a major contributing factor.

With diet playing a major contributory role in the early onset of colorectal cancer in the young population, it is important to mention that having a correct diet can significantly help in the prevention and development of colorectal cancer. Based on a meta-analysis conducted by Chan and his team, it is now well established that there is a strong correlation between meat consumption and the onset of colorectal cancer. Therefore, excessive consumption of red meat and processed meat poses a high risk of developing colorectal, rectal, and colon cancers.

In addition, some research trials have also successfully concluded that prebiotics in conjunction with probiotics, and even the combination of both known as synbiotics, also have a pivotal role to play in the chemoprevention of colorectal cancer (Geier et al., 2006). Prebiotics, probiotics, and the combination of both are thought to be a new, interesting therapeutic option. Experts say and research supports that prebiotics and probiotics in conjunction help in improving the beneficial bacteria of the gut, such as *Bifidobacteria* and *Lactobacillus*. This process, in turn, helps significantly reduce the pathogenic bacterial load of the gut. These series of events are capable enough of inhibiting the development as well as the progression of neoplasms. Prebiotics such as inulin and oligofructose have special inhibitory properties that help against the development of colonic aberrant crypt foci (ACF). ACF are preneoplastic lesions, which may later progress towards the development of carcinomas as well as adenomas in the colon. The mechanisms by which the development and progression of neoplasia are inhibited include –

- decreased intestinal inflammation
- improved immune function
- enhanced anti – tumorigenic activity
- reduction in bacterial enzymes that hydrolyze pre – carcinogenic compounds like beta – glucuronidase
- enhanced activity that enables binding of the potential carcinogens and toxins found in food

Several research studies conducted on milk and other dairy products have shown that the lactic cultures primarily used in the fermentation of milk and its products also possess anticarcinogenic and antimutagenic properties.

Additionally, there have been studies proving that cultures of *Bifidobacteria* improve and enhance host immunity.

Furthermore, certain research trials have concluded that prebiotics such as jaboticaba seed extract contain active phenolic compounds like ellagic acid, which, when it reaches the gut, is degraded to urolithins. This compound, urolithin, is also known to effectively impede the spread of cancer cells by decreasing the stimulation of proinflammatory cytokines in the gut (Loo et al., 2020).

12.9 HEPATOCELLULAR CARCINOMA AND PREBIOTICS

Hepatocellular carcinoma is one of the most fatal malignancies and is also a leading cause of cancer-related deaths globally. Time and again, it has been repeatedly proven that prebiotics have a pivotal role to play when it comes to cancer prevention. Changing the gut microflora, which would have more of the beneficial bacteria, would bring about a massive decline in the population of pathogenic bacteria, which would eventually help in achieving a healthy gut. Achieving a healthier gut microflora has immense benefits to its credit. It helps in improving intestinal inflammation, along with improvement in mucosal barrier functions, which in turn helps in effectively improving cirrhotic conditions (Fatima et al., 2017).

12.10 PREBIOTICS AND CANCER IN A NUTSHELL – KEY HIGHLIGHTS

Using prebiotics to improve gut health is a promising and safe therapy. Making modifications to the diet to include high-fiber foods along with prebiotics can stimulate the gut to produce more SCFA-producing bacteria and subsequently increase the level of short-chain fatty acid metabolites. Various research studies have shown that an increase in the level of short-chain fatty acid metabolites results in a significant decrease in tumor loads. Additionally, prebiotics have a host of other benefits, including protection against pathogenic bacteria, modulation of the immune system, antimicrobial activity against gut pathogens, reduction of colitis and gut inflammation, and prevention of colorectal cancers.

It can be safely concluded that making conscious efforts to enrich our gut microbiota with prebiotic-rich foods will not only help strengthen gut immunity but also work towards improving the overall immunity of our body, which in turn can go a long way in preventing various types of cancers.

KEYWORDS

- **bacteria**
- **cancerous diseases**
- **fatty acids**
- **food culture**
- **food habits**
- **food production**
- **gastrointestinal microbiota**
- **global problems**
- **health**
- **lifestyle habits**

REFERENCES

Arnold, M., Sierra, M. S., Laversanne, M., Soerjomataram, I., Jemal, A., & Bray, F. (2017). Global patterns and trends in colorectal cancer incidence and mortality. *Gut, 66*(4), 683–691. https://doi.org/10.1136/gutjnl-2015-310912.

Carlson, J. L., Erickson, J. M., Lloyd, B. B., & Slavin, J. L. (2018). Health effects and sources of prebiotic dietary fiber. *Current Developments in Nutrition, 2*(3). https://doi.org/10.1093/cdn/nzy005.

Casari, I., & Falasca, M. (2015). Diet and pancreatic cancer prevention. *Cancers, 7*(4), 2309. https://doi.org/10.3390/cancers7040892.

Cheng, W. Y., Wu, C. Y., & Yu, J. (2020). The role of gut microbiota in cancer treatment: Friend or foe? *Gut, 69*(10), 1867–1876. https://doi.org/10.1136/gutjnl-2020-321153.

Fatima, N., Akhtar, T., & Sheikh, N. (2017). Prebiotics: A novel approach to treat hepatocellular carcinoma. *Canadian Journal of Gastroenterology and Hepatology, 2017*. https://doi.org/10.1155/2017/6238106.

Geier, M. S., Butler, R. N., & Howarth, G. S. (2006). Probiotics, prebiotics and symbiotic: A role in chemoprevention for colorectal cancer? *Cancer Biology & Therapy, 5*(10), 1265–1269. https://doi.org/10.4161/cbt.5.10.3296.

Jemal, A., Bray, F., Center, M. M., Ferlay, J., Ward, E., & Forman, D. (2011). Global cancer statistics. *CA: A Cancer Journal for Clinicians, 61*(2), 69–90. https://doi.org/10.3322/caac.20107.

Jumpertz, R., Le, D. S., Turnbaugh, P. J., Trinidad, C., Bogardus, C., Gordon, J. I., & Krakoff, J. (2011). Energy-balance studies reveal associations between gut microbes, caloric load, and nutrient absorption in humans. *The American Journal of Clinical Nutrition, 94*(1), 58–65. https://doi.org/10.3945/ajcn.110.010132.

Koumbi, L. (2017). Dietary factors can protect against liver cancer development. *World Journal of Hepatology, 9*(3), 119. https://doi.org/10.4254/wjh.v9.i3.119.

Li, Y., Go, V. L. W., & Sarkar, F. H. (2015). The role of nutraceuticals in pancreatic cancer prevention and therapy: Targeting cellular signaling, MicroRNAs, and Epigenome. *Pancreas, 44*(1), 1–10. https://doi.org/10.1097/mpa.0000000000000257.

Liong, M. T. (2008). Roles of probiotics and prebiotics in colon cancer prevention: Postulated mechanisms and in-vivo evidence. *International Journal of Molecular Sciences, 9*(5), 854. https://doi.org/10.3390/ijms9050854.

Loo, Y. T., Howell, K., Chan, M., Zhang, P., & Ng, K. (2020). Modulation of the human gut microbiota by phenolics and phenolic fiber-rich foods. *Comprehensive Reviews in Food Science and Food Safety, 19*(4), 1268–1298. https://doi.org/10.1111/1541-4337.12563.

National Cancer Institute. Understanding cancer. Retrieved from: https://www.cancer.gov/about-cancer/understanding (accessed on 19 July 2024).

Niittynen, L., Kajander, K., & Korpela, R. (2016). Galacto-oligosaccharides and bowel function. *Nutritional Therapy & Metabolism, 51*(2), 62–66. https://doi.org/10.1080/17482970701414596.

Pool-Zobel, B. L. (2005). Inulin-type fructans and reduction in colon cancer risk: Review of experimental and human data. *British Journal of Nutrition, 93*(S1), S73–S90. https://doi.org/10.1079/bjn20041349.

Pouncey, A. L., Scott, A. J., Alexander, J. L., Marchesi, J., & Kinross, J. (2018). Gut microbiota, chemotherapy and the host: The influence of the gut microbiota on cancer treatment. *Ecancermedicalscience, 12*. https://doi.org/10.3332/ecancer.2018.868.

Pujari, R., & Banerjee, G. (2021). Impact of prebiotics on immune response: From the bench to the clinic. *Immunology and Cell Biology, 99*(3), 255–273. https://doi.org/10.1111/imcb.12409.

Samanta, S. (2022). Potential impacts of prebiotics and probiotics on cancer prevention. *Anti-Cancer Agents in Medicinal Chemistry, 22*(4). https://doi.org/10.2174/1871520621999201210220442.

Scott, K. P., Tuohy, K. M., Mach, F. E., San, A., All'adige, M., Gibson, G. R., Rastall, R. A., Hotchkiss, A., Dubert-Ferrandon, A., Gareau, M., Murphy, E. F., Saulnier, D., Loh, G., Macfarlane, S., Delzenne, N., Ringel, Y., Kozianowski, G., Dickmann, R., Lenoir-Wijnkoop, I., & Buddington, R. (2010). Dietary prebiotics: Current status and new definition. *Food Science and Technology Bulletin: Functional Foods, 7*(1), 1–19. https://doi.org/10.1616/1476-2137.15880.

Viaud, S., Saccheri, F., Mignot, G., Yamazaki, T., Daillère, R., Hannani, D., Enot, D. P., Pfirschke, C., Engblom, C., Pittet, M. J., Schlitzer, A., Ginhoux, F., Apetoh, L., Chachaty, E., Woerther, P. L., Eberl, G., Bérard, M., Ecobichon, C., Clermont, D., & Zitvogel, L. (2013). The intestinal microbiota modulates the anticancer immune effects of cyclophosphamide. *Science, 342*(6161), 971–976. https://doi.org/10.1126/science.1240537.

Watzl, B., Girrbach, S., & Roller, M. (2005). Inulin, oligofructose and immunomodulation. *British Journal of Nutrition, 93*(S1), S49–S55. https://doi.org/10.1079/bjn20041357.

World Health Organization (2022). Cancer. Retrieved from: https://www.who.int/news-room/fact-sheets/detail/cancer (accessed on 19 July 2024).

Zitvogel, L., Galluzzi, L., Viaud, S., Vétizou, M., Daillère, R., Merad, M., & Kroemer, G. (2015). Cancer and the gut microbiota: An unexpected link. *Science Translational Medicine, 7*(271), 271ps1. https://doi.org/10.1126/scitranslmed.3010473.

CHAPTER 13

Human Milk Oligosaccharides – Earliest Prebiotics and Their Impact on Infant Health and Immunity

VIJAYATA SANEGAR

Department of Foods and Nutrition, The Maharaja Sayajirao University of Baroda, Gujarat, India

Human Milk Oligosaccharides (HMOs) act as antimicrobial and antiviral agents, inhibiting pathogen binding and promoting beneficial gut bacteria growth. Research is ongoing to understand how HMOs in breast milk protect newborns and influence gut microbiota development. Studies are needed to explore supplementing nursing mothers with pre- and probiotics to modify milk composition.

13.1 HUMAN MILK OLIGOSACCHARIDES IN HUMAN MILK: A HISTORICAL PERSPECTIVE

A newborn needs mother's milk soon after birth and nothing else until 6 months of age. Several components in human milk not only provide nutritional benefits but also provide immunological benefits too. Breastfeeding influences the gut microbiota directly through the neonate's exposure to the milk microbiota and indirect influences on gut microbiota are observed through

Prebiotic Foods for a Healthy Gut: Scientific Studies to Manage Metabolic Disorders.
Mini K. Sheth and Shruti Dwivedi (Eds.)

maternal milk components like HMOs, antimicrobial agents, and secretory IgA that influence bacterial growth and metabolism (Elsen et al., 2019).

Several research studies were initiated towards the end of the nineteenth century in order to ascertain the health benefits of breastfeeding. Escherich (1886) established associations between gut microflora and the physiology of digestion in infants. Following his work, Morro and Tissier (1900) in separate studies revealed significant differences in gut microflora of bottle-fed and breastfed infants.

Simultaneously around this time, Eschbach (1888) found that lactose found in human milk was different from bovine milk. These findings were reviewed by Montreuil in 1992. However, almost four decades later in 1930, Polonowski and Lespagnol, after a series of researches, categorized this carbohydrate fraction and named it Gyno lactose. Further studies conducted by Polonowski and Montreuil led to the separation of individual oligosaccharides in Gyno lactose. Still further research was required to understand the structures and functions of these oligosaccharides (Polonowski and Montreuil, 1954).

In 1926, Schonfeld proposed that human milk contained a growth-promoting factor for *Lactobacillus bifidus* in the whey fraction. Kuhn and Gyorgy connected the work done by Moro and Tissier on bacteria and Polonowski and Lespagnol's work on Gyno lactose and concluded that the bifidus factor was indeed made up of oligosaccharides (Gauhe et al., 1954; György, Hoover, et al., 1954; György et al., 1954a, 1954b; Rose et al., 1954). Figure 13.1 shows the timeline with key events in HMO discovery until the end of the 20th century. Several studies since then have identified more than 200 different HMOs and have explained not only biological but significant nutritional benefits of HMOs too (Kunz et al., 1996; Rudloff et al., 1996, 2002, 2006, 2011; Gnoth et al., 2000, 2001; Bode, Kunz, et al., 2004; Bode, Rudloff, et al., 2004; Kunz et al., 2008, 2009; Morozov et al., 2018).

13.2 STRUCTURE AND BIOGENESIS OF HMOS

Followed by lactose and lipids, HMOs are the third largest solid component found in human milk. HMOs are slightly hydrolyzed and thus accumulate in the intestinal tract of infants. Being complex carbohydrates, HMOs act as a substrate for gut microbes due to their indigestibility and thus serve the role of prebiotics. HMOs are composed of five monosaccharides, namely glucose (Glc), galactose (Gal), fucose (Fuc), N-acetylglucosamine (GlcNAc), and the sialic acid (Sia) derivative called N-acetylneuraminic acid (Neu5Ac). Most of the HMO structures have a lactose (Galβ1–4Glc) core, barring a

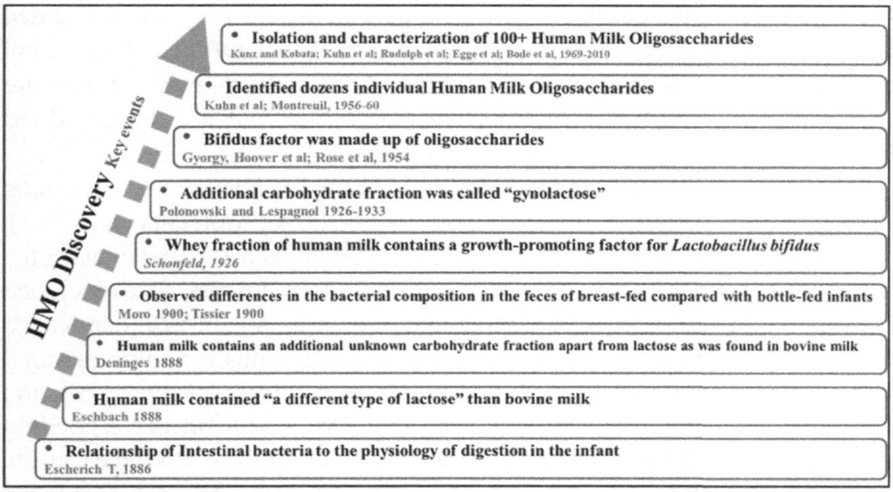

FIGURE 13.1 Key Events in the discovery of HMOs.

few exceptions (Kunz et al., 2000). This core can be elongated at β1–3- or β1–6-linkage by two different disaccharides, either Galβ1-3 glcNAc (type 1 chain) or Galβ1–4GlcNAc (type 2 chain).

Sialylation leads to the addition of Sia in α2–3- or α2–6-linkage to the HMO backbone. Variations in the expression of the enzyme sialyltransferases responsible for sialylation lead to variations in HMOs. However, sialylation is more subtle compared to fucosylation (Bode 2015).

Several genetic and environmental factors are believed to contribute to HMO biosynthesis. However, these factors remain mostly unknown and are currently being researched.

Despite the fact that human milk composition varies in each individual, HMOs are the most important nutrients for newborns' growth. Around 20% of the total carbohydrate in colostrum is accounted for by HMOs (Xu et al., 2017; Chen et al., 2015). The amount of HMOs in human milk is determined by the lactation stage, ranging from 20–24 g/L in early human milk to 10–15 g/L in mature milk on average (Andreas et al., 2015; Thurl et al., 2017; Xu et al., 2017; Chen et al., 2015).

13.3 METABOLISM OF HMOS

HMOs in milk are not significantly affected by acidic pH or by the enzymes of the brush border or those secreted by the pancreas of a breastfed infant.

Around 99% of the HMOs reach the intestine where they are fermented by the action of microbes in the intestinal microbiota. Of this, 40–50% are excreted in feces and approximately 1% of the total ingested HMOs are excreted unchanged in urine (Gnoth et al., 2000; Engfer et al., 2000; Andreas et al., 2015; Kulinich et al., 2016; Vazquez et al., 2017).

Ingested HMOs are either excreted as they are with feces or are acted upon by microbes in the gut of infants (Albrecht et al., 2010; Albrecht et al., 2011).

Recent studies have revealed that HMO metabolism is dependent on the age, blood group, and feeding patterns of the infant. Infants' feces included sialylated and non-sialylated HMOs that were similar but not identical to corresponding milk samples until they were two months old. In the second stage, the feces primarily included products derived from HMO degradation and processing that were distinct from the HMOs seen in corresponding milk samples. HMOs completely disappear from the infant's feces in the third stage, which initiates when foods apart from breastmilk are introduced (Albrecht et al., 2011).

Excretion of HMO in feces showed variations depending on the infant's intestinal microbiota while excretion in urine depended upon the mother's secretor or non-secretor status. Post-microbial degradation, some novel HMOs were found in urine and feces indicating variations in the metabolism of HMOs (Dotz et al., 2015).

13.4 FUNCTIONS OF HMOS

HMOs were initially recognized for their "bifidogenic" or prebiotic properties since they were first identified as the "bifidus factor" in human milk. Over the years, researchers have proven that HMOs have several beneficial effects not only for the infant but also for the breastfeeding mother.

13.4.1 HMOS AS PREBIOTICS

By definition, a prebiotic is "a selectively fermented ingredient that allows specific changes, both in the composition and/or activity in the gastrointestinal microflora that confers benefits upon host well-being and health" (Gibson et al., 2004; Roberfroid, 2007).

Before serving as nourishment for colon bacteria, intact HMOs rinse the infant's esophagus, stomach, and small intestine since they escape the low pH in the stomach and also the degrading enzymes in the pancreas and brush border membrane (Gnoth et al., 2000; Engfer et al., 2000). HMOs aid in the growth of beneficial microbes such as *Bifidobacterium*, which accounts for 50–90% of the entire bacterial population found in the feces of breastfed children (Jost et al., 2015; Hao et al., 2019).

Bifidobacterium also affects the composition of other microorganisms. When HMOs are the only carbohydrate supply, *B. infantis*, i.e., *Bifidobacterium* longum subspecies infantis (*B. infantis*, JCM1222), thrives well (LoCascio et al., 2007; Marcobal et al., 2010; Asakuma et al., 2011). *B. infantis* utilizes all HMOs, including mono- and disaccharide breakdown products, over time (Asakuma et al., 2011). Dominance of *B. infantis* mediated through HMOs may keep potentially harmful bacteria under control as they strive for limited nutrient supply. Furthermore, *B. infantis* and numerous other bacteria associated with infants produce SCFAs (short-chain fatty acids) and other metabolites (known as post-biotics), which establish an environment that favors commensal development over pathogen growth (Gibson & Wang, 1994).

B. bifidum can secrete glycosyl hydrolases to break down HMOs extracellularly and then leave degraded sugars outside the cells for other bacteria to use to synthesize SCFAs like butyrate and propionate (Asakuma et al., 2011). Because butyric and propionic acids interact with epithelial cells in the hosts' body by inducing mucin secretion, boosting mucosal blood flow, and controlling immunity and thus are vital for gut health (Ploger et al., 2012; Reichardt et al., 2014).

In addition to modulating the composition of *Bifidobacteria* in the intestinal tract, HMOs have a direct interaction with other bacteria by affecting the distribution of gut microbes. In conclusion, despite their minimal capacity, gut microorganisms, with the exception of *Bifidobacteria*, have an integral part in maintaining intestinal homeostasis in breastfed newborns (Walsh et al., 2020).

HMOs can also cross the intestinal barrier, causing effects outside of the intestines (Rudolf et al., 2012). Research has shown that G protein-coupled receptors (GPCRs) are activated by HMOs, which can influence practically every physiological function, including development, taste, olfaction, heart rate regulation, hormone signaling, and neurotransmission (Wettschureck et al., 2005).

13.4.2 HMOS AND COLONIZATION OF BIFIDOBACTERIA

Extent to which probiotic strains attach to the intestinal epithelium has a big impact on the gut microbiota's makeup. Surface-associated populations of microbes can outcompete other species simply by being more adherent (McLoughlin et al., 2016).

Several studies have revealed that HMOs could improve probiotic strain adherence by modulating bacterial adhesins action mechanism (Morrin & Hickey, 2020).

Improved adherence may extend the probiotic's gut transit time and/or boost host-microbe and/or microbe-microbe interactions, resulting in higher health advantages for the host (Monteagudo et al., 2019).

Thus, HMOs may help in increasing the duration of residence for probiotic bacteria in the gut.

13.4.3 ANTI-INFECTIVE PROPERTIES OF HMO

As can be seen in Figure 13.2, HMOs as prebiotic agents provide a selection edge to beneficial microorganisms over pathogenic microbes, thereby protecting the infant from infectious diseases. Due to the incapability of pathogenic bacteria to metabolize HMOs, symbionts can thrive and outdo harmful microbes by competitively excluding them (Hoeflinger et al., 2015).

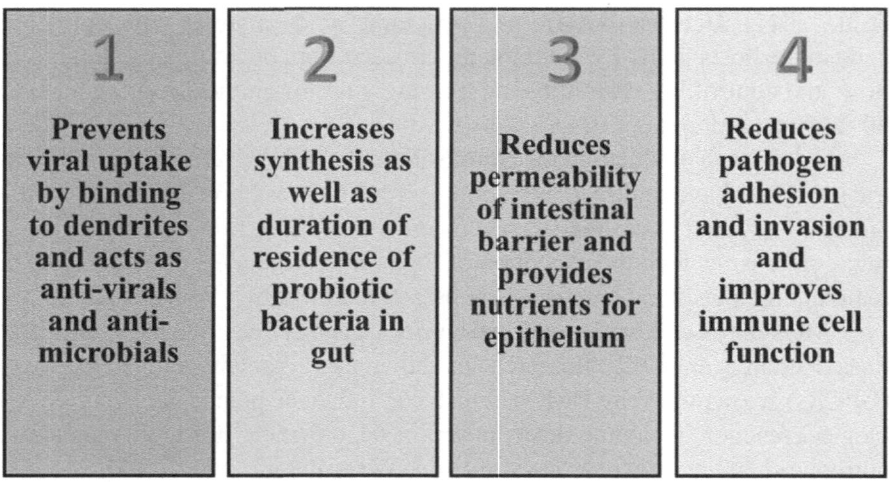

FIGURE 13.2 Functions of HMOs in breast fed infants.

Organic acids produced by *Bifidobacterial* HMO metabolism generate an acidic environment that inhibits harmful bacterial growth (Tan et al., 2014). Along with indirect pathogenic rebound, HMOs also function as soluble receptor decoys, restricting pathogen entry directly (Figure 13.2).

In order to infect the host and cause disease, many viruses, bacteria, and protozoa must attach to the 'glycocalyx' (Costerton, Irvin, & Cheng, 1981). The glycocalyx is a carbohydrate-rich layer that coats epithelial cells and is made up of glycans that have been conjugated with lipids or proteins (Kavanaugh et al., 2015). HMOs have chemical structures that are comparable to the glycans which pathogenic bacteria use to adhere to the epithelial cell surface (Newburg & Grave, 2014). Consequently, harmful pathogens and toxins adhering with HMOs rather than cell-surface glycans will pass through the gastrointestinal tract without causing infections (Craft & Townsend, 2018). Apart from reducing pathogen adhesion and invasion, HMOs can also change epithelial surface gene expression and inhibit pathogen growth, thus reducing infection (Chen et al., 2017).

HMOs are known to improve infants' resistance to rotavirus and norovirus, two deadly gastrointestinal illness-causing viruses (Laucirica et al., 2017). HMOs were shown to provide protection against viral infections by preventing viruses from accessing host cells (by imitating receptor sites) and thus promoting immunity, as observed in a few mechanistic studies (Ramani et al., 2018).

HMOs have been found to protect against a wide range of pathogens, including those linked to diarrheal, respiratory, and urinary infections, as well as the human immunodeficiency virus (HIV). These investigations shed light on HMOs' potential for future pathogenic infection treatment developments (Craft & Townsend, 2019).

13.4.4 MODULATION OF EPITHELIAL CELL RESPONSES

HMOs can function as prebiotics or antiadhesive antimicrobials to disrupt host-microbial interactions, but they can also modulate infants' intestinal epithelial cell responses directly (Angeloni et al., 2005). The intestinal epithelium lining is the initial defense in innate immunity as it acts as a rate-limiting barrier between the intestinal lumen and the circulatory system (Maldonado-Contreras & McCormick, 2011). The epithelium's permeability is essentially determined by the form and expression of tight junctions. Nutrients are absorbed either trans-epithelially (through cells) or paracellularly (via the epithelium) between tight junctions. Irregular immune responses, such as gastrointestinal infections, inflammatory bowel disorders, and allergies, are caused by disruptions in barrier integrity (Vancamelbeke & Vermeire, 2017). HMOs in several research studies have been revealed to induce tight junction protein expression and differentiation along the crypt-villus axis as shown in Figure 13.2 (Holscher, Davis, & Tappenden, 2014). However, findings from *in vivo*, *in vitro*, and ex vivo research suggest that HMOs improve intestinal barrier function and, as a result, may help to avoid gut dysfunction by improving the expression of mucins along with tight junction proteins and glycocalyx components (Cheng et al., 2019; Wu et al., 2019).

13.5 HMOS AS NUTRIENT FOR BRAIN DEVELOPMENT

Better developmental scores have been observed in breastfed preterm infants at 18 months and higher intellectual quotients at 7 years (Lucas et al., 1990, 1992). A growing amount of evidence suggests that Sia-containing gangliosides and poly-Sia-containing glycoproteins have a role in brain development and cognition (Wang et al., 2009).

Ganglioside- and protein-bound Sia concentrations were found to be significantly higher in the brains of breastfed infants than in the brains of formula-fed infants after postmortem analysis (Wang et al., 2003). Most of the Sia in human milk is obtained from sialylated HMOs. The amount of Sia in HMO is 2–3 times more than that in glycoproteins, and only around 1% of Sia in human milk comes from Sia in glycolipids (Wang et al., 2001). Further research is required to ascertain whether HMOs are the principal Sia carrier, supplying this seemingly critical nutrient to the growing brain and contributing to improved developmental scores and cognitive quotients in breastfed children (Lars, 2012).

13.6 EFFECTS OF HMOS ON MOTHERS HEALTH

So far, all of the HMOs' potential benefits have been focused on the breastfed baby. HMOs may, however, affect breastfeeding mothers as well, and the underlying mechanisms may be comparable (Lars, 2012). Human milk is not sterile in the sense that it contains complex bacterial populations that are often highly tailored to the breastfeeding mother (Martin et al., 2007; Hunt et al., 2011). HMOs may affect the bacterial composition in milk in the mammary gland by acting as prebiotics or antiadhesive antimicrobials, or by a direct influence on responses of mammary gland epithelial cells or immunological responses, similar to what has been shown in the neonate (Barbosa-Cesnik et al., 2003; Delgado et al., 2009). HMOs were detected in urine samples of pregnant women just before delivery (Date, 1964; Hallgren et al., 1977; Hallgren and Lundblad, 1977a, 1977b). This shows that there may be a retrograde "leakage" into the circulation, which could have systemic consequences not just for the breastfed infant but also for the breastfeeding mother (Lars, 2012).

13.7 CONCLUSION

HMOs function as antimicrobial or antiviral agents by binding to pathogens and toxins or by binding to receptors in the epithelium, thus blocking adhesion and infection. As a prebiotic agent, HMOs enrich intestinal bifidogenic bacteria, providing an advantage over pathogens and protecting from infections. The end products of microbial fermentation of HMOs play a crucial role in the physiological development of infants. Maturation of intestinal epithelial cells (IEC) depends on short-chain fatty acids produced after intestinal fermentation of HMOs.

Various factors contribute to the involvement of HMOs in controlling mucosal immunological and barrier function, although the exact mechanisms are unknown. Further research is needed to understand how different structures of HMOs in breast milk contribute to protecting newborns against diseases (Ayechu et al., 2018).

The significance of the HMOs along with the microbiological composition of breastmilk in the development of the neonatal gut microbiota has lately become increasingly apparent. Research is needed on ways of modifying breastfeeding elements that shape the microbiota. The administration of probiotics to nursing mothers has yielded mixed findings in terms of the ability to alter the milk microbiota (Wickens et al., 2018; Jimenez et al., 2008), and there is currently no indication of the ability to alter the content of HMOs in human milk.

Thus, there is an urgent need for randomized intervention trials to investigate the effects of supplementing nursing mothers with milk-modulating substances like pre- and probiotics in order to change the infant's microbiome and, as a result, program the immune response (Elsen et al., 2019). Several research studies are underway across the world to create and include HMOs into newborn meals. As stated, nature's ways are tough to emulate; therefore, HMOs development and imitation of mother's milk appear to be a long journey.

KEYWORDS

- **babies**
- **bacteria**
- **fatty acids**
- **gastrointestinal microbiota**
- **health benefits**
- **HIV infection**
- **immunology**
- **mothers**
- **oligosaccharides**
- **pathogens**

REFERENCES

Albrecht, S., Schols, H. A., van den Heuvel, E. G., Voragen, A. G., & Gruppen, H. (2010). CE-LIF-MSn profiling of oligosaccharides in human milk and feces of breast-fed babies. *Electrophoresis, 31*(7), 1264–1273.

Albrecht, S., Schols, H. A., van den Heuvel, E. G., Voragen, A. G., & Gruppen, H. (2011). Occurrence of oligosaccharides in feces of breast-fed babies in their first six months of life and the corresponding breast milk. *Carbohydrate Research, 346*(16), 2540–2550.

Andreas, N. J., Kampmann, B., & Mehring Le-Doare, K. (2015). Human breast milk: A review on its composition and bioactivity. *Early Human Development, 91*(11), 629–635. https://doi.org/10.1016/j.earlhumdev.2015.08.013.

Asakuma, S., Hatakeyama, E., Urashima, T., Yoshida, E., Katayama, T., Yamamoto, K., Kumagai, H., Ashida, H., Hirose, J., & Kitaoka, M. (2011). Physiology of consumption of human milk oligosaccharides by infant gut-associated bifidobacteria. *Journal of Biological Chemistry, 286*(40), 34583–34592. https://doi.org/10.1074/jbc.M111.248138.

Ayechu-Muruzabal, V., van Stigt, A. H., Mank, M., Willemsen, L. E. M., Stahl, B., Garssen, J., & van't Land, B. (2018). Diversity of human milk oligosaccharides and effects on early life immune development. *Frontiers in Pediatrics, 6*, 239. https://doi.org/10.3389/fped.2018.00239.

Barbosa-Cesnik, C., Schwartz, K., & Foxman, B. (2003). Lactation mastitis. *JAMA, 289*(13), 1609–1612.

Bode, L. (2012). Human milk oligosaccharides: Every baby needs a sugar mama–A review. *Glycobiology, 22*(9), 1147–1162.

Bode, L. (2015). The functional biology of human milk oligosaccharides. *Early Human Development, 91*(11), 619–622. https://doi.org/10.1016/j.earlhumdev.2015.09.001.

Chaturvedi, P., Warren, C. D., Altaye, M., Morrow, A. L., Ruiz-Palacios, G., Pickering, L. K., et al. (2001). Fucosylated human milk oligosaccharides vary between individuals and over the course of lactation. *Glycobiology, 11*, 365–372.

Chen, P., Reiter, T., Huang, B., Kong, N., & Weimer, B. C. (2017). Prebiotic oligosaccharides potentiate host protective responses against *L. monocytogenes* infection. *Pathogens, 6*(4), Article 68. https://doi.org/10.3390/pathogens6040068.

Chen, X. (2015). Human milk oligosaccharides (HMOS): Structure, function, and enzyme-catalyzed synthesis. *Advances in Carbohydrate Chemistry and Biochemistry, 72*, 113–190. https://doi.org/10.1016/bs.accb.2015.08.002.

Cheng, L., Kong, C., Walvoort, M. T. C., Faas, M. M., & de Vos, P. (2019). Human milk oligosaccharides differently modulate goblet cells under homeostatic, proinflammatory conditions and ER stress. *Molecular Nutrition & Food Research, 64(5)*, 1900976. https://doi.org/10.1002/mnfr.201900976.

Costerton, J. W., Irvin, R. T., & Cheng, K. J. (1981). The bacterial glycocalyx in nature and disease. *Annual Review of Microbiology, 35*(1), 299–324. https://doi.org/10.1146/annurev.mi.35.100181.001503.

Craft, K. M., & Townsend, S. D. (2018). The human milk glycome as a defense against infectious diseases: Rationale, challenges, and opportunities. *ACS Infectious Diseases, 4*(2), 77–83. https://doi.org/10.1021/acsinfecdis.7b00209.

Craft, K. M., & Townsend, S. D. (2019). Mother knows best: Deciphering the antibacterial properties of human milk oligosaccharides. *Accounts of Chemical Research, 52*(3), 760–768. https://doi.org/10.1021/acs.accounts.8b00630.

Date, J. W. (1964). The isolation of some oligosaccharides from the urine of pregnant and lactating women. *Scandinavian Journal of Clinical and Laboratory Investigation, 16*, 597–603.

Delgado, S., Arroyo, R., Jimenez, E., Marin, M. L., del Campo, R., Fernandez, L., & Rodriguez, J. M. (2009). *Staphylococcus epidermidis* strains isolated from breast milk of women suffering infectious mastitis: Potential virulence traits and resistance to antibiotics. *BMC Microbiology, 9*, 82. https://doi.org/10.1186/1471-2180-9-82.

Dotz, V., Rudlof, S., Meyer, C., Lochnit, G., & Kunz, C. (2015). Metabolic fate of neutral human milk oligosaccharides in exclusively breast-fed infants. *Molecular Nutrition & Food Research, 59*(2), 355–364. https://doi.org/10.1002/mnfr.201400160.

Engfer, M. B., Stahl, B., Finke, B., Sawatzki, G., & Daniel, H. (2000). Human milk oligosaccharides are resistant to enzymatic hydrolysis in the upper gastrointestinal tract. *American Journal of Clinical Nutrition, 71*(6), 1589–1596.

Gibson, G. R., & Wang, X. (1994). Regulatory effects of bifidobacteria on the growth of other colonic bacteria. *Journal of Applied Bacteriology, 77*(4), 412–420.

Gibson, G. R., Probert, H. M., Loo, J. V., Rastall, R. A., & Roberfroid, M. B. (2004). Dietary modulation of the human colonic microbiota: Updating the concept of prebiotics. *Nutrition Research Reviews, 17*(2), 259–275.

Gnoth, M. J., Kunz, C., Kinne-Saffran, E., & Rudloff, S. (2000). Human milk oligosaccharides are minimally digested in vitro. *Journal of Nutrition, 130*(12), 3014–3020.

Hallgren, P., & Lundblad, A. (1977a). Structural analysis of nine oligosaccharides isolated from the urine of a blood group O, nonsecretor, woman during pregnancy and lactation. *Journal of Biological Chemistry, 252*(4), 1014–1022.

Hallgren, P., & Lundblad, A. (1977b). Structural analysis of oligosaccharides isolated from the urine of a blood group A, secretor, woman during pregnancy and lactation. *Journal of Biological Chemistry, 252*(4), 1023–1033.

Hallgren, P., Lindberg, B. S., & Lundblad, A. (1977). Quantitation of some urinary oligosaccharides during pregnancy and lactation. *Journal of Biological Chemistry, 252*(4), 1034–1040.

Hao, H., Zhu, L., & Faden, H. S. (2019). The milk-based diet of infancy and the gut microbiome. *Gastroenterology Report, 7*(4), 246–249. https://doi.org/10.1093/gastro/goz031.

Hoeflinger, J. L., Davis, S. R., Chow, J., & Miller, M. J. (2015). In vitro impact of human milk oligosaccharides on Enterobacteriaceae growth. *Journal of Agricultural and Food Chemistry, 63*(12), 3295–3302. https://doi.org/10.1021/jf505721p.

Hunt, K. M., Foster, J. A., Forney, L. J., Schutte, U. M., Beck, D. L., Abdo, Z., Fox, L. K., Williams, J. E., McGuire, M. K., & McGuire, M. A. (2011). Characterization of the diversity and temporal stability of bacterial communities in human milk. *PLoS One, 6*(6), e21313. https://doi.org/10.1371/journal.pone.0021313.

Jimenez, E., Fernandez, L., Maldonado, A., Martin, R., Olivares, M., Xaus, J., et al. (2008). Oral administration of *Lactobacillus* strains isolated from breast milk as an alternative for the treatment of infectious mastitis during lactation. *Applied and Environmental Microbiology, 74*(15), 4650–4655. https://doi.org/10.1128/AEM.02599-07.

Jost, T., Lacroix, C., Braegger, C., & Chassard, C. (2015). Impact of human milk bacteria and oligosaccharides on neonatal gut microbiota establishment and gut health. *Nutrition Reviews, 73*(7), 426–437. https://doi.org/10.1093/nutrit/nuu016.

Kavanaugh, D., O'Callaghan, J., Kilcoyne, M., Kane, M., Joshi, L., & Hickey, R. M. (2015). The intestinal glycome and its modulation by diet and nutrition. *Nutrition Reviews, 73*(6), 359–375. https://doi.org/10.1093/nutrit/nuu019.

Kulinich, A., & Liu, L. (2016). Human milk oligosaccharides: The role in the fine-tuning of innate immune responses. *Carbohydrate Research, 432*, 62–70. https://doi.org/10.1016/j.carres.2016.07.009.

Kunz, C., Rudloff, S., Baier, W., Klein, N., & Strobel, S. (2000). Oligosaccharides in human milk: Structural, functional, and metabolic aspects. *Annual Review of Nutrition, 20*, 699–722.

Laucirica, D. R., Triantis, V., Schoemaker, R., Estes, M. K., & Ramani, S. (2017). Milk oligosaccharides inhibit human rotavirus infectivity in MA104 cells. *Journal of Nutrition, 147*(9), 1709–1714. https://doi.org/10.3945/jn.116.246090.

Lo Cascio, R. G., Ninonuevo, M. R., Freeman, S. L., Sela, D. A., Grimm, R., Lebrilla, C. B., Mills, D. A., & German, J. B. (2007). Glycoprofiling of bifidobacterial consumption of human milk oligosaccharides demonstrates strain-specific, preferential consumption of small-chain glycans secreted in early human lactation. *Journal of Agricultural and Food Chemistry, 55*(22), 8914–8919.

Marcobal, A., Barboza, M., Froehlich, J. W., Block, D. E., German, J. B., Lebrilla, C. B., & Mills, D. A. (2010). Consumption of human milk oligosaccharides by gut-related microbes. *Journal of Agricultural and Food Chemistry, 58*(9), 5334–5340.

Martin, R., Heilig, H. G. H., Zoetendal, E. G., Jimenez, E., Fernandez, L., Smidt, H., & Rodriguez, J. M. (2007). Cultivation-independent assessment of the bacterial diversity of breast milk among healthy women. *Research in Microbiology, 158*(1), 31–37.

McLoughlin, K., Schluter, J., Rakoff-Nahoum, S., Smith, A. L., & Foster, K. R. (2016). Host selection of microbiota via differential adhesion. *Cell Host & Microbe, 19*(4), 550–559. https://doi.org/10.1016/j.chom.2016.02.021.

Monteagudo-Mera, A., Rastall, R. A., Gibson, G. R., Charalampopoulos, D., & Chatzifragkou, A. (2019). Adhesion mechanisms mediated by probiotics and prebiotics and their potential impact on human health. *Applied Microbiology and Biotechnology, 103*(16), 6463–6472. https://doi.org/10.1007/s00253-019-09978-7.

Morozov, V., Hansman, G., Hanisch, F. G., Schroten, H., & Kunz, C. (2018). Human milk oligosaccharides as promising antivirals. *Molecular Nutrition & Food Research, 62*(6), e1700679. https://doi.org/10.1002/mnfr.201700679.

Morrin, S. T., & Hickey, R. M. (2020). New insights on the colonization of the human gut by health-promoting bacteria. *Applied Microbiology and Biotechnology, 104*(4), 1511–1515. https://doi.org/10.1007/s00253-019-10336-w.

Newburg, D. S., & Grave, G. (2014). Recent advances in human milk glycobiology. *Pediatric Research, 75*(5), 675–679. https://doi.org/10.1038/pr.2014.24.

Ploger, S., Stumpf, F., Penner, G. B., Schulzke, J. D., Gabel, G., Martens, H., Shen, Z., Gunzel, D., & Aschenbach, J. R. (2012). Microbial butyrate and its role for barrier function in the gastrointestinal tract. *Annals of the New York Academy of Sciences, 1258*(1), 52–59. https://doi.org/10.1111/j.1749-6632.2012.06553.x.

Ramani, S., Stewart, C. J., Laucirica, D. R., Ajami, N. J., Robertson, B., Autran, C. A., Shinge, D., Rani, S., Anandan, S., Hu, L., Ferreon, J. C., Kuruvilla, K. A., Petrosino, J. F., Venkataram Prasad, B. V., Bode, L., Kang, G., & Estes, M. K. (2018). Human milk oligosaccharides, milk microbiome, and infant gut microbiome modulate neonatal rotavirus infection. *Nature Communications, 9*(1), 5010. https://doi.org/10.1038/s41467-018-07476-4.

Reichardt, N., Duncan, S. H., Young, P., Belenguer, A., McWilliam Leitch, C., Scott, K. P., Flint, H. J., & Louis, P. (2014). Phylogenetic distribution of three pathways for propionate production within the human gut microbiota. *ISME Journal, 8*(6), 1323–1335. https://doi.org/10.1038/ismej.2014.14.

Roberfroid, M. (2007). Prebiotics: The concept revisited. *Journal of Nutrition, 137*(3), 830S–837S.

Rudloff, S., Pohlentz, G., Borsch, C., Lentze, M. J., & Kunz, C. (2012). Urinary excretion of in vivo (13)C-labelled milk oligosaccharides in breastfed infants. *British Journal of Nutrition, 107*(7), 957–963. https://doi.org/10.1017/S0007114511004016.

Stahl, B., Thurl, S., Henker, J., Siegel, M., Finke, B., & Sawatzki, G. (2001). Detection of four human milk groups with respect to Lewis-blood-group-dependent oligosaccharides by serologic and chromatographic analysis. *Advances in Experimental Medicine and Biology, 501*, 299–306.

Tan, J., McKenzie, C., Potamitis, M., Thorburn, A. N., Mackay, C. R., & Macia, L. (2014). The role of short-chain fatty acids in health and disease. *Advances in Immunology, 121*, 91–119. https://doi.org/10.1016/b978-0-12-800100-4.00003-9.

Thurl, S., Munzert, M., Boehm, G., Matthews, C., & Stahl, B. (2017). Systematic review of the concentrations of oligosaccharides in human milk. *Nutrition Reviews, 75*(11), 920–933. https://doi.org/10.1093/nutrit/nux044.

Van den Elsen, L. W. J., Garssen, J., Burcelin, R., & Verhasselt, V. (2019). Shaping the gut microbiota by breastfeeding: The gateway to allergy prevention? *Frontiers in Pediatrics, 7*, 47. https://doi.org/10.3389/fped.2019.00047.

Vazquez, E., Santos-Fandila, A., Buck, R., Rueda, R., & Ramirez, M. (2017). Major human milk oligosaccharides are absorbed into the systemic circulation after oral administration in rats. *British Journal of Nutrition, 117*(2), 237–247. https://doi.org/10.1017/S0007114516004554.

Walsh, C., Lane, J. A., van Sinderen, D., & Hickey, R. M. (2020). Human milk oligosaccharides: Shaping the infant gut microbiota and supporting health. *Journal of Functional Foods, 72*, 104074. https://doi.org/10.1016/j.jf.2020.104074.

Wang, B. (2009). Sialic acid is an essential nutrient for brain development and cognition. *Annual Review of Nutrition, 29*, 177–222.

Wang, B., Brand-Miller, J., McVeagh, P., & Petocz, P. (2001). Concentration and distribution of sialic acid in human milk and infant formulas. *American Journal of Clinical Nutrition, 74*(4), 510–515.

Wang, B., McVeagh, P., Petocz, P., & Brand-Miller, J. (2003). Brain ganglioside and glycoprotein sialic acid in breastfed compared with formula-fed infants. *American Journal of Clinical Nutrition, 78*(5), 1024–1029.

Wettschureck, N., & Offermanns, S. (2005). Mammalian G proteins and their cell type-specific functions. *Physiological Reviews, 85*(4), 1159–1204. https://doi.org/10.1152/physrev.00003.2005.

Wickens, K., Barthow, C., Mitchell, E. A., Stanley, T. V., Purdie, G., Rowden, J., et al. (2018). Maternal supplementation alone with *Lactobacillus rhamnosus* HN001 during pregnancy and breastfeeding does not reduce infant eczema. *Pediatric Allergy and Immunology, 29*(3), 296–302. https://doi.org/10.1111/pai.12874.

Wu, R. Y., Li, B., Koike, Y., Maattanen, P., Miyake, H., Cadete, M., & Sherman, P. M. (2019). Human milk oligosaccharides increase mucin expression in experimental necrotizing enterocolitis. *Molecular Nutrition & Food Research, 63*(3), Article e1800658. https://doi.org/10.1002/mnfr.201800658.

Xu, G., Davis, J. C., Goonatilleke, E., Smilowitz, J. T., German, J. B., & Lebrilla, C. B. (2017). Absolute quantitation of human milk oligosaccharides reveals phenotypic variations during lactation. *Journal of Nutrition, 147*(1), 117–124. https://doi.org/10.3945/jn.116.238279.

CHAPTER 14

Techniques to Evaluate Prebiotic Properties

V. H. PATEL AND URVASHI MALL

P.G. Department of Home Science, Sardar Patel University, Anand, Gujarat, India

Prebiotic ability in vivo (animal studies and human trials) can be considered a gold standard due to the most accurate results but they are costly and time-consuming. In vitro methods are an alternative to in vivo methods, which facilitate the research and reduce cost. Therefore, a huge effort is given to in vitro techniques.

14.1 GENOMIC TECHNIQUES

Various *in vitro* and *in vivo* technologies were employed for assessing carbohydrate prebiotic ability (Gibson et al., 2000; Hur et al., 2011; de Souza Aquino et al., 2017). Hundreds of years, *in vitro* methods like (i) Batch fermentation model, (ii) Dynamic Fermentation models, (iii) The TNO *in vitro* model of the colon (TIM-2), (iv) The simulator of the human intestinal microbial ecosystem (SHIME®), (v) Mucosal Simulator of the Human Intestinal Microbial Ecosystem (M-SHIME®), (vi) The SIMulator gastro-intestinal: SIMGI, (vii) Polyfermentor intestinal model (PolyFermS), (viii) Mini-Bio *in vitro* model, and (ix) The smallest intestine *in vitro* model (TSI), and (x) Gas production technique (SCFAs) built to a degree model the gut has been widely used by researchers. To classify bacteria involved

Prebiotic Foods for a Healthy Gut: Scientific Studies to Manage Metabolic Disorders.
Mini K. Sheth and Shruti Dwivedi (Eds.)

in prebiotic activity, genomic techniques generally used such as technique based on culture (bacteria isolated from feces), quantitative polymerase chain reactions (qPCR), sequencing methods, DNA sequencing, Denaturing gradient gel electrophoresis (DGGE), Terminal restriction fragment length polymorphism (TRFLP), Fluorescence in situ hybridization (FISH), and Next Generation Sequencing (NGS) (Bajury et al., 2018; Nissen et al., 2020) (Figure 14.1).

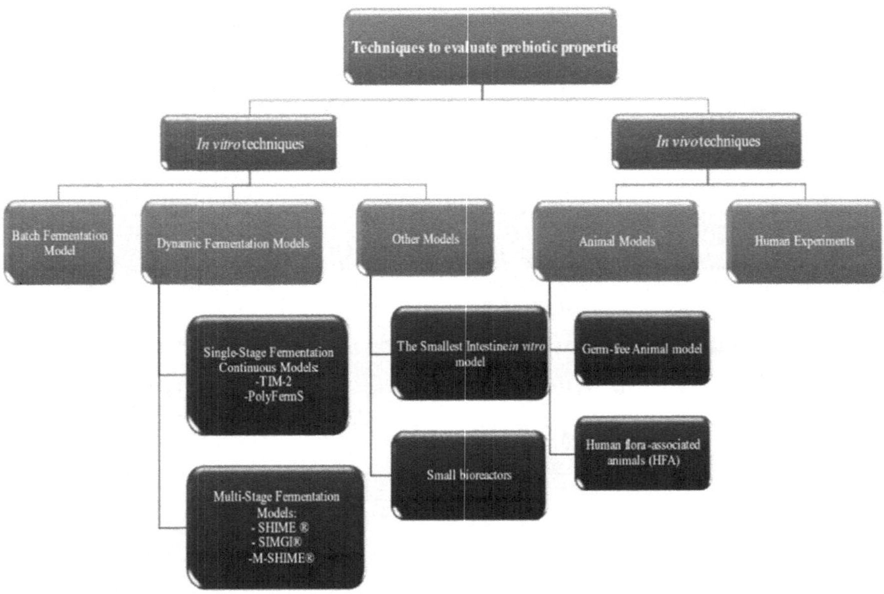

FIGURE 14.1 Overview of techniques to evaluate prebiotic properties.

14.2 *IN VITRO* TECHNIQUES

14.2.1 BATCH FERMENTATION MODEL

Batch cultures are the most basic *in vitro* methods for studying human gut bacteria. They are mostly used to evaluate the metabolization of particular substrates by chosen strains or complicated fecal microbiota (Venema & Van den Abbeele, 2013). Batch cultures are often performed in single fermenter vessels that contain basal medium supplemented with a particular substrate or a growth inhibitor. Constant physiological temperature (37°C) and anaerobic environment (usually achieved by bubbling nitrogen) are used as working conditions. Tests are usually done for short periods of time (24–48 hours, with a maximum of 72 hours) (Rumney & Rowland, 1992; El Oufir et al., 2000). To initiate fermentation, bacterial species, pure cultures, or fecal suspensions are introduced into sterile media, with or without pH control and without extra nutrient supply (Roupar et al., 2021). The quantity of fermentation is often determined by the breakdown of non-starch polysaccharides degradation, SCFAs or gas production, or loss of dry matter (DM)/organic matter (OM) (Coles et al., 2005). The batch fermentation model is used to investigate the effects of various substrates on the physiology and biodiversity of gastrointestinal microbes. The effect on the microflora is measured using quantitative as well as qualitative molecular techniques, while the effect on metabolic activity is examined using an assessment of the SCFAs released. Batch methods are important for assessing cross-variation in response to a certain bioactive molecule or agent, as well as comparing the effects of exposure to different compound types or concentrations. Furthermore, they give a brief analysis of microbial metabolites produced and aid in the elucidation of metabolic pathways involved (Verhoeckx et al., 2015). Because of the simplicity of the whole setup design, batch models have the benefits of rapid assembly, low operating costs, simple operation, plus repeatability. Because of their high throughput and potential for automation, they are virtually a first-line method for studying intestinal microbiome composition, metabolism, and control by probiotics and dietary substances. These processes are usually convenient and cost-effective so that large quantities of substrates and/or fecal samples can be easily checked (Gagnon et al., 2011). As these systems are readily scaled down, they require less quantity of test substances to scan novel substrates on a smaller scale and analyze within 24 hours (Rycroft et al., 2001). Before doing more complicated multi-vessel continuous fermentation investigations, batch models are often used (Williams et al., 2015). Figure 14.2 shows the presentation of *in vitro* digestion.

14.2.1.1 METHOD FOR IN VITRO BATCH-CULTURE FERMENTATIONS (COLONIC FERMENTATION)

In an *in vitro* batch-culture fermentation experiment intended to mimic colonic fermentation, a pre-digested food sample is placed in batch culture fermentation containers containing a sterile basal nutritional solution. These containers are subsequently placed in an anaerobic chamber at 37°C

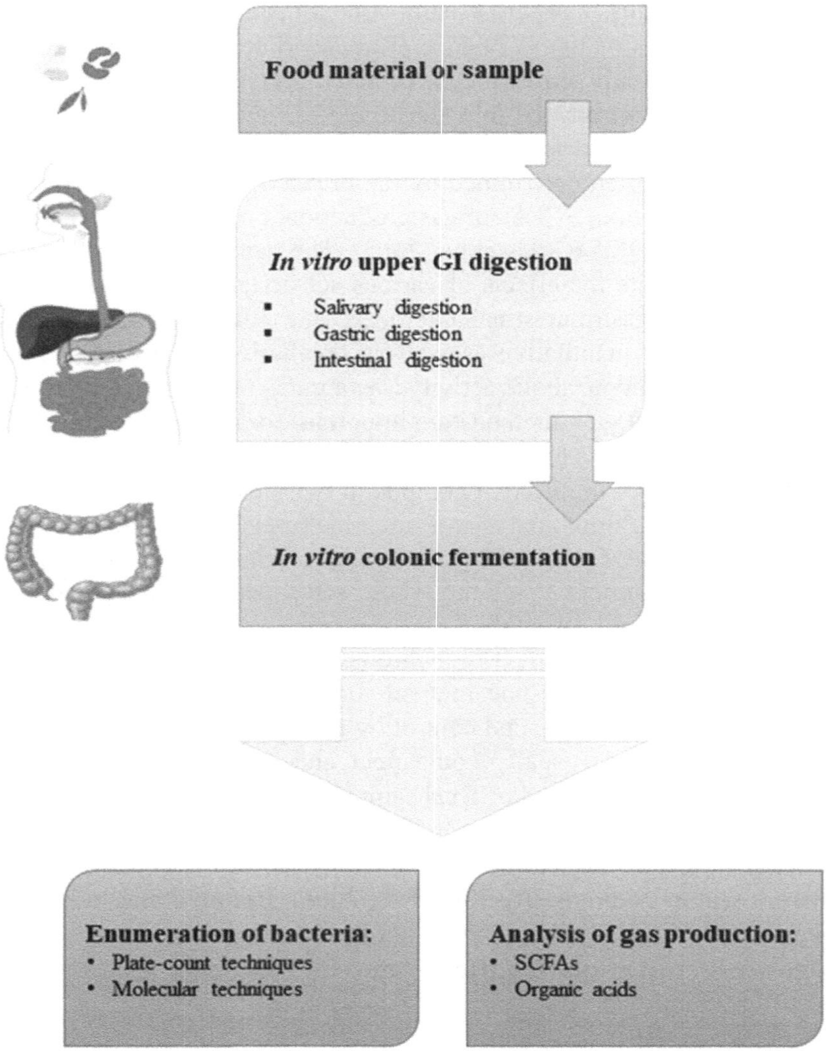

FIGURE 14.2 Presentation of in *vitro* digestion.

overnight to allow the medium to equilibrate in an anaerobic environment. Fermentation is started the next day by inoculating the medium with a known probiotic strain or a faecal slurry. After inoculation, the vessels are placed back to the anaerobic chamber and incubated at 37°C for 48 hours to simulate colonic fermentation conditions.

During the fermentation process, samples are collected from the containers at regular intervals, i.e., 0, 6, 24, 30, and 48 hours. These samples are analyzed to determine bacterial growth as well as activity. Bacterial populations are quantified and identified using plate-count techniques and sophisticated molecular approaches such as fluorescence in situ hybridization (FISH) and quantitative polymerase chain reaction (qPCR). Furthermore, the samples are examined for short-chain fatty acids (SCFAs), which constitute key metabolic by-products of gut fermentation. This approach provides important details on the fermentation process and the interactions between dietary components and gut flora (Rycroft et al., 2001; Parra-Matadamas et al., 2015; Liu et al., 2016).

14.2.2 DYNAMIC FERMENTATION MODELS

In recent years, several dynamic fermentation models have become well-known with the aim of establishing a relatively stable microbial ecosystem *in vitro* under colon conditions that are physiologically important. Long-term studies are necessary to test the spatial and temporal adaptation of food compounds by intestinal bacteria and the microbial metabolism of dietary compounds, which are useful for continuous multi-phase and single-fermentation modeling. Gibson et al. (1988) defined that the majority of models of dynamic multistage fermentation are based on the Reading model, consisting of three vessels of increasing scale aligned in series to produce a sequential feed on the growing medium that matches proximal, transverse, and distal colon conditions with distinct pH and fermentation activity. The authors have also established a nutrient medium widely used in batch and dynamic models of fermentation. The medium includes protein substrates, complex carbohydrates (pectin, xylan, arabinogalactan, and resistant starch) that are not digested by gastric enzymes, as well as combination salts and vitamins. These models are subject to environmental parameters such as pH, retention period, temperature, anaerobiosis, and the medium flow rate of the *in vivo* obtained by human gut microbiota, which have adapted, survived, and proliferated (Payne et al., 2012).

14.2.3 THE TNO IN VITRO MODEL OF THE COLON (TIM-2)

The TNO Gastro-Intestinal Model (TIM) is a multi-compartmental model intended to accurately mimic conditions in the lumen of the gastrointestinal tract. TIM is widely used to research the gastrointestinal activity of a large range of feed, dietary, and medicinal drugs. Human digestive functions that are represented in the TIM-1 upper gastrointestinal tract model comprise bile secretion, motility, pH, and absorption capacities of the small intestine (Minekus et al., 1995). Other host mechanisms such as peristaltic mixing, water and metabolite absorption are used in proximal colon simulator models like TIM-2 (Minekus et al., 1999). The combination of the TIM-1 and TIM-2 models resulted in the formation of an artificial human digestive system that has been used in pharmaceutical research of drug delivery as well as modern nutritional research (Blanquet-Diot et al., 2009; Souliman et al., 2007; Souliman et al., 2006). In the TIM-2 system, the following standardized conditions were simulated: body temperature, pH in the lumen, secretion rate and composition, delivery of a pre-digested substrate from the "ileum" (SIEM; Standard Ileal Efflux Medium), mixing and transportation of the gastric tract contents, water absorption, the presence of a highly complex, high-density, metabolically active, anaerobic microbes of human origin, and microbial metabolite absorption through a semipermeable membrane inside the colon model. The SIEM simulates substances entering the human colon (Reimer et al., 2014). At a concentration of 10 g/dL, the system was used to investigate the effects of lactulose (Venema et al., 2003), inulin (van Nuenen et al., 2003), and GOS (Maathuis et al., 2012). Interestingly, research using TIM-2 inoculated with fecal suspension from lean and obese subjects demonstrated that GOS, lactulose, apple fiber, and sugar beet pectin are fermented diversely by lean and obese microbiotas (Aguirre et al., 2014). Interesting findings were obtained when antibiotics as well as probiotics were added in TIM-2 together, leading to higher *Bifidobacteria*l and *Lactobacillus* counts (Rehman et al., 2012).

14.2.4 SIMULATOR OF THE HUMAN INTESTINAL MICROBIAL ECOSYSTEM (SHIME®)

The SHIME is among the few gastrointestinal systems that simulate the whole digestive system, including the gastric, small bowel, and various parts of the large intestines. The SHIME (Stimulator of the Human Intestinal Microbial Ecosystem) is a proprietary device developed by ProDigest and the University of Ghent (Van den Abbeele et al., 2010). It is comprised of five attached units that recreate the entire gastrointestinal tract, from the stomach to the distal colon. The first pair of units, which simulate the gut and small bowel, use a fill-and-draw method to add a predetermined quantity of nutritional mixture, pancreatic enzymes, and bile to the simulated gut and small bowel containers. The other three bowel chambers are constantly stirred systems of steady volume (0.5, 0.8, as well as 0.6 liters, respectively) and pH regulation between 5.6 and 5.9, 6.15–6.4, as well as 6.6–6.9 (Venema et al., 2013). The retention period ranges from 24 to 72 hours. Following fecal microbial inoculums, a couple of weeks of stabilization phase help these fecal bacteria to adjust to an *in vitro* environment and develop from fecal to colon region-specific flora (Venema et al., 2013; Van den Abbeele et al., 2010; Molly et al., 1994 and Molly et al., 1993). Giuliani et al. (2019) conducted research on the impact of vegetable waste products on the gut microbiome. Terpend et al. (2013) as well as Van de Wiele et al. (2004) investigated the impact of arabinogalactan FOS natural chicory and inulin on the Gut microbiome using the SHIME method. This type of equipment is time-consuming and labor-intensive, and it is not ideal for small laboratories, but it has the highest *in vivo* similarities among many other versions.

14.2.5 MUCOSAL SIMULATOR OF THE HUMAN INTESTINAL MICROBIAL ECOSYSTEM (M-SHIME®)

Macfarlane and his team investigated the impact of using mucin in their *in vitro* model to discover how gut microbiota colonize mucus and to investigate the physiologic and enzymatic aspects responsible for the degradation of the glycoprotein. The colonization of mucin gels was observed under environments of nutrient supply and restriction similar to those found in the distal and proximal colons, respectively. Mucin gels were quickly colonized and microbes developing in nutrient-restricted environments destroyed the most mucin in the gels (Macfarlane et al., 2005). Van den Abbeele et al. (2013) conducted the above study in an optimized SHIME model for the luminal flora by adding a mucosal system including mucin-covered subareas (mucosal SHIME or M-SHIME) (Van den Abbeele et al., 2012; Van den Abbeele et al., 2013). To begin, the mucosal condition enabled the colonization of specific *Lactobacilli* (i.e., *L. mucosae* and *L. rhamnosus*) through *in vivo* processes (Payne et al., 2012). Furthermore, detailed analysis of the entire microbiome showed that it is compatible with *in vivo* research (Shen et al., 2010 and Nava et al., 2011). The main discovery of this M-SHIME, or mucosal SHIME, was the colonization of the mucosal atmosphere marked by a greater proportion of *Clostridium* clusters (IV and XIVa) that produce butyrate (Van den Abbeele et al., 2013). Giuliani et al. (2016) used the

M-SHIME to investigate the impact of a grape extract on gut microbiota. This method was often used to investigate the impact of long-chain arabinoxylans (LC-AX) on microbiota in the luminal and mucosal lining. The findings of PCR-based microbial testing by MiSeq (Illumina, Hayward, CA) and gas chromatographic assay were performed for fatty acids, indicating that LC-AX could have a prebiotic advantage because it was beneficial in enriching some gut bacteria (Truchado et al., 2017).

14.2.6 SIMULATOR OF THE GASTROINTESTINAL TRACT (SIMGI®)

The Simulator of the Gastrointestinal Tract (SimGI®) model is a completely automatic digestive multichambered system recently built at the CIAL Institute of Food Science Research (CSIC-UAM, Madrid, Spain) (Gil-Sánchez et al., 2020). The SimGI model is an artificial gastrointestinal *in vitro* method developed to effectively mimic the physiological mechanisms that occur throughout digestion in the gastrointestinal tract, as well as to recreate the fecal microbiome that is essential for metabolic bioconversions throughout the large bowel (Barroso et al., 2015). The system is made up of five units that are operated remotely by a computer running special software and are linked by pumps (Pereira-Caro et al., 2015; Marín et al., 2015; Cueva et al., 2019). The first two modules represent the stomach section as well as the small intestine, while the remaining three comprise the large intestine's three major parts. The first device consists of two methacrylate tubes located over a tank and split by a jacket. The tank works as a container for stomach acid, and the portable jacket mixes the contents using water sprays (Cueva et al., 2015; Thuenemann et al., 2015). Other four downstream systems, on the other hand, are glass reactors with magnetic stirrers to keep the media agitated. The environmental factors for each container are different, with unique pH and environment specifications being set. Based on simulation analysis, the experimental cycles will last up to six days (Barroso et al., 2015 and Nissen et al., 2020). Tamargo et al. (2018) analyzed the benefits of chia seed polysaccharide using SIMGI experiments and obtained their findings using culture-based microbiology and solid-phase microextraction accompanied by gas chromatography-mass spectrometry (SPME-GC-MS) analysis of SCFAs. Aside from the fact that substantial increases in SCFA activity, no significant differences were observed in the bacteria species of gut microbes. SIMGI has the disadvantages of being time-intensive and less practical and reproducible due to less stringent atmospheric conditions, but the advantages of being cost-effective, logistically scalable, and operator-oriented.

14.2.7 POLY FERMENTOR INTESTINAL MODEL (POLYFERMS)

The gastrointestinal dynamic fermentation models mentioned earlier require a fecal suspension to replicate the gut microbes. Even though a flexible approach is beneficial, it has been called into doubt, mostly since it does not allow for the various skills and techniques employed by microbes to colonize and prevent in the host gastrointestinal epithelia. The simulated population is dysfunctional, and those microbiome groups capable of overcoming others, specifically by tolerance to hunger, ecological displacement, and rapid development, enforce their power. As a result, critical species that are much less violent in their fight for bowel material become unviable, limiting the model's similarity to fact. A further explanation of a failure is the absence of bacterial diversity found in a biofilm in feces samples (Macfarlane & Macfarlane, 2007; Macfarlane & Dillon, 2007; and Nissen et al., 2020). To overcome this limitation by maintaining a certain amount of microbiota variety and abundance in deep research, an immobilization mechanism was created that involves retaining fecal bacteria using a micro-encapsulation method (Le Blay et al., 2010; Zihler et al., 2011; Payne et al., 2012; Dostal et al., 2013). The PolyFermS system, a dynamic fermentation model developed at ETH Zurich through a joint European effort, was the first to use this concept (Berner et al., 2013). PolyFermS is a model that simulates the proximal colon region. It helps to overcome biological reliability issues associated with gastrointestinal microbe control by allowing the analysis of various impacts of interventions on flora, such as atmospheric factors, nutritional supplements, and medicines (Poeker et al., 2013). The effluent from a first-stage ongoing inoculum reactor (IR) comprising immobilized fecal microbes and mimicking the upper proximal colon was used to constantly supply a group of second-stage monitoring (CR) and test (TR) reactors connected in parallel with proximal colon environments (Berner et al., 2013). Fermented effluents are spread evenly throughout the units (volume of 0.3L), with an average processing period of 7 hours and 30 minutes (Roupar et al., 2021). All of the model's five processors have their specific microbiome suspension, and the system will maintain a balanced microbial population throughout the length of a 38-day test phase. Furthermore, the simulated bacteria closely match the fecal microbiome of a normal donor, maintaining main taxonomical microbial communities for both variety and abundance during the research. Poeker et al. (2018) tested this system to assess the impact of fermentable dietary fibers on intestinal flora. Doo et al. (2017) investigated the effect of certain ingredients of baby formula that resemble the quality of human milk on infant intestinal flora. This model requires time and dedication to develop, but it appears to be extremely consistent over time, developing *in vivo* equivalents (Nissen et al., 2020).

14.2.8 SMALL BIOREACTORS

The Mini-Bioreactor *in vitro* system is a licensed device newly implemented in experimental science by Applikon Biotechnology in Delft, Netherlands, which can perform as a dynamic fermentation system, generating high-speed and large datasets that are monitored, examined, and analyzed by specialized software. The requirements for the different culture conditions are stored in a server and loaded directly into the device in order to accurately reproduce the tests that have already been run (O'Donnell et al., 2018). The software's capability to directly monitor fermentation and supervise simultaneous research containing up to 32 bioreactors is the key innovation. The operating capacity within each glass bioreactor can even be modified to meet the needs of the experiment, up to a maximum of 50 ml. Furthermore, various systems enable excellent efficiency such as: (i) a group of mechanical pumps rotating up to 2000 rpm, providing for the cultivation of high-density cultured cells or thick media; (ii) a defoam unit combined with scale detectors; (iii) gas spargers controlled via four mass flow control systems; (iv) six numerical speed control valves; (v) customizable autotuning adaptive proportional-integral-derivative control to manage system parameter changes while fermentation; (vi) a liquid-free perfused peltier system for thermoregulation and rapid adjustment; (vii) an electronic condenser capable of maintaining a pure environment; (viii) an optical small camera capable of monitoring cell culture confluency (Lattermann and Büchs, 2015). This bioreactor also has the disadvantage of being less affordable; however, the advantages include being modular, operationally fitting inside a laboratory, and having good reproducibility due to strong parameter selection and monitoring (Nissen et al., 2020).

14.2.9 THE SMALLEST INTESTINE IN VITRO MODEL

The Smallest Intestine *in vitro* model (TSI) mimics human small bowel passage by adjusting the pH and amount of bile salts, pancreases, and degradations to simulate digestion. Every The Smallest Intestine *in vitro* model device is made up of five batch-like processors, each of which mimics the

SI of a single organism and allows for the testing of 5 samples at the same time. The operating capacity of the units is 12 ml (Wiese et al., 2018). The main device of The Smallest Intestine *in vitro* model is a composite environment unit in which the temperature (37°C) is regulated by the movement of water from a rotating water bath via a copper/aluminum heat exchanger located within the box. The temperature is continuously monitored by the outside temperature controller attached to the water bath and installed within the main device. An individual temperature dataset recorder records the temperature throughout the procedure. Each container comprises five stirred fused quartz glass units, one for each individual's passage through the SI. To create anaerobic environments, reactors are individually sealed and continuously bubbled by nitrogen. Anaerobic environments can also be achieved by utilizing anaerobic pouches. Compartments are installed on a five-unit electromagnetic stirrer plate that ensures that materials are mixed evenly within the units. The units are sealed with a septa lid that comprises a pH probe inlet, needle inlets for enzyme and bile salt distribution, and processing, input, and output for the dialysis tank. Using dialysis cassettes, bile is ingested during the dialysis stage (Cieplak et al., 2018). The key parameters are managed by software, for example, the pH treatment is monitored and acid or base are dosed instantly by a code in Matlab software. To mimic the electrolytic structure and osmotic force that develop *in vivo*, distinct synthetic gastrointestinal fluids are utilized (Minekus et al., 2014). Cieplak et al. (2018) investigated the activity of three probiotic *Lactobacillus* species in the TSI system and examined the strains' persistence moving through the upper GI tract using culture-dependent microbiology. It has the disadvantage of being less realistic and reproducible due to less restrictive environmental settings, but it has the advantage of being operationally scalable and compatible.

14.3 *IN VIVO* TECHNIQUES

In vivo research, whether on human beings or animals, is the ultimate result for prebiotic efficiency before any significant judgments can be made. No doubt, the ideal method for evaluating prebiotic activity is human research, during which the gastrointestinal microbiome of the human subject can be directly investigated. Yet, this technique has downsides since it must adhere to strong ethical norms of practice, not to mention that gastrointestinal content, tissues, and organs are more easily obtainable in animal research (Bajury et al., 2018).

14.3.1 ANIMAL MODELS

Traditional animals such as mice, rats, and pigs have been commonly employed as models of gut microbiota. Mice are an excellent animal model since their gastrointestinal anatomy is quite close to human beings. Rats, on the other hand, have the advantage of providing a greater sample amount for examination as compared to mice. Pigs, which have the equivalent gastrointestinal physiology and anatomy as humans, are also employed as a monogastric animal model. These animals were commonly utilized in research on the effects of weight change, toxicity, and gas generation in diet samples. However, owing to existing variations between animal and human microbiota, comparable outcomes may emerge (Bajury et al., 2018). Mice and rats were already reported to be granivores, whereas pigs are omnivorous comparable to humans. Both of the animal models contain gut flora that is comparable to the human digestive system, with *Firmicutes* and Bacteroidetes phyla dominating. Moreover, the huge quantity of the dominating species in each of these animal studies differs significantly from that of humans (Nguyen et al., 2015). Nonetheless, mice and rats remain the most widely utilized animal model for understanding the human intestine. When compared to pigs, they have cheaper maintenance expenses and a short lifespan, yet pigs have the benefit of being bigger animal models that enable scientists to gather significant volumes of organs. Amazingly, the gastrointestinal systems of rodents and pigs are made of organs that, with minor changes, are physically comparable. This could be influenced by their food habits, genetic makeups, physiological needs, and divergent diets. Since their gastrointestinal structure and functionality are comparable to humans, the pig is a useful animal model for human digestive research. The sole disadvantage of utilizing pigs versus rats is money, since it is definitely more expensive to maintain pig housing and husbandry (Jimenez et al., 2015).

14.3.2 GERM-FREE ANIMAL MODEL

Germ-free (GF) animal models are essential for learning how gut bacteria may alter host anatomy and metabolism (Grover & Kashyap, 2014; Bhattarai & Kashyap, 2016; Fiebiger et al., 2016; Kubelkova et al., 2016; Kennedy et al., 2018; Qv et al., 2020). GF animals specifically colonized by one or more microbial strains are considered gnotobiotic (Fritz et al., 2013; Smith et al., 2007). Furthermore, gnotobiotic is often used as a similar term for

Germ-free (Umesaki et al., 2014). Germ-free (GF) animals are an extremely useful testing tool for studying interconnections among a host and its microbiome. A germ-free (axenic) animal is one that is devoid of germs such as bacteria, viruses, fungi, protozoa, and parasites throughout its lifespan. Germ-free animals are kept in isolation to avoid contamination, and they are supplied sanitized meals to limit their contact with microbes. Generally, it is not recommended to utilize the first-born GF animals in investigations as they may still retain viruses, bacteria, and microbial metabolites passed from mothers that are not GF animals (Al-Asmakh et al., 2015).

Inoculation of human gastrointestinal microbes into GF mice, humanized gnotobiotic models, enables phylogenetic recapitulation of the gut microbiome (Goodman et al., 2011). These systems are a great technique for evaluating the source and impact of intestinal microbiota in a human-like system (Al-Asmakh et al., 2015). Germ-free mice are sometimes used to research the function between intestinal flora and body mass since it is simple to continually track the mice's diet and body weight, which may result in any long-term influence on body mass and organ growth. Using interleukin 10-null mice, researchers investigated the influence of dietary synbiotics on body mass. The research discovered that prebiotics through probiotic supplements had a minimal influence on food consumption, resulting in an inconsequential impact on body weight (Kuo et al., 2013).

14.3.3 HUMAN FLORA-ASSOCIATED ANIMALS (HFA)

As axenic animals have a diverse gut microbiome compared to humans, human flora-associated (HFA) animals are most suitable for examining the ecosystem as well as the metabolic system of human intestinal bacteria because they offer the specific function of the human gastrointestinal flora. Scientists report that by inoculating gnotobiotic mice and rats with fecal slurry from healthy volunteers, they may transmit the fecal germs of human volunteers (Bajury et al., 2018). The method revealed that the dominating microbial species in these HFA mice remain stable. However, there have been circumstances when *Bifidobacteria* were eliminated from HFA mice, and these results are being related to the nature of the microbiota in the inoculation samples (Hirayama, 2010). Therefore, it is considered that HFA mice may be employed for a long period of time since their offspring can create similar content and metabolic activity as human gastrointestinal microbiota (Hirayama et al., 1995). Furthermore, by shifting GF mice into cages holding HFA mice, they can mimic the characteristics of human gastrointestinal flora, permitting them to be colonized by the human intestinal flora (Bajury et al., 2018). Swine, in addition to rodents, have been utilized to obtain human intestinal flora after a fecal slurry of a young 10-year-old child was inoculated into a piglet (Pang et al., 2007). The outcomes reveal that the microflora of the HFA piglet is more similar to that of the human donor than that of usual swine (Shen, 2010). HFA piglets were used in research on the prebiotic effect of short-chain fructooligosaccharide (scFOS). As a result, HFA pig models may be useful for evaluating prospective prebiotics.

14.4 HUMAN EXPERIMENTS

The definitive evaluation of prebiotic effects is to feed candidate materials or food to human subjects and measure microbial changes in feces. Because fecal matter is the only easily accessed part of gastrointestinal materials. The randomized, double-blind, placebo-controlled trial is the most often used clinical trial to examine potential prebiotics. Randomized means that individual group participants' variables, like age, gender, lifestyles, and socioeconomic disparities, are assigned at random. In alternative terms, randomization may lower the chances that the detected connection among treatment and result is due to a third component. A double-blind trial is one where both the therapist and client know which therapy the client receives, while a placebo trial is wherein an "inert" alternative of a substance with no active components that may alter the conclusion is being used (Misra, 2012). A beneficial risk analysis must be performed to explain the usage of placebo in such a clinical trial, as well as the client must be adequately aware of the risks associated. Human candidates are taken based on particular rejection and acceptance requirements which include being among the ages of 18–60, being in excellent overall health, and not being allergic to milk products for lactose-related compounds like galactooligosaccharides (GOS) (Depeint et al., 2008).

The avoidance of digestive diseases and not ingested either prebiotics, probiotics, or antibiotics in the two months prior to the experiment. Furthermore, donors must refrain from using the aforementioned substances throughout the research and not vary their regular meal or water consumption. Subjects are selected at random to eat various experimental materials. Then, fecal samples are taken on the same day that the materials are consumed as well as at the end of each intervention (Molis et al., 1995). Normally, multiple human trials are used to confirm stability and to eliminate incidental observations for only one prebiotic substance (Roberfroid et al., 2010; Bajury et al., 2018). The indirect technique of studying bacterial fermentation may be done by collecting exhaled breath at periodic durations and measuring the amount of gases, mostly hydrogen or methane gas, after a single oral intake of carbs in subjects (Christl et al., 1992). Anaerobic microorganisms metabolize hydrogen to create methane, which may contribute to a variety of disorders like irritable bowel syndrome (IBS) as well as persistent constipation (Pimentel, 2012). Short-chain fatty acids (SCFAs) in venous plasma or stool are an indirect approach to evaluate fermentation in human beings, such as acetate, propionate, as well as butyrate. Acetate is the most abundant SCFA in blended venous blood while propionate as well as butyrate ratios are very minimal as quantification is challenging unless a significant test quantity (Bajury et al., 2018).

Rodrigues et al. (2016) stated that the addition of 5 g/day of fructooligosaccharide (FOS) and trans-galacto-oligosaccharides (TGOS) may lead to beneficial impacts on the intestinal flora. previous research done by Letexier et al. (2003) on inulin showed strong effects in the short term after ingesting 10 grams/day of inulin in a high-carbohydrate, low-fat diet, with 5 g before breakfast and 5 g before the evening meal. Overdosing on the substrate resulted in prebiotic adverse effects, specifically gastrointestinal irritation due to gas generation. In conclusion, it was shown that if a substrate consumption of up to 20 g/day causes gas distension, the material cannot be classified as a "good" prebiotic. This is because the dose is too high, and the prebiotic impact is compromised since microbes other than desired species are involved in the fermentation (Probert & Gibson, 2002).

14.5 GAS PRODUCTION TECHNIQUE [SHORT-CHAIN FATTY ACIDS (SCFAS)]

Generally, the gut microbes get energy through fermented polysaccharides that were not absorbed in the upper GI system. Additional metabolic compounds include electron sink molecules such as lactate, pyruvate, ethanol, as well as succinate, and also gases like hydrogen, carbon dioxide, methane, and hydrogen sulfide (Sarbini and Rastall, 2011). It gives several advantages to the host's health by lowering luminal pH and inhibiting the development of intestinal pathogens (Blaut, 2002). Furthermore, SCFAs influence intestinal motility via colonic mucosa absorption as well as energy sources for the body. Certain protein-fermented components, including phenols, hydrogen sulfide, and ammonia, are harmful to humans (Dass et al., 2007; Bajury et al., 2018). Acetate is primarily metabolized in human muscle, kidney, heart, and brain tissues (Xiong et al.,

2004) as it is the most abundantly generated SCFA and it is most likely fermented through *Bacteroides, Lactobacillus, Clostridium,* and *Bifidobacterium* (Cook and Sellin, 1998). Similarly, propionate aids in the suppression of cholesterol production in the liver and the management of adipose tissue deposition (Sarbini and Rastall, 2011). Furthermore, butyrate is the main essential acid produced by *Clostridium* in intestinal metabolism. Butyrate has a role in the control of cell growth and differentiation (Williams et al., 2003).

SCFAs are often evaluated using gas chromatography (GC) and liquid chromatography (LC) such as high-performance liquid chromatography (HPLC), nuclear magnetic resonance (NMR), capillary electrophoresis (CE), and enzymatic detection. Figure 14.3 shows an overview of gas production (SCFAs) estimation. Since SCFAs are volatile and feces carry high numbers of microorganisms, it is crucial to store biological samples in proper conditions after collection to avoid sample degradation. Generally, samples are maintained at -80 °C, while numerous studies have effectively stored them at -20 °C as well. When evaluating SCFAs, two crucial pre-treatment procedures must be done: extraction and derivatization. Because SCFAs are quite hydrophilic, accurate extraction to hydrophobic organic solvents is more challenging. Therefore, sample acidification is often used to retain acids protonated and lower hydrophilicity, allowing for improved extraction (Zhao et al., 2006).

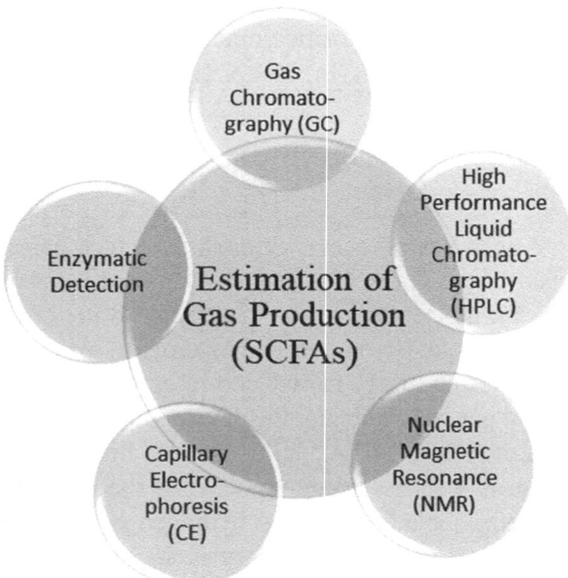

FIGURE 14.3 Overview of gas production (SCFAs) estimation.

The direct analysis of fatty acids (FA) using gas chromatography was first disclosed in 1952 by James & Martin. The GC concept is based on a carrier gas functioning as a mobile phase in which sample components are separated through differential interactions with the column stationary phase. Compounds' qualitative and quantitative evaluations are based on their retention period upon reaching a detector and its response, which depends on the chemical composition (Guthrie, 1988). A suitable pre-treatment chromatographic column as well as detector are required for a successful gas chromatographic determination of SCFAs. It should be noted that some gas chromatographic techniques can cause thermal deterioration and structural alteration of fatty acids during the methyl esterification procedure or they might easily destroy the sample, making re-analysis impossible (Lima & Abdalla, 2002). The GC is most often found in combination with flame-ionization detection (Richardson et al., 1989; Ribeiro et al., 2018) or mass spectrometric detection based on the derivation of non-volatile SCFA to enhance chromatographic performance and improve molecular weight (Bianchi et al., 2011).

High-performance liquid chromatography (HPLC) is another technique for analyzing SCFAs and a feasible replacement for GC. The most popular method is reverse-phase HPLC (RP-HPLC), in which the stationary solid phase (column) is hydrophobic and the mobile liquid phase is hydrophilic. The mobile phase containing analytes moves due to increased pressure, and the small stationary phase particles with a greater area allow for improved interaction with the analytes. The utilization of lower operating temperatures is the most significant benefit of the HPLC process over GC technology (Gutnikov, 1995). For effective SCFAs assessment, particular sets of Pretreatment, columns, operating conditions, and detectors must be adjusted, just like GC.

The Nuclear Magnetic Resonance (NMR) technique is dependent on the magnetic properties of certain atomic cores that receive and release electromagnetic waves in the presence of a magnetic field. The generated energy emerges at certain resonant frequencies that are determined by the magnetic field and magnetic properties of individual isotopes. Isotope cores ^{1}H (Saric et al., 2008; Ndagijimana et al., 2009; Jacobs et al., 2008; Le Gall et al., 2011) and ^{13}C (Monleon et al., 2009) were used for SCFAs analysis in stool samples. NMR is a non-destructive, measurable, reproducible, and objective approach. It has mostly been applied for chemical component and concentration tracking of several metabolites in a complicated biological sample. It is a complete and sufficiently sensitive technique with detection limits of metabolite concentrations reaching the μM range (Lindon et al., 2000; Smolinska et al., 2012). It requires minimum sample preparation. Furthermore, pretreatment technology requires improvement to allow for the extraction of a wide range of metabolites from a complicated biological sample. Moreover, deuterated solvents should be used in NMR studies to adjust the field strength and prevent the large solvent absorbance that otherwise would damage the ^{1}H-NMR spectra (Hoye & Koltun, 1998). It might result in a costly, time-consuming process with considerable analyte wastage, particularly when dealing with large volumes of materials on a regular basis. At the same time, the technique must be non-destructive and create only small material variations (Jacobs et al., 2008). Indeed, variances and contradictions in NMR findings indicate that the probable reason for poor sensitivity (less than that reported with GC-MS or LC-MS) may be related to differing pretreatment techniques and data gathering methods (Smolinska et al., 2012).

Capillary electrophoresis (CE) is a method that uses an electrical field to separate charged particles in a capillary. CE has been recognized as a useful analytical method for the identification of many organic molecules such as SCFAs over the last 30 years. CE is based on the electrophoretic mobility of ions across the gel, which is affected by the particle's molecular weight, structure, and charge (Whatley, 2001). CE is particularly useful for the separation of water-soluble substances. The quickness and low sample preparation technique of CE make it ideal for routine evaluation. However, CE has also been effectively used in the quantitative measurement of SCFAs in a variety of biological samples, such as fecal samples (Garcia et al., 2008). Unfortunately, it has certain drawbacks, including limited repeatability and reproducibility and the requirement for high analyte quantity. The majority of these limits are due to the small volume of injected material (Toyo'oka, 1995). Many difficulties in CE detection have been overcome by using a combined detection technique with MS, increasing the technique's efficiency in identifying small compounds in the μM to nM range (Primec et al., 2017).

Enzymatic detection, which depends on spectrophotometric assessment of enzymatic compounds derived using SCFAs as substrates, is among the various SCFAs measurement techniques in stools. In their metabolic processes, enzymes favor stereoselective isomers and may therefore distinguish among optical isomers. Lactate may be present in both D- and L-forms, and various research groups have confirmed D- and L-lactate enzymatically (Mohan et al., 2008; Thompson et al., 2011; Bakke et al., 2005; Reimer et al., 2012; Gostner, 2006). Even though the technique is quick, limitations include the possibility of cross-enzymatic interactions with other endogenous substances and the occurrence of enzymatic blockers in analyzed materials. Furthermore, in aspects of lactate analysis, it is vital to remember that total lactate has been effectively measured by GC (Woodmansey et al., 2004; Macfarlane et al., 2013) as well as HPLC (Huda-Faujan et al., 2010).

14.6 GENOMIC TECHNIQUES

Microbial species reported in the intestinal bacteria that ferment non-digestible carbohydrates are largely from the genera *Bacteroides*, *Bifidobacterium*, *Ruminococcus*, *Lactobacillus*, *Clostridium*, and *Eubacterium* (Roberfroid et al., 2010). The first plating method to describe intestinal flora was a culture-dependent method in the 1880s, in which the gut microbiota was isolated from human fecal mass and cultured in selective media to obtain their biochemical as well as physiological characteristics based on nutrient utilization, inhibitory substance resistance, enzyme production, and motility recognized by Gram staining (Hiergeist et al., 2015). The methodology had switched to DNA-based culture-independent techniques in the 1990s, using the DNA of the intestinal microbiota extracted from *in vitro* fermentation. Since then, quantitative polymerase chain reactions (qPCR), sequencing techniques, fingerprinting techniques, DNA microarrays, Denaturing gradient gel electrophoresis (DGGE), Terminal restriction fragment length polymorphism (TRFLP), Next Generation Sequencing (NGS), and fluorescent in situ hybridization (FISH) have all been used to identify and quantify particular DNA or RNA (Figure 14.4).

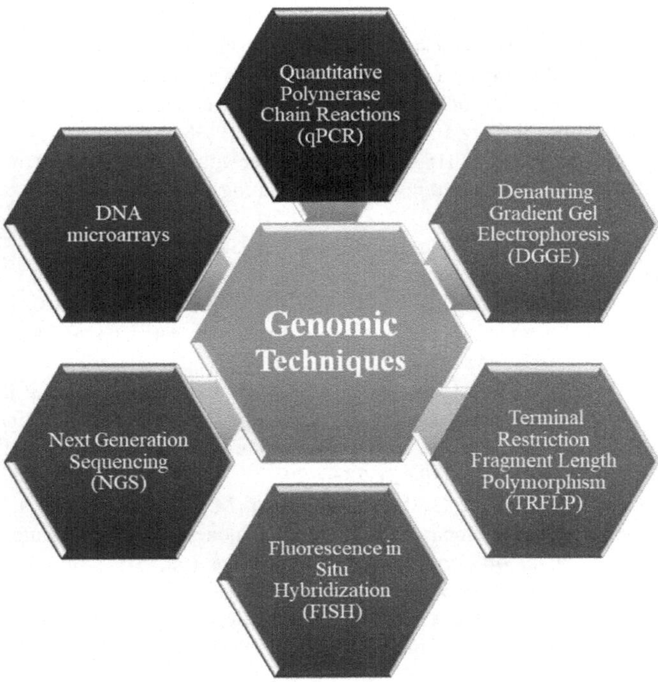

FIGURE 14.4 Outline of genomic techniques.

KEYWORDS

- **bacteria**
- **evaluation**
- **gas production**
- **isotopes**
- **metabolism**
- **mice**
- **prebiotics**
- **research methods**
- **simulation**
- **techniques**

REFERENCES

Aguirre, M., Jonkers, D. M., Troost, F. J., Roeselers, G., & Venema, K. (2014). In vitro characterization of the impact of different substrates on metabolite production, energy extraction, and composition of gut microbiota from lean and obese subjects. *PLoS One, 9*(11), e113864. https://doi.org/10.1371/journal.pone.0113864.

Al-Asmakh, M., & Zadjali, F. (2015). Use of germ-free animal models in microbiota-related research. *Journal of Microbiology and Biotechnology, 25*(10), 1583–1588. https://doi.org/10.4014/jmb.1508.08008.

Bajury, D. M., Nashri, S. M., King Jie Hung, P., & Sarbini, S. R. (2018). Evaluation of potential prebiotics: A review. *Food Reviews International, 34*(7), 639–664.

Bajury, D. M., Nashri, S. M., King Jie Hung, P., & Sarbini, S. R. (2018). Evaluation of potential prebiotics: A review. *Food Reviews International, 34*(7), 639–664. https://doi.org/10.1080/87559129.2018.1452691.

Bakker-Zierikzee, A. M., Alles, M. S., Knol, J., Kok, F. J., Tolboom, J. J., & Bindels, J. G. (2005). Effects of infant formula containing a mixture of galacto- and fructo-oligosaccharides or viable *Bifidobacterium animalis* on the intestinal microflora during the first 4 months of life. *British Journal of Nutrition, 94*(5), 783–790.

Barroso, E., Cueva, C., Peláez, C., Martínez-Cuesta, M. C., & Requena, T. (2015). The computer-controlled multicompartmental dynamic model of the gastrointestinal system SIMGI. In *The Impact of Food Bioactive on Health* (pp. 319). Springer. https://doi.org/10.1007/978-3-319-15903-6_18.

Berner, A. Z., Fuentes, S., Dostal, A., Payne, A. N., Gutierrez, P. V., Chassard, C., & Lacroix, C. (2013). Novel Polyfermentor intestinal model (PolyFermS) for controlled ecological studies: Validation and effect of pH. *PLOS ONE, 8*(10), e77772. https://doi.org/10.1371/journal.pone.0077772.

Bhattarai, Y., & Kashyap, P. C. (2016). Germ-free mice model for studying host–microbial interactions. In *Mouse Models for Drug Discovery* (pp. 123–135). Springer. https://doi.org/10.1007/978-1-4939-2901-7_8.

Bianchi, F., Dall'Asta, M., Del Rio, D., Mangia, A., Musci, M., & Scazzina, F. (2011). Development of a headspace solid-phase microextraction gas chromatography–mass spectrometric method for the determination of short-chain fatty acids from intestinal fermentation. *Food Chemistry, 129*(1), 200–205.

Blanquet-Diot, S., Soufi, M., Rambeau, M., Rock, E., & Alric, M. (2009). Digestive stability of xanthophylls exceeds that of carotenes as studied in a dynamic in vitro gastrointestinal system. *The Journal of Nutrition, 139*(5), 876–883.

Blaut, M. (2002). Relationship of prebiotics and food to intestinal microflora. *European Journal of Nutrition, 41*(1), i11–i16.

Christl, S. U., Murgatroyd, P. R., Gibson, G. R., & Cummings, J. H. (1992). Production, metabolism, and excretion of hydrogen in the large intestine. *Gastroenterology, 102*(4), 1269–1277.

Cieplak, T., Wiese, M., Nielsen, S., Van de Wiele, T., van den Berg, F., & Nielsen, D. S. (2018). The smallest intestine (TSI)—A low volume in vitro model of the small intestine with increased throughput. *FEMS Microbiology Letters, 365*(21), fny231. https://doi.org/10.1093/femsle/fny231.

Coles, L. T., Moughan, P. J., & Darragh, A. J. (2005). In vitro digestion and fermentation methods, including gas production techniques, as applied to nutritive evaluation of foods in the hindgut of humans and other simple-stomached animals. *Animal Feed Science and Technology, 123*, 421–444.

Cook, S. I., & Sellin, J. H. (1998). Short chain fatty acids in health and disease. *Alimentary Pharmacology & Therapeutics, 12*(6), 499–507.

Cueva, C., Gil-Sánchez, I., Tamargo, A., Miralles, B., Crespo, J., Bartolomé, B., & Moreno-Arribas, M. V. (2019). Gastrointestinal digestion of food-use silver nanoparticles in the dynamic SIMulator of the Gastrointestinal tract (simgi®): Impact on human gut microbiota. *Food and Chemical Toxicology, 132*, 110657.

Cueva, C., Jiménez-Girón, A., Muñoz-González, I., Esteban-Fernández, A., Gil-Sánchez, I., Dueñas, M., & Moreno-Arribas, M. V. (2015). Application of a new dynamic gastrointestinal simulator (SIMGI) to study the impact of red wine in colonic metabolism. *Food Research International, 72*, 149–159. https://doi.org/10.1016/j.foodres.2015.03.007.

Cueva, C., Jiménez-Girón, A., Muñoz-González, I., Esteban-Fernández, A., Gil-Sánchez, I., Dueñas, M., & Moreno-Arribas, M. V. (2015). Application of a new dynamic gastrointestinal simulator (SIMGI) to study the impact of red wine in colonic metabolism. *Food Research International, 72*, 149–159.

Dass, N. B., John, A. K., Bassil, A. K., Crumbley, C. W., Shehee, W. R., Maurio, F. P., & Sanger, G. J. (2007). The relationship between the effects of short-chain fatty acids on intestinal motility in vitro and GPR43 receptor activation. *Neurogastroenterology & Motility, 19*(1), 66–74.

de Souza Aquino, J., Batista, K. S., Menezes, F. N. D. D., Lins, P. P., de Sousa Gomes, J. A., & da Silva, L. A. (2017). Models to evaluate the prebiotic potential of foods. In *Functional Food: Improve Health Through Adequate Food* (pp. 235–256).

Depeint, F., Tzortzis, G., Vulevic, J., I'anson, K., & Gibson, G. R. (2008). Prebiotic evaluation of a novel galactooligosaccharide mixture produced by the enzymatic activity of *Bifidobacterium bifidum* NCIMB 41171, in healthy humans: A randomized, double-blind, crossover, placebo-controlled intervention study. *The American Journal of Clinical Nutrition, 87*(3), 785–791.

Doo, E. H., Chassard, C., Schwab, C., & Lacroix, C. (2018). Dietary nucleosides and yeast extracts on composition and metabolic activity of infant gut microbiota in PolyFermS colonic fermentation models. In *Impacts of Dietary Nucleosides, Nucleotides and Yeast Extract on Human and Animal Gut Microbial Composition and Metabolic Activity Studied with In Vitro Fermentation Models* (Vol. 93, pp. 73–84). Springer. https://doi.org/10.1007/978-3-319-61255-6_7.

Dostal, A., Fehlbaum, S., Chassard, C., Zimmermann, M. B., & Lacroix, C. (2013). Low iron availability in continuous in vitro colonic fermentations induces strong dysbiosis of the child gut microbial consortium and a decrease in main metabolites. *FEMS Microbiology Ecology, 83*(1), 161–175. https://doi.org/10.1111/1574-6941.12024.

El Oufir, L., Barry, J. L., Flourie, B., Cherbut, C., Cloarec, D., Bornet, F., & Galmiche, J. P. (2000). Relationships between transit time in man and in vitro fermentation of dietary fiber by fecal bacteria. *European Journal of Clinical Nutrition, 54*(8), 603–609.

Fiebiger, U., Bereswill, S., & Heimesaat, M. M. (2016). Dissecting the interplay between intestinal microbiota and host immunity in health and disease: Lessons learned from germfree and gnotobiotic animal models. *European Journal of Microbiology and Immunology, 6*(4), 253–271. https://doi.org/10.1556/1886.2016.00035.

Fritz, J. V., Desai, M. S., Shah, P., Schneider, J. G., & Wilmes, P. (2013). From meta-omics to causality: Experimental models for human microbiome research. *Microbiome, 1*(1), 1–15. https://doi.org/10.1186/2049-2618-1-14.

Gagnon, M., Zihler, A., Chassard, C., & Lacroix, C. (2011). Ecology of probiotics and enteric protection. In *Probiotic Bacteria and Enteric Infections* (pp. 65–85). Springer, Dordrecht.

Garcia, A., Olmo, B., Lopez-Gonzalvez, A., Cornejo, L., Rupérez, F. J., & Barbas, C. (2008). Capillary electrophoresis for short chain organic acids in faeces: Reference values in a Mediterranean elderly population. *Journal of Pharmaceutical and Biomedical Analysis, 46*(2), 356–361.

Gibson, G. R., & Fuller, R. (2000). Aspects of in vitro and in vivo research approaches directed toward identifying probiotics and prebiotics for human use. *The Journal of Nutrition, 130*(2), 391S–395S.

Gibson, G. R., Cummings, J. H., & Macfarlane, G. T. (1988). Use of a three-stage continuous culture system to study the effect of mucin on dissimilatory sulfate reduction and methanogenesis by mixed populations of human gut bacteria. *Applied and Environmental Microbiology, 54*(11), 2750–2755.

Gil-Sánchez, I., Cueva, C., Tamargo, A., Quintela, J. C., de la Fuente, E., Walker, A. W., & Bartolomé, B. (2020). Application of the dynamic gastrointestinal simulator (simgi®) to assess the impact of probiotic supplementation in the metabolism of grape polyphenols. *Food Research International, 129*, 108790.

Giuliani, C., Marzorati, M., Daghio, M., Franzetti, A., Innocenti, M., Van de Wiele, T., & Mulinacci, N. (2019). Effects of olive and pomegranate by-products on human microbiota: A study using the SHIME® in vitro simulator. *Molecules, 24*(20), 3791. https://doi.org/10.3390/molecules24203791.

Giuliani, C., Marzorati, M., Innocenti, M., Vilchez-Vargas, R., Vital, M., Pieper, D. H., & Mulinacci, N. (2016). Dietary supplement based on stilbenes: A focus on gut microbial metabolism by the in vitro simulator M-SHIME®. *Food & Function, 7*(11), 4564–4575. https://doi.org/10.1039/C6FO00895F.

Goodman, A. L., Kallstrom, G., Faith, J. J., Reyes, A., Moore, A., Dantas, G., & Gordon, J. I. (2011). Extensive personal human gut microbiota culture collections characterized and manipulated in gnotobiotic mice. *Proceedings of the National Academy of Sciences, 108*(15), 6252–6257. https://doi.org/10.1073/pnas.1017713108.

Gostner, A., Blaut, M., Schäffer, V., Kozianowski, G., Theis, S., Klingeberg, M., & Scheppach, W. (2006). Effect of isomalt consumption on faecal microflora and colonic metabolism in healthy volunteers. *British Journal of Nutrition, 95*(1), 40–50.

Grover, M., & Kashyap, P. C. (2014). Germ-free mice as a model to study effect of gut microbiota on host physiology. *Neurogastroenterology & Motility, 26*(6), 745–748. https://doi.org/10.1111/nmo.12372.

Guthrie, F. A. (1988). Introduction of instrumental analysis (Braun, Robert D.).

Gutnikov, G. (1995). Fatty acid profiles of lipid samples. *Journal of Chromatography B: Biomedical Sciences and Applications, 671*(1–2), 71–89.

Hiergeist, A., Gläsner, J., Reischl, U., & Gessner, A. (2015). Analyses of intestinal microbiota: Culture versus sequencing. *ILAR Journal, 56*(2), 228–240.

Hirayama, K. (2010). Human flora-associated animals as a model for studying probiotics and prebiotics. In *Bioactive Foods in Promoting Health* (pp. 531–540). Academic Press. https://doi.org/10.1016/B978-0-12-374628-3.00052-3.

Hirayama, K., Miyaji, K., Kawamura, S., Itoh, K., Takahashi, E., & Mitsuoka, T. (1995). Development of intestinal flora of human-flora-associated (HFA) mice in the intestine of their offspring. *Experimental Animals, 44*(3), 219–222. https://doi.org/10.1538/expanim1995.44.219.

Hoye, T. R., & Koltun, D. O. (1998). An NMR strategy for determination of configuration of remote stereogenic centers: 3-methylcarboxylic acids. *Journal of the American Chemical Society, 120*(19), 4638–4643.

Huda-Faujan, N., Abdulamir, A. S., Fatimah, A. B., Anas, O. M., Shuhaimi, M., Yazid, A. M., & Loong, Y. Y. (2010). The impact of the level of the intestinal short chain fatty acids in inflammatory bowel disease patients versus healthy subjects. *The Open Biochemistry Journal, 4*, 53.

Hur, S. J., Lim, B. O., Decker, E. A., & McClements, D. J. (2011). In vitro human digestion models for food applications. *Food Chemistry, 125*(1), 1–12.

Jacobs, D. M., Deltimple, N., van Velzen, E., van Dorsten, F. A., Bingham, M., Vaughan, E. E., & van Duynhoven, J. (2008). 1H NMR metabolite profiling of feces as a tool to assess the impact of nutrition on the human microbiome. *NMR in Biomedicine: An International Journal Devoted to the Development and Application of Magnetic Resonance In vivo, 21*(6), 615–626.

James, A. T., & Martin, A. J. (1952). Gas-liquid partition chromatography: The separation and micro-estimation of volatile fatty acids from formic acid to dodecanoic acid. *Biochemical Journal, 50*(5), 679.

Jiminez, J. A., Uwiera, T. C., Inglis, G. D., & Uwiera, R. R. (2015). Animal models to study acute and chronic intestinal inflammation in mammals. *Gut Pathogens, 7*(1), 1–31. https://doi.org/10.1186/s13099-015-0106-x.

Kennedy, E. A., King, K. Y., & Baldridge, M. T. (2018). Mouse microbiota models: Comparing germ-free mice and antibiotics treatment as tools for modifying gut bacteria. *Frontiers in Physiology, 9*, 1534. https://doi.org/10.3389/fphys.2018.01534.

Kubelkova, K., Benuchova, M., Kozakova, H., Sinkora, M., Krocova, Z., Pejchal, J., & Macela, A. (2016). Gnotobiotic mouse model's contribution to understanding host–pathogen interactions. *Cellular and Molecular Life Sciences, 73*(20), 3961–3969. https://doi.org/10.1007/s00018-016-2220-1.

Kuo, S. M., Merhige, P. M., & Hagey, L. R. (2013). The effect of dietary prebiotics and probiotics on body weight, large intestine indices, and fecal bile acid profile in wild type and IL10−/− mice. *PLOS ONE, 8*(3), e60270. https://doi.org/10.1371/journal.pone.0060270.

Lattermann, C., & Büchs, J. (2015). Microscale and miniscale fermentation and screening. *Current Opinion in Biotechnology, 35*, 1–6. https://doi.org/10.1016/j.copbio.2015.07.010.

Le Blay, G., Chassard, C., Baltzer, S., & Lacroix, C. (2010). Set up of a new in vitro model to study dietary fructans fermentation in formula-fed babies. *British Journal of Nutrition, 103*(3), 403–411. https://doi.org/10.1017/S0007114509992149.

Le Gall, G., Noor, S. O., Ridgway, K., Scovell, L., Jamieson, C., Johnson, I. T., & Narbad, A. (2011). Metabolomics of fecal extracts detects altered metabolic activity of gut microbiota in ulcerative colitis and irritable bowel syndrome. *Journal of Proteome Research, 10*(9), 4208–4218.

Letexier, D., Diraison, F., & Beylot, M. (2003). Addition of inulin to a moderately high-carbohydrate diet reduces hepatic lipogenesis and plasma triacylglycerol concentrations in humans. *The American Journal of Clinical Nutrition, 77*(3), 559–564.

Lima, E. S., & Abdalla, D. S. P. (2002). High-performance liquid chromatography of fatty acids in biological samples. *Analytica Chimica Acta, 465*(1–2), 81–91.

Lindon, J. C., Nicholson, J. K., Holmes, E., & Everett, J. R. (2000). Metabonomics: Metabolic processes studied by NMR spectroscopy of biofluids. *Concepts in Magnetic Resonance: An Educational Journal, 12*(5), 289–320.

Liu, Y., Gibson, G. R., & Walton, G. E. (2016). An in vitro approach to study effects of prebiotics and probiotics on the faecal microbiota and selected immune parameters relevant to the elderly. *PLoS One, 11*(9), e0162604.

Maathuis, A. J., van den Heuvel, E. G., Schoterman, M. H., & Venema, K. (2012). Galacto-oligosaccharides have prebiotic activity in a dynamic in vitro colon model using a 13C-labeling technique. *The Journal of Nutrition, 142*(7), 1205–1212.

Macfarlane, G. T., & Macfarlane, S. (2007). Models for intestinal fermentation: Association between food components, delivery systems, bioavailability and functional interactions in the gut. *Current Opinion in Biotechnology, 18*(2), 156–162. https://doi.org/10.1016/j.copbio.2007.03.001.

Macfarlane, S., & Dillon, J. F. (2007). Microbial biofilms in the human gastrointestinal tract. *Journal of Applied Microbiology, 102*(5), 1187–1196. https://doi.org/10.1111/j.1365-2672.2007.03353.x.

Macfarlane, S., Cleary, S., Bahrami, B., Reynolds, N., & Macfarlane, G. T. (2013). Symbiotic consumption changes the metabolism and composition of the gut microbiota in older people and modifies inflammatory processes: A randomized, double-blind, placebo-controlled crossover study. *Alimentary Pharmacology & Therapeutics, 38*(7), 804–816.

Macfarlane, S., Woodmansey, E. J., & Macfarlane, G. T. (2005). Colonization of mucin by human intestinal bacteria and establishment of biofilm communities in a two-stage continuous culture system. *Applied and Environmental Microbiology, 71*(11), 7483–7492. https://doi.org/10.1128/AEM.71.11.7483-7492.2005.

Marín, L., Miguélez, E. M., Villar, C. J., & Lombó, F. (2015). Bioavailability of dietary polyphenols and gut microbiota metabolism: Antimicrobial properties. *BioMed Research International, 2015*.

Minekus, M., Alminger, M., Alvito, P., Ballance, S., Bohn, T. O. R. S. T. E. N., Bourlieu, C., & Brodkorb, A. (2014). A standardized static in vitro digestion method suitable for food—An international consensus. *Food & Function, 5*(6), 1113–1124. https://doi.org/10.1039/C4FO00186K.

Minekus, M., et al. (1999). A computer-controlled system to simulate conditions of the large intestine with peristaltic mixing, water absorption, and absorption of fermentation products. *Applied Microbiology and Biotechnology, 53*(1), 108–114.

Minekus, M., Marteau, P., Havenaar, R., & Veld, J. H. H. I. T. (1995). A multicompartmental dynamic computer-controlled model simulating the stomach and small intestine. *Alternatives to Laboratory Animals, 23*(2), 197–209.

Misra, S. (2012). Randomized double blind placebo control studies, the "Gold Standard" in intervention-based studies. *Indian Journal of Sexually Transmitted Diseases and AIDS, 33*(2), 131.

Mohan, R., Koebnick, C., Schildt, J., Mueller, M., Radke, M., & Blaut, M. (2008). Effects of *Bifidobacterium lactis* Bb12 supplementation on body weight, fecal pH, acetate, lactate, calprotectin, and IgA in preterm infants. *Pediatric Research, 64*(4), 418–422.

Molis, C., Flourie, B., Ouarne, F., Gailing, M. F., Lartigue, S., Guibert, A., & Galmiche, J. P. (1996). Digestion, excretion, and energy value of fructooligosaccharides in healthy humans. *The American Journal of Clinical Nutrition, 64*(3), 324–328.

Molly, K., Woestyne, M. V., & Verstraete, W. (1993). Development of a 5-step multi-chamber reactor as a simulation of the human intestinal microbial ecosystem. *Applied Microbiology and Biotechnology, 39*(2), 254–258. https://doi.org/10.1007/BF00164190.

Molly, K., Woestyne, M. V., Smet, I. D., & Verstraete, W. (1994). Validation of the simulator of the human intestinal microbial ecosystem (SHIME) reactor using microorganism-associated activities. *Microbial Ecology in Health and Disease, 7*(4), 191–200. https://doi.org/10.3109/08910609409149166.

Monleon, D., Morales, J. M., Barrasa, A., Lopez, J. A., Vazquez, C., & Celda, B. (2009). Metabolite profiling of fecal water extracts from human colorectal cancer. *NMR in Biomedicine: An International Journal Devoted to the Development and Application of Magnetic Resonance In vivo, 22*(3), 342–348.

Nava, G. M., Friedrichsen, H. J., & Stappenbeck, T. S. (2011). Spatial organization of intestinal microbiota in the mouse ascending colon. *The ISME Journal, 5*(4), 627–638. https://doi.org/10.1038/ismej.2010.177.

Ndagijimana, M., Laghi, L., Vitali, B., Placucci, G., Brigidi, P., & Guerzoni, M. E. (2009). Effect of a symbiotic food consumption on human gut metabolic profiles evaluated by 1H nuclear magnetic resonance spectroscopy. *International Journal of Food Microbiology, 134*(1–2), 147–153.

Nguyen, T. L. A., Vieira-Silva, S., Liston, A., & Raes, J. (2015). How informative is the mouse for human gut microbiota research? *Disease Models & Mechanisms, 8*(1), 1–16. https://doi.org/10.1242/dmm.018122.

Nissen, L., Casciano, F., & Gianotti, A. (2020). Intestinal fermentation in vitro models to study food-induced gut microbiota shift: An updated review. *FEMS Microbiology Letters, 367*(12), fnaa097. https://doi.org/10.1093/femsle/fnaa097.

O'Donnell, M. M., Rea, M. C., Shanahan, F., & Ross, R. P. (2018). The use of a mini-bioreactor fermentation system as a reproducible, high-throughput ex vivo batch model of the distal colon. *Frontiers in Microbiology, 9*, 1844. https://doi.org/10.3389/fmicb.2018.01844

Pang, X., Hua, X., Yang, Q., Ding, D., Che, C., Cui, L., & Zhao, L. (2007). Inter-species transplantation of gut microbiota from human to pigs. *The ISME Journal, 1*(2), 156–162.

Parra-Matadamas, A., Mayorga-Reyes, L., & Pérez-Chabela, M. L. (2015). In vitro fermentation of agro-industrial by-products: Grapefruit albedo and peel, cactus pear peel and pineapple peel by lactic acid bacteria. *International Food Research Journal, 22*(2), 859–865.

Payne, A. N., Chassard, C., Banz, Y., & Lacroix, C. (2012). The composition and metabolic activity of child gut microbiota demonstrate differential adaptation to varied nutrient loads in an in vitro model of colonic fermentation. *FEMS Microbiology Ecology, 80*(3), 608–623. https://doi.org/10.1111/j.1574-6941.2012.01323.x.

Payne, A. N., Chassard, C., Banz, Y., & Lacroix, C. (2012). The composition and metabolic activity of child gut microbiota demonstrate differential adaptation to varied nutrient loads in an in vitro model of colonic fermentation. *FEMS Microbiology Ecology, 80*(3), 608–623. https://doi.org/10.1111/j.1574-6941.2012.01323.x.

Payne, A. N., Zihler, A., Chassard, C., & Lacroix, C. (2012). Advances and perspectives in in vitro human gut fermentation modeling. *Trends in Biotechnology, 30*(1), 17–25.

Pereira-Caro, G., Oliver, C. M., Weerakkody, R., Singh, T., Conlon, M., Borges, G., & Augustin, M. A. (2015). Chronic administration of a microencapsulated probiotic enhances the bioavailability of orange juice flavanones in humans. *Free Radical Biology and Medicine, 84*, 206–214.

Pimentel, M., Gunsalus, R. P., Rao, S. S., & Zhang, H. (2012). Methanogens in human health and disease. *The American Journal of Gastroenterology Supplements, 1*(1), 28.

Poeker, S. A., Geirnaert, A., Berchtold, L., Greppi, A., Krych, L., Steinert, R. E., & Lacroix, C. (2018). Understanding the prebiotic potential of different dietary fibers using an in vitro continuous adult fermentation model (PolyFermS). *Scientific Reports, 8*(1), 1–12. https://doi.org/10.1038/s41598-018-28331-8.

Primec, M., Mičetić-Turk, D., & Langerholc, T. (2017). Analysis of short-chain fatty acids in human feces: A scoping review. *Analytical Biochemistry, 526*, 9–21.

Probert, H. M., & Gibson, G. R. (2002). Investigating the prebiotic and gas-generating effects of selected carbohydrates on the human colonic microflora. *Letters in Applied Microbiology, 35*(6), 473–480.

Qv, L., Yang, Z., Yao, M., Mao, S., Li, Y., Zhang, J., & Li, L. (2020). Methods for establishment and maintenance of germ-free rat models. *Frontiers in Microbiology, 11*, 1148. https://doi.org/10.3389/fmicb.2020.01148.

Rehman, A., Heinsen, F. A., Koenen, M. E., Venema, K., Knecht, H., Hellmig, S., & Ott, S. J. (2012). Effects of probiotics and antibiotics on the intestinal homeostasis in a computer-controlled model of the large intestine. *BMC Microbiology, 12*(1), 1–10. https://doi.org/10.1186/1471-2180-12-62.

Reimer, R. A., Maathuis, A. J., Venema, K., Lyon, M. R., Gahler, R. J., & Wood, S. (2014). Effect of the novel polysaccharide PolyGlycopleX® on short-chain fatty acid production in a computer-controlled in vitro model of the human large intestine. *Nutrients, 6*(3), 1115–1127.

Reimer, R. A., Pelletier, X., Carabin, I. G., Lyon, M. R., Gahler, R. J., & Wood, S. (2012). Faecal short chain fatty acids in healthy subjects participating in a randomized controlled trial examining a soluble highly viscous polysaccharide versus control. *Journal of Human Nutrition and Dietetics, 25*(4), 373–377.

Ribeiro, W. R., Vinolo, M. A. R., Calixto, L. A., & Ferreira, C. M. (2018). Use of gas chromatography to quantify short chain fatty acids in the serum, colonic luminal content and feces of mice. *Bio-protocol, 8*(22), e3089–e3089.

Richardson, A. J., Calder, A. G., Stewart, C. S., & Smith, A. (1989). Simultaneous determination of volatile and non-volatile acidic fermentation products of anaerobes by capillary gas chromatography. *Letters in Applied Microbiology, 9*(1), 5–8.

Roberfroid, M., Gibson, G. R., Hoyles, L., McCartney, A. L., Rastall, R., Rowland, I., & Meheust, A. (2010). Prebiotic effects: Metabolic and health benefits. *British Journal of Nutrition, 104*(S2), S1–S63.

Roberfroid, M., Gibson, G. R., Hoyles, L., McCartney, A. L., Rastall, R., Rowland, I., & Meheust, A. (2010). Prebiotic effects: Metabolic and health benefits. *British Journal of Nutrition, 104*(S2), S1–S63.

Rodrigues, D., Walton, G., Sousa, S., Rocha-Santos, T. A., Duarte, A. C., Freitas, A. C., & Gomes, A. M. (2016). In vitro fermentation and prebiotic potential of selected extracts from seaweeds and mushrooms. *LWT, 73*, 131–139.

Roupar, D., Berni, P., Martins, J. T., Caetano, A. C., Teixeira, J. A., & Nobre, C. (2021). Bioengineering approaches to simulate human colon microbiome ecosystem. *Trends in Food Science & Technology, 115*, 512–529. https://doi.org/10.1016/j.tifs.2021.07.001.

Roupar, D., Berni, P., Martins, J. T., Caetano, A. C., Teixeira, J. A., & Nobre, C. (2021). Bioengineering approaches to simulate human colon microbiome ecosystem. *Trends in Food Science & Technology, 112, 808–822.*

Rumney, C. J., & Rowland, I. R. (1992). In vivo and in vitro models of the human colonic flora. *Critical Reviews in Food Science & Nutrition, 31*(4), 299–331.

Rycroft, C. E., Jones, M. R., Gibson, G. R., & Rastall, R. A. (2001). A comparative in vitro evaluation of the fermentation properties of prebiotic oligosaccharides. *Journal of Applied Microbiology, 91*(5), 878–887.

Sarbini, S. R., & Rastall, R. A. (2011). Prebiotics: Metabolism, structure, and function. *Functional Food Reviews, 3*(3), 93–106.

Saric, J., Wang, Y., Li, J., Coen, M., Utzinger, J., Marchesi, J. R., & Holmes, E. (2008). Species variation in the fecal metabolome gives insight into differential gastrointestinal function. *Journal of Proteome Research, 7*(01), 352–360.

Shen, X. J., Rawls, J. F., Randall, T. A., Burcall, L., Mpande, C., Jenkins, N., & Keku, T. O. (2010). Molecular characterization of mucosal adherent bacteria and associations with colorectal adenomas. *Gut Microbes, 1*(3), 138–147. https://doi.org/10.4161/gmic.1.3.12197.

Shen, Y. T. (2010). Primate models for cardiovascular drug research and development. *Current Opinion in Investigational Drugs (London, England: 2000), 11*(9), 1025–1029.

Smith, K., McCoy, K. D., & Macpherson, A. J. (2007). Use of axenic animals in studying the adaptation of mammals to their commensal intestinal microbiota. In *Seminars in Immunology* (Vol. 19, No. 2, pp. 59–69). Academic Press. https://doi.org/10.1016/j.smim.2007.02.005.

Smolinska, A., Blanchet, L., Buydens, L. M., & Wijmenga, S. S. (2012). NMR and pattern recognition methods in metabolomics: From data acquisition to biomarker discovery: A review. *Analytica Chimica Acta, 750*, 82–97.

Smolinska, A., Blanchet, L., Buydens, L. M., & Wijmenga, S. S. (2012). NMR and pattern recognition methods in metabolomics: From data acquisition to biomarker discovery: A review. *Analytica Chimica Acta, 750*, 82–97.

Souliman, S., Beyssac, E., Cardot, J. M., Denis, S., & Alric, M. (2007). Investigation of the biopharmaceutical behavior of theophylline hydrophilic matrix tablets using USP methods and an artificial digestive system. *Drug Development and Industrial Pharmacy, 33*(4), 475–483.

Souliman, S., Blanquet, S., Beyssac, E., & Cardot, J. M. (2006). A level A in vitro/in vivo correlation in fasted and fed states using different methods: Applied to solid immediate release oral dosage forms. *European Journal of Pharmaceutical Sciences, 27*(1), 72–79.

Tamargo, A., Cueva, C., Laguna, L., Moreno-Arribas, M. V., & Muñoz, L. A. (2018). Understanding the impact of chia seed mucilage on human gut microbiota by using the dynamic gastrointestinal model SIMGI®. *Journal of Functional Foods, 50*, 104–111. https://doi.org/10.1016/j.jff.2018.09.021.

Terpend, K., Possemiers, S., Daguet, D., & Marzorati, M. (2013). Different colonic fermentation behaviour of FOS and AG. *Environmental Microbiology Reports, 5*(4), 595–603. https://doi.org/10.1111/1758-2229.12039.

Thompson-Chagoyan, O. C., Fallani, M., Maldonado, J., Vieites, J. M., Khanna, S., Edwards, C., & Gil, A. (2011). Faecal microbiota and short-chain fatty acid levels in faeces from infants with Cow's milk protein allergy. *International Archives of Allergy and Immunology, 156*(3), 325–332.

Thuenemann, E. C., Mandalari, G., Rich, G. T., & Faulks, R. M. (2015). The impact of food bioactive on gut health: In vitro and ex vivo models.

Toyo'oka, T. (1995). Use of derivatization to improve the chromatographic properties and detection selectivity of physiologically important carboxylic acids. *Journal of Chromatography B: Biomedical Sciences and Applications, 671*(1–2), 91–112.

Truchado, P., Hernandez-Sanabria, E., Salden, B. N., Van den Abbeele, P., Vilchez-Vargas, R., Jauregui, R., & Van de Wiele, T. (2017). Long chain arabinoxylans shift the mucosa-associated microbiota in the proximal colon of the simulator of the human intestinal microbial ecosystem (M-SHIME). *Journal of Functional Foods, 32*, 226–237. https://doi.org/10.1016/j.jff.2017.03.037.

Umesaki, Y. (2014). Use of gnotobiotic mice to identify and characterize key microbes responsible for the development of the intestinal immune system. *Proceedings of the Japan Academy, Series B, 90*(9), 313–332. https://doi.org/10.2183/pjab.90.313.

Van de Wiele, T., Boon, N., Possemiers, S., Jacobs, H., & Verstraete, W. (2004). Prebiotic effects of chicory inulin in the simulator of the human intestinal microbial ecosystem. *FEMS Microbiology Ecology, 51*(1), 143–153. https://doi.org/10.1016/j.femsec.2004.07.007.

Van den Abbeele, P., Belzer, C., Goossens, M., Kleerebezem, M., De Vos, W. M., Thas, O., & Van de Wiele, T. (2013). Butyrate-producing *Clostridium* cluster XIVa species specifically colonize mucins in an in vitro gut model. *The ISME Journal, 7*(5), 949–961. https://doi.org/10.1038/ismej.2012.174.

Van den Abbeele, P., Grootaert, C., Marzorati, M., Possemiers, S., Verstraete, W., Gérard, P., & Van de Wiele, T. (2010). Microbial community development in a dynamic gut model is reproducible, colon region specific, and selective for *Bacteroidetes* and *Clostridium* cluster IX. *Applied and Environmental Microbiology, 76*(15), 5237–5246. https://doi.org/10.1128/AEM.03026-09.

Van den Abbeele, P., Roos, S., Eeckhaut, V., MacKenzie, D. A., Derde, M., Verstraete, W., ... & Van de Wiele, T. (2012). Incorporating a mucosal environment in a dynamic gut model results in a more representative colonization by lactobacilli. *Microbial Biotechnology, 5*(1), 106–115. https://doi.org/10.1111/j.1751-7915.2011.00306.x.

van Nuenen, M. H., Meyer, P. D., & Venema, K. (2003). The effect of various inulin's and *Clostridium difficile* on the metabolic activity of the human colonic microbiota in vitro. *Microbial Ecology in Health and Disease, 15*(2–3), 137–144.

Venema, K., & Van den Abbeele, P. (2013). Experimental models of the gut microbiome. *Best Practice & Research Clinical Gastroenterology, 27*(1), 115–126. https://doi.org/10.1016/j.bpg.2013.01.011.

Venema, K., & Van den Abbeele, P. (2013). Experimental models of the gut microbiome. *Best Practice & Research Clinical Gastroenterology, 27*(1), 115–126.

Venema, K., van Nuenen, M. H., van den Heuvel, E. G., Pool, W., & van der Vossen, J. M. (2003). The effect of lactulose on the composition of the intestinal microbiota and short-chain fatty acid production in human volunteers and a computer-controlled model of the proximal large intestine. *Microbial Ecology in Health and Disease, 15*(2–3), 94–105.

Verhoeckx, K., Cotter, P., López-Expósito, I., Kleiveland, C., Lea, T., Mackie, A., & Wichers, H. (2015). The impact of food bioactive on health: In vitro and ex vivo models (p. 338). Springer Nature.

Whatley, H. (2001). Basic principles and modes of capillary electrophoresis. In *Clinical and Forensic Applications of Capillary Electrophoresis* (pp. 21–58). Humana Press, Totowa, NJ.

Wiese, M., Khakimov, B., Nielsen, S., Sørensen, H., van den Berg, F., & Nielsen, D. S. (2018). CoMiniGut—A small volume in vitro colon model for the screening of gut microbial fermentation processes. *PeerJ, 6*, e4268. https://doi.org/10.7717/peerj.4268.

Williams, C. F., Walton, G. E., Jiang, L., Plummer, S., Garaiova, I., & Gibson, G. R. (2015). Comparative analysis of intestinal tract models. *Annual Review of Food Science and Technology, 6*, 329–350.

Williams, E. A., Coxhead, J. M., & Mathers, J. C. (2003). Anti-cancer effects of butyrate: Use of micro-array technology to investigate mechanisms. *Proceedings of the Nutrition Society, 62*(1), 107–115.

Woodmansey, E. J., McMurdo, M. E., Macfarlane, G. T., & Macfarlane, S. (2004). Comparison of compositions and metabolic activities of fecal microbiotas in young adults and in antibiotic-treated and non-antibiotic-treated elderly subjects. *Applied and Environmental Microbiology, 70*(10), 6113.

Xiong, Y., Miyamoto, N., Shibata, K., Valasek, M. A., Motoike, T., Kedzierski, R. M., & Yanagisawa, M. (2004). Short-chain fatty acids stimulate leptin production in adipocytes through the G protein-coupled receptor GPR41. *Proceedings of the National Academy of Sciences, 101*(4), 1045–1050.

Zhao, G., Nyman, M., & Åke Jönsson, J. (2006). Rapid determination of short-chain fatty acids in colonic contents and faeces of humans and rats by acidified water-extraction and direct-injection gas chromatography. *Biomedical Chromatography, 20*(8), 674–682.

Zihler, A., Gagnon, M., Chassard, C., & Lacroix, C. (2011). Protective effect of probiotics on *Salmonella* infectivity assessed with combined in vitro gut fermentation-cellular models. *BMC Microbiology, 11*(1), 1–13. https://doi.org/10.1186/1471-2180-11-10.

CHAPTER 15

Xylooligosaccharide: A Potential Prebiotic

ABNITA THAKURIA

Assistant Professor, Department of Food Science and Nutrition, College of Community Science, Assam Agricultural University, Jorhat, Assam, India

It has been determined that XOS may be a novel prebiotic with enormous health advantages. Enzymatic, chemo-enzymatic, partial hydrolysis of xylan from a variety of sources, including barley hulls, rice hulls, corn cobs, peanut pods, sugarcane baggase, wheat straw, cotton stalks, orange peels, mango peels, etc., can be used to produce XOS.

15.1 INTRODUCTION

Nutrition and health highlight great potential to improve our health. Functional foods are foods that can promote our health in addition to basic nutrition.

One of the major areas of research in food science and nutrition is the discovery of natural ingredients with biological activities (antioxidant, antihypertensive, hypoglycemic, etc.) that can improve human health as part of new functional foods (Plaza et al., 2008). Human clinical trials reported side-effect-free benefits with XOS doses up to 12 g per day (XOS GRAS Notice, January, 2013).

In order to produce XOS from agricultural wastes, technology transfer can be done to the industries that produce fruits, vegetables, nuts, and oilseeds.

Prebiotic Foods for a Healthy Gut: Scientific Studies to Manage Metabolic Disorders.
Mini K. Sheth and Shruti Dwivedi (Eds.)

This will not only improve human health but will also strengthen our nation's economy by making XOS available in markets both locally and globally.

15.2 GLOBAL XOS MARKET SIZE AND ITS EXTRACTION

The global Xylo oligosaccharide (XOS) market size will grow to US$130 million by 2025, from US$94 million in 2018, at a CAGR of 4.1% during the forecast period. Xylo oligosaccharides (XOS) Market has reported an in-depth analysis of the competitive landscape of this business, including companies such as HBTX, Henan, Shengtai, Longlive, etc. (Marketwatch, 2019).

Thakuria and Sheth (2018) determined the XOS content of various agricultural wastes such as corncob, raw green banana peels, green pea shells, and orange peels (Figure 15.1). These agricultural wastes were extracted using an alkaline method, and XOS was extracted using enzymatic hydrolysis and quantified using HPLC. Corncob, orange peels, raw green banana, and green pea shells yielded 91.11% (1.8 g), 92.36% (1.41 g), 95.05% (1.01 g), and 92.40% (0.79 g) of XOS, respectively ($p<0.01$). As a result of the research, it was discovered that XOS could be extracted from various agro waste. The formula below was used to calculate the true yield of xylan.

$$\text{True yield (\%)} = \frac{\text{Dry weight of extracted xylan (g)}}{\text{Weight of the sample (g)}} \times 100$$

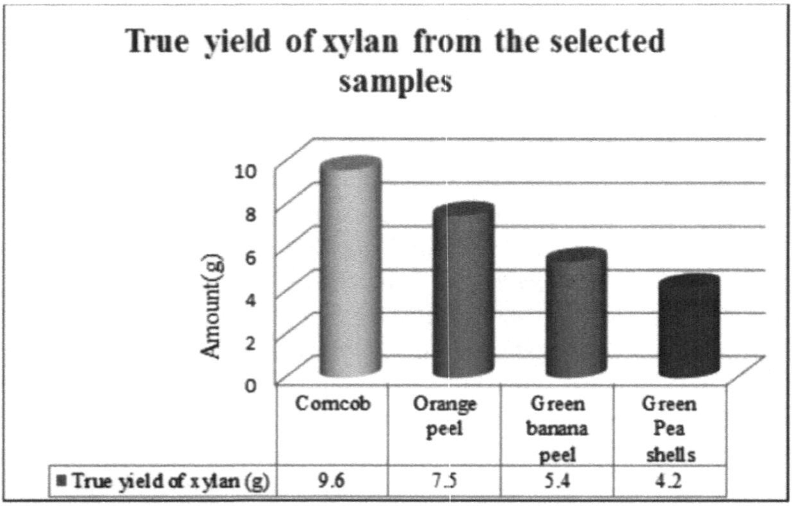

FIGURE 15.1 True yield of xylan from green pea shells, orange peel, and corn cob.

Source: Reprinted from Thakuria and Sheth et al., 2018. https://creativecommons.org/licenses/by/4.0/

15.4 PREBIOTIC POTENTIAL OF XOS

XOS have been identified as a potential novel prebiotic, and its benefits include enhancement of the biological availability of minerals such as calcium, iron, etc. by improving their absorption, reduction in colon cancer risk, cytotoxic effects on human leukemia cells, positive effects on diabetes, and improvement in bowel function and gut health (Jain et al., 2015). The prebiotic mechanism of Xylooligosaccharides (XOS) was observed in a randomized clinical trial (RCT), in which XOS administration for 2 and 4 weeks was determined. Both cases reported an increase in butyrate fecal concentrations and *Bifidobacteria*. The selective growth of *Bifidobacterium* adolescentis and *Bifidobacterium* lactis in pure cultures was determined by XOS. In a colon model, the bifidogenic effects of Xylo oligosaccharides on *Bifidobacterium* lactis appeared to be more efficient than those of Fructooligosaccharides alone, implying that the association of XOS and *B. lactis* could result in a successful symbiotic product. A 6-week randomized controlled trial on 20 healthy subjects found that those who ate 150 g of rice porridge with 1.2 g Xylo oligosaccharide supplementation daily had higher fecal counts of *Lactobacilli* and *Bifidobacterium*; lower *Clostridium* counts, but no change in total anaerobic bacterial counts, as compared to the control group having only rice porridge without Xylo oligosaccharide supplementation (Guarino et al., 2020).

Normal, healthy Americans who had 2 g of XOS daily for 8 weeks saw an abundant increase in *Bifidobacteria* without experiencing any gastrointestinal side effects (Finegold et al., 2014). According to a study, total cholesterol, LDL cholesterol, oxidized low-density lipoprotein (ox-LDL), apolipoprotein B, glucose, HbA1c, and fructosamine levels were reduced by XOS supplementation. The results of the study showed that taking XOS as a dietary supplement for 8 weeks significantly improved type 2 diabetic subjects' blood sugar and lipid levels, suggesting that XOS-containing diets may be advantageous to those with type II diabetes mellitus (Sheu et al., 2008).

In streptozotocin-induced diabetic rats, oligosaccharide (XOS and FOS) supplementation at a 10% dietary level had positive effects on plasma glucose, cholesterol, and a number of other metabolic parameters. By lowering hyperglycemia and hypercholesterolemia, XOS and FOS helped diabetic rats' nephromegaly and decreased the amount of AGE products in their renal tissue. According to the study, XOS and FOS at a 10% dietary level could be incorporated as an addition to dietary therapy to benefit diabetics and postpone complications (Gobinath et al., 2010) (Table 15.1).

TABLE 15.1 Clinical Trial Summary for XOS (Saville et al., 2019)

Study authors	XOS purity / dose	Study design and duration	Endpoints
Childs et al. [34]	8 g per day with or without B. lactis	Randomized, double-blind, placebo-controlled, factorial cross-over study with 41 healthy adults; 3 weeks dosing, with a 4 week washout between dosing periods	Increased HDL levels by 0.07 mM ($P = 0.005$) Increase of 0.10 mM in associated with 10% risk reduction in coronary artery disease Increased # of bowel movements per day ($P = 0.009$) Increased Bifidobacteria ($P = 0.008$) Reduced IL-10 production ($P = 0.049$)
Chung et al. [35]	95% pure XOS, 4 g per day	Randomised, placebo-controlled trial with 22 healthy adults (9 control, 13 XOS); includes a 1 week run-in, 3 weeks of dosing, and a 3 week washout	Statistically significant increase in fecal moisture (alleviates constipation) and fecal pH, with return to baseline post-washout ($P < 0.05$) Statistically significant increase in Bifidobacteria content of the stool ($P < 0.05$)
Finegold et al. [36]	70% pure; 1.4 or 2.8 g per day	Randomized double blind placebo-controlled trial with 32 healthy adjusts; 2 week run-in period, followed by 8 weeks of treatment and a 2 week washout	Statistically significant increase in Bifidobacteria ($P = 0.007$; 2.4 g per day dose) and B. fragilis ($P = 0.001$; 2.4 g per day dose) relative to placebo/baseline
lino et al. [40]	0.40 g per day	4 Weeks with 40 healthy adults	Improved stool frequency ($P < 0.05$)
Kajihara et al. [37]	3 g per day	Baseline data in patient population comprised of 14 adults with cirrhosis, followed by 2 weeks of dosing with 3 g per day of XOS.	Increased Bifidobacteria content Reduced Bacteriodes Reduced ammonia in serum (statistical data not presented)
Lecerf et al. [38]	5 g per day WitaXOS (80% Pure); 1 g per day XOS + 3 g per day inulin	Randomized, placebo-controled, doubled blind trial with 60 healthy adults (20 in each of three groups: (i) XOS, (ii) XOS and inulin, (iii) placebo); 4 weeks	Increased Bifidobacteria ($P = 0.002$) Increased butyrate ($P = 0.002$) Increased butyrate ($P = 0.036$) Reduced cresol ($P = 0.020$), acetate ($P = 0.011$) and fecal pH ($P = 0.033$)

TABLE 15.1 *(Continued)*

Study authors	XOS purity / dose	Study design and duration	Endpoints
Na and Kim [39]	1.4 and 2.8 g per day	Randomized dose-dependent trial without a placebo with 14 healthy adults (7 per test group); 4 weeks	Reduced triglycerides, cholesterol and glucose in 2.8 g per day group (all $P < 0.05$) Increased Bifidobacteria content after 14 days ($P < 0.05$) Increased lactic acid concentration ($P < 0.05$) hile study lacked placebo, subjects saw effect of dose response over multiple doses
Sheu et al. [42]	4 g per day	Randomized, double blind placebo-controlled trial in 26 adults with type II diabetes; 8 weeks	Statistically significant reduction in: Blood glucose ($P < 0.05$) (described as a point measurement of blood glucose, with the sample obtained after fasting) LDL ($P < 0.01$) Total cholesterol ($P < 0.01$) HbA1c ($P < 0.05$) Apolipoprotein B ($P < 0.05$)
Tateyama et al. [41]	4.2 g per day in 8 g syrup	29 Adult Women (pregnant with constipation); all received product, at the same dose, for 4 weeks	Statistically significant increase in stool frequency and reduction in constipation Improvement increased every week over the duration of treatment Normalized stool consistency
Yang et al. [43]	2.8 g per day of 70% XOS	8 week trial with 13 healthy adults and 16 prediabetic (increased risk of diabetes) adults	Decreased Firmicutes Attenuated changes in Howardella, Slackia, and B. Hydrogenotrophica Tendency to reduce OGTT insulin No effect on blood gulucose, triglycerides Limited by small study numbers and lack of statistical analysis
Cloetens, L. et al. [44]	AXOS, 10 grams per day (63% XOS, 17% arabinan, 12% glucan)	Randomized, placebo-controlled cross-over study with 20 healthy adults; 3 weeks with AXOS or placebo, with 4-week washout between treatments	Increased Bifidobacteria ($P = 0.012$) Increased B. adolescentis ($P = 0.013$) After 3 weeks, reduced urinary p-cresol ($P = 0.011$)

TABLE 15.1 *(Continued)*

Study authors	XOS purity / dose	Study design and duration	Endpoints
Francois, l. et al. [45]	Fugeia wheat bran extract, 3 grams per day, 10 grams per day (comprised of 79% AXOS, which, in turn, consists of 49% XOS and 10% glucan) [1]	Double blind randomized placebo controlled cross-over trial with 20 healthy adults; 3-week treatment periods interspersed with 2-week washout periods	At 10 grams per day: Increased acetate, propionate, and butyrate ($P = 0.009$, $P = 0.05$, $P = 0.001$, respectively) Reduced p-cresol ($P = 0.039$) Lower fecal pH ($P < 0.039$) Increased Bifidobacteria ($P < 0.001$) Increased stool frequency ($P = 0.258$) Reduced LDL ($P = 0.168$)
Francois, I. et al. [46]	Fugeia Wheat Bran Extract, 5 grams per day (comprised of 79% AXOS, which, in turn, consists of 49% XOS and 10% glucan)	Double blind randomized placebo controlled cross-over trial with 29 healthy children; 3 weeks	Increased Bifidobacteria ($P = 0.069$) Reduced isovaleric and isobutyric acid ($P < 0.01$) No impact on stool frequency or consistency
Maki, K.C., et al. [47]	2.2 grams or 4.8 grams AXOS per day added to cereal	Randomized, double blind, placebo-controlled cross-over trial with 65 healthy adults; 3weeks treatment with 2 weeks washout between treatments	Dose-dependent increase in Bifidobacteria observed; statistically significant increase in the 4.8 grams per day AXOS group ($P < 0.001$) No statistically significant impact on fasting blood glucose, total cholesterol, HDL, or triglycerides
Walton, G. et al. [48]	180 grams per day fo Bread, with or without enrichment with 2.2 g AXOS	Double-blind, randomized, placebo-controlled cross-over trial with 40 healthy adults; five periods of 3 weeks each	Increased Bifidobacteria relative to baseline ($P = 0.0011$) Increased butyrate ($P = 0.041$) and propionate ($P = 0.045$) Other results confounded by XOS/fructans in the placebo bread, which increased Bifidobacteria, short chain fatty acids, and bowel frequency relative to the baseline. The trend to further improve these characteristics via the AXOS-enriched bread were thus masked.

A comparative study suggested that 4 g of XOS for 3 weeks per day improves the gut microbiota in elderly people over 65 years old. Several reports were found on the utilization of XOS by beneficial bacteria like *Bifidobacteria* and other lactic acid bacteria. These bacteria produce SCFAs by fermenting Xylo oligosaccharides resulting in various health benefits (Jain et al., 2015).

Lactate is produced as an intermediary product of XOS in the beginning of fermentation, and is further metabolized into acetate, butyrate, and propionate by intestinal microflora. Prebiotic consumption is related to a reduction in fecal and caecal pH; decreased growth of C. perfringens, *E. coli*, and many more harmful microflorae by switching the growth stimulation of beneficial gut flora (Samanta et al., 2015).

A 21-day intervention trial revealed that 8 g/day XOS supplementation was given to healthy adults (25–65 years). XOS supplementation improved mean bowel movements per day, but did not show any change in abdominal pain, flatulence, or bloating. XOS supplementation significantly increased participant-reported vitality and happiness. There was a significant increase in fecal *Bifidobacteria*l counts and High Density Lipoprotein (HDL) concentrations. In healthy adults, it was found that XOS induced bifidogenesis, modified immune function markers, and enhanced certain aspects of the plasma lipid profile (Childs et al., 2014).

XOS also has immense potential in improving the quality of many foods by food flavor modification and improvement in their physicochemical characteristics because of its stability and broad range of pH and temperature. It could be added to soy milk, soft drinks, cocoa drinks, tea, nutritive preparations, dairy products like milk, milk powder, yogurt, candies, jellies, jam, and honey products to formulate health foods or value-added food products. The cost of manufacturing XOS is a big hurdle in incorporating them into daily meals, which could be reduced by some economical substrates and efficient processes (Jain et al., 2015).

A study reported the effects of XOS incorporated rice porridge consumption on the intestinal tract's ecosystem of 20 healthy subjects in a 6-week intervention trial. Ten subjects received rice porridge incorporated with XOS, and the rest received placebo rice porridge. According to their findings, consuming rice porridge with XOS added daily for six weeks significantly increased fecal bacterial counts of *Bifidobacterium* spp., and *Lactobacillus* spp., while inhibiting the growth of *Clostridium* perfringens. The study concluded that after daily consumption of 150 g of XOS incorporated rice porridge with 1.5 g XOS for 6 weeks, the intestinal microbiota balance improved, indicating the prebiotic potential of XOS incorporated foods (Lin et al., 2016).

There is an emerging market for functional oligosaccharides to be used in food. XOS is considered a potential ingredient in functional foods. Several factors are responsible for the incorporation of XOS as functional ingredients, such as scientific research-based evidence supporting that such foods help to improve and maintain overall health and wellbeing, consumer demand, the role of XOS for prevention of disease, and self-medication. Xylo oligosaccharides have advantages over other nondigestible oligosaccharides in terms of health and technological properties (Gupta et al., 2016).

It can be concluded that XOS has many health benefits and is known for its interesting physicochemical properties. It has a moderately sweet taste, is stable over a wide pH and temperature range, and has sensory properties that make it suitable for use in a variety of foods. XOS escapes the low pH and enzymes of the stomach to the colon where it selectively stimulates specific bacterial populations and acts as an indigestible food component (Gupta et al., 2016). These properties of XOS make it a novel ingredient in new functional foods. In the food industry, XOS has been found to produce low-calorie sweeteners such as xylitol and antioxidant compounds, and is also recognized as a flavor enhancer in beverage formulations (Moure et al., 2006).

XOS has been shown to be highly effective in maintaining and improving a balanced gut microbiota to improve host health and well-being. *In vivo* studies can also be performed to demonstrate the clinical efficacy of XOS intake in treating various non-communicable diseases.

KEYWORDS

- **diabetes**
- **food preparation**
- **functional foods**
- **health**
- **health benefits**
- **health effects**
- **health risks**
- **national economy**
- **nutrition**
- **oligosaccharides**

REFERENCES

Childs, C. E., Röytiö, H., Alhoniemi, E., Fekete, A. A., Forssten, S. D., Hudjec, N., & Rastall, R. A. (2014). Xylo-oligosaccharides alone or in symbiotic combination with *Bifidobacterium animalis* subsp. *lactis* induce bifidogenesis and modulate markers of immune function in healthy adults: A double-blind, placebo-controlled, randomized, factorial cross-over study. *British Journal of Nutrition, 111*(11), 1945–1956.

De Sousa, V. M. C., dos Santos, E. F., & Sgarbieri, V. C. (2011). The importance of prebiotics in functional foods and clinical practice. *Food and Nutrition Sciences, 2*(2), 133–144.

Finegold, S. M., Li, Z., Summanen, P. H., Downes, J., Thames, G., Corbett, K., et al. (2014). Xylooligosaccharide increases bifidobacteria but not lactobacilli in human gut microbiota. *Food Function, 5*, 436–445.

Gibson, G. R., Probert, H. M., Van Loo, J. A. E., Rastall, R. A., & Roberfroid, M. B. (2004). Dietary modulation of the human colonic microbiota: Updating the concept of prebiotics. *Nutrition Research Reviews, 17*, 259–275.

Gobinath, D., Madhu, A., Prashant, G., Srinivasan, K., & Prapulla, S. (2010). Beneficial effect of xylo-oligosaccharides and fructo-oligosaccharides in streptozotocin-induced diabetic rats. *British Journal of Nutrition, 104*(1), 40–47.

GRAS (Generally Recognised as Safe). (2013). GRN No. 458: Notification for Shandong Longlive Biotechnology, Ltd.'s Xylo oligosaccharides.

Guarino, M., Altomare, A., Emerenziani, S., Di Rosa, C., Ribolsi, M., Balestrieri, P., Iovino, P., Rocchi, G., & Cicala, M. (2020). Mechanisms of action of prebiotics and their effects on gastrointestinal disorders in adults. *Nutrients, 12*(4), 1037.

Gupta, P. K., Agrawal, P., Hegde, P., Shankarnarayan, N., Vidyashree, S., Singh, S. A., & Ahuja, S. (2016). Xylooligosaccharide: A valuable material from waste to taste: A review. *Journal of Environmental Research and Development, 10*(3), 555–563.

Jain, I., Kumar, V., & Satyanarayana, T. (2015). Xylooligosaccharides: An economical prebiotic from agro residues and their health benefits. *Indian Journal of Experimental Biology, 53*, 131–142.

Jang, S. K., Jung, C. D., Yu, J. H., & Kim, H. (2020). Environmentally friendly approach for the production of glucose and high-purity xylooligosaccharides from edible biomass byproducts. *Applied Sciences, 10*(22), 8119.

Lin, S. H., Chou, L. M., Chien, Y. W., Chang, J. S., & Lin, C. I. (2016). Prebiotic effects of Xylo oligosaccharides on the improvement of microbiota balance in human subjects. *Gastroenterology Research and Practice, 2016*, 1–6.

Marketwatch. (2019). At 4.1% CAGR, xylooligosaccharides (XOS) market size will reach 130 million USD by 2025. Retrieved from: https://www.marketwatch.com/press-release/at-41-cagr-xylooligosaccharides-xosmarket-size-will-reach-130-million-usd-by-2025-2019-03-22 (accessed on 12 March 2020).

Moure, A., Gullón, P., Domínguez, H., & Parajó, J. C. (2006). Advances in the manufacture, purification and applications of xylo-oligosaccharides as food additives and nutraceuticals. *Process Biochemistry, 41*(9), 1913–1923.

Plaza, M., Cifuentes, A., & Ibáñez, E. (2008). In the search of new functional food ingredients from algae. *Trends in Food Science & Technology, 19*, 31–39.

Samanta, A. K., Jayapal, N., Kolte, A. P., Senani, S., Sridhar, M., Dhali, A., Suresh, K. P., Jayaram, C., & Prasad, C. S. (2015). Process for enzymatic production of xylooligosaccharides from the xylan of corn cobs. *Journal of Food Processing and Preservation, 39*(6), 729–736.

Saville, B. A., & Saville, S. H. (2019). Functional attributes and health benefits of novel prebiotic oligosaccharides derived from xylan, arabinan, and mannan. In E. Franco-Robles & J. Ramírez-Emiliano (Eds.), *Prebiotics and Probiotics–Potential Benefits in Nutrition and Health*. IntechOpen. Retrieved from: https://www.intechopen.com/chapters/69385 (accessed on 19 July 2024).

Sheu, W., Lee, I. T., Chen, W., & Chan, Y. C. (2008). Effects of xylooligosaccharides in type 2 diabetes mellitus. *Journal of Nutritional Science and Vitaminology, 54*, 396–401. https://doi.org/10.3177/jnsv.54.396.

Thakuria, A., & Sheth, M. (2018). Alkaline Extraction of Xylan from Agro Waste and Determining Xylooligosaccharide (XOS) Using Enzymatic Hydrolysis. *Paripex–Indian Journal of Research*, *7*(11).

CHAPTER 16

Role of Prebiotics in Synbiotic Fermented Dairy Foods

A. H. JANA[1] and HATI SUBROTA[2]

[1]*Dairy Processing and Operations, SMC College of Dairy Science, Anand, Gujarat, India*

[2]*Department of Dairy Microbiology, SMC College of Dairy Science, Anand, Gujarat, India*

Prebiotics, probiotics, and symbiotic dairy foods are gaining importance in our daily diet and have preventive and therapeutic interventions in certain types of human health disorders. Prebiotics are essentially indigestible food substances that have beneficial physiological effects on the host by stimulating the growth or activity of a small number of naturally occurring bacteria, especially probiotics, in the gut. Prebiotics play a significant role in the fermentation of difficult-to-digest carbohydrates and promote the growth of probiotic bacteria such as Bifidobacteria and some Gram-positive bacteria. Novel synbiotic products are developed using these prebiotics with probiotics for the management of certain health problems like hypertension, diabetes, tumors, cancer, etc.

Prebiotic Foods for a Healthy Gut: Scientific Studies to Manage Metabolic Disorders.
Mini K. Sheth and Shruti Dwivedi (Eds.)

16.1 INTRODUCTION

Lactulose, galacto-oligosaccharides (GOS), fructo-oligosaccharides (FOS), inulin and their hydrolysates, malto-oligosaccharides, and resistant starches are the commonly used prebiotics in the manufacture of synbiotic dairy foods. The salient end products of carbohydrate metabolism include short-chained fatty acids (SCFAs) which are further used by the host. Prebiotic compounds favorably influence the technological properties, physico-chemical, rheological, sensory profile, economics and even the survivability of probiotics until the products' consumption.

Foods may inherently contain 'prebiotics' or food may be fortified with prebiotics during its manufacture with a view to improve the functional efficacy of probiotics. The prebiotics favorably influence the intestinal gut microflora by multiplying the numbers of beneficial anaerobic bacteria and retarding the population of pathogenic microbes. Probiotics, on the other hand, affect the intestinal ecosystem by stimulating the mucosal immune systems as well as non-immune mechanisms via antagonism and competition with the pathogens (WGO Global guidelines, 2011).

16.2 PREBIOTICS – AN ESSENTIAL COMPONENT OF SYNBIOTIC FOOD

Prebiotics, when used, or when used in conjunction with probiotic bacteria, produce 'synbiotics'; the latter tends to promote and enhance human gastrointestinal health (Tuohy et al., 2003). From the viewpoint of technological aspects, the amalgamation of both probiotics and prebiotics is vital for the proper development of flavor and texture of functional fermented dairy products (Martin et al., 2011); they are held responsible for the variations in the quantity of organic acids, SCFAs, volatile compounds, and exopolysaccharides released during the manufacture of such food products (Ong and Shah, 2009).

The term "synbiotic" describes a product that combines prebiotics and probiotics to serve as a nutritional supplement and ultimately improve the host's health (Schrezenmeir and de Vrese, 2001). Since probiotics can survive in the digestive system when prebiotics are present and can tolerate oxygen, low pH, and unfavorable temperature, synbiotics are particularly preferred (Nagpal et al., 2007; Pandey et al., 2015). Prebiotics are now a common ingredient in food formulations, particularly dairy products, to aid in the growth of probiotics and enhance nutrient quality. Table 16.1 shows the utility of some prevalent prebiotics in the manufacture of some fermented dairy foods.

TABLE 16.1 Examples of Prebiotic Ingredients in Dairy Based Fermented Foods

Fermented Dairy Foods	Prebiotic	Impact on Product	Reference
Fermented milk	Lupin seeds	Reduction in fermentation period Increased probiotic bacterial count	Martinez-Villaluenga et al., 2006
	Lemon and orange fibers	Promoted bacterial growth and led to superior survival of probiotic bacteria Superior sensory traits	Sendra et al., 2008
	Apple, banana and grape flours	Increased the viability of probiotic starter strains Alternative method to produce functional dairy food	Casarotti and Penna, 2015
Yogurt	Lentil flour	Improved the physical and rheological properties of product	Zare et al., 2011
	Pea protein, chickpea flour, lentil flour, pea fiber, soy flour and soy protein concentrate	Improved the growth and metabolic activity of probiotic as well as of yogurt starter cultures	Zare et al., 2012
	Apple, banana and passion fruit fibers	Improved the fatty acid profile of product Development of value-added fermented dairy food	Do Espirito-Santo et al., 2012
	Grain mixture (wheat, rye, oat, barley and millet)	Acceptable sensory acceptance of fiber-enriched yogurt Could serve as partial sugar replacer in product	Hoppert et al., 2013
	Raw and roasted almonds	Both type of almonds promoted the growth of *L. acidophilus* La-14 and *B. breve* JCM 1192	Liu et al., 2016
	Oligosaccharides from honey pomegranate peels	Increased growth and population of *Bifidobacterial* and *Lactobacilli.* Fermentation of pomegranate peel flour by colonic bacteria generated propionic, acetic and butyric acids	Sanz et al., 2005; Gullon et al., 2015
	Oligosaccharides from Pitaya (*Hylocereus undatus* (Haw.))	Increased resistance to gastric acidity and growth of both *Lactobacillus* and *Bifidobacterium*	Wichienchot et al., 2010

Compiled from several sources.

Use of probiotic microbes in milk products leads to improved mucosal surfaces of the host and possible production of antimicrobial compounds that have the capacity to inhibit pathogens culminating in bio-preservation of dairy foods (Parvez et al., 2015; Sharma et al., 2017).

The efficacy of prebiotics in synbiotic food products is dependent on several factors which include (a) type of prebiotic, (b) prebiotic purity, (c) chain length of prebiotics, (d) its usage level, type of probiotic microorganisms, (f) characteristics of products (viz., pH, titratable acidity, milk solids not fat (MSNF), fat content), and (g) storage temperature and time, (h) processing conditions during synbiotic product manufacture (temperature, duration, packaging) (Mohammadi and Mortazavian, 2011).

Prebiotics are not digested by the enzymes inherent in the small intestine, but can be fermented by probiotic bacteria colonized in the large intestine. As a result of fermentation, SCFAs, primarily acetate, propionate, and butyrate, are produced, and a possible reduction in pathogenic microbes is caused by a fall in the medium's pH (Mussatto and Manchilha, 2007; McFarlane et al., 2008). These are utilized by the host organisms as a source of energy. The prebiotics upon fermentation by the probiotic bacteria improve the host's health by enhancing the absorption of minerals such as Ca, Mg, and Fe and producing compounds that are purported to combat colon cancer (Gibson and Roberfroid, 1995; Ozcan et al., 2016).

Prebiotic foods are classified in the category of functional foods and represent a significant part of this market (Davani-Davari et al., 2019). Nevertheless, the quantity of prebiotics in such sources is usually quite low to exert any marked effect on the composition of intestinal microflora. The synbiotic effects of probiotics and prebiotics (such as inulin, lactobionic acid, lactulose) are strain-dependent in food medium (Adebola et al., 2014).

Recent data suggest that the world demand for prebiotics is estimated at 167,000 metric tons, generating €390 million (Siro et al., 2008). In fact, the 180 new food products launched onto the market in 2006 demonstrate the status of prebiotics (Sveje, 2007).

16.3 CLASSES OF PREBIOTIC SUBSTANCES

The classification of prebiotic substances is based on (a) chemical nature, (b) chain length, (c) mode of utility, and (d) the degree of polymerization. Based on the chemical nature, the prebiotic substances are classified into three types *viz.*, (i) saccharide derivatives (disaccharides, oligosaccharides, polysaccharides), (ii) proteins or peptides, and (iii) lipids (Ishwarya and Prabhasankar, 2014).

Prebiotics that are used as food additives typically contain derivatives of saccharides that are primarily derived from plants. Carbohydrates such as oligofructose, inulin, FOS, GOS, SOS, transgalacto-oligosaccharides, gluco-oligosaccharides, gentio-oligosaccharides, xylo-oligosaccharides (XOS), lactulose, isomalto-oligosaccharides (IOS), and polysaccharides (viz., pectin, starch) are considered to be efficient prebiotic substances (Macfarlane et al., 2008; De Vrese and Schrezenmeir, 2008; Fuller and Gibson, 2008; Sharma et al., 2017).

The prebiotic carbohydrates of plant origin include cereals and legume crops such as barley, oats, wheat, chickpeas, and lentils; vegetables such as chicory root, Jerusalem artichoke, jicama, yacon, onion, garlic, asparagus, okra, and leek (Ozcan et al., 2016); and fruits such as dragon fruit, jackfruit, palm fruit, nectarine, and mushroom (Manning and Gibson, 2004; Fazilah et al., 2012; Dwivedi et al., 2014).

Gulewicz et al. (2003) have dealt with three methods to produce prebiotic oligosaccharides, namely, (i) extraction from plant materials, (ii) microbiological/enzymatic synthesis, and (iii) enzymatic hydrolysis of polysaccharides.

Fermented foods may naturally contain prebiotics or they may be fortified with prebiotics during their manufacture in order to improve the functional efficacy of probiotics. The complex carbohydrates pass into the small intestine, followed by the lower gut area where they become available to the colonic bacteria, but remain unutilized by the majority of bacteria residing in the colon.

The ideal characteristics of prebiotic substances can be listed as follows (Maftei, 2019):

- Neither hydrolyzed nor absorbed by the mammalian tissue or enzymes.
- Selectively promotes the growth of some specific probiotic species.
- Capacity to modify the intestinal microbial flora and its activity.
- Ability to alter the systemic aspects of host defense system.

The selection of prebiotics is based on the requirement that the substrate be used only by the microorganisms in the large intestine and not hydrolyzed or absorbed in the stomach or small intestine (Ozcan et al., 2016).

16.4 PREBIOTICS IN THE GLOBAL MARKET

Some products made utilizing prebiotics in the global market involve Vitalinea containing inulin (Yogurt and meal; Danone, UK), Taillefine containing oligofructose (Yogurt; Danone, Belgium), low sugar sorbet containing oligofructose (Sorbet; Thiriet, France), Aviva containing oligofructose (chocolate drink, Novartis, Switzerland), Jour Apres Jout containing oligofructose (Milk, Lactel, France) among others (Kumar et al., 2005).

16.5 PREBIOTIC INDEX

The prebiotic index measures how well a particular substrate supports an organism's growth in comparison to that of other cells (Huebner et al., 2007). Many quantitative methodologies have been attempted to assess the functional activity of prebiotics under *in vitro* fermentation conditions (Jimenez-Vera et al., 2008; Figueroa-Gonzalez et al., 2010). These include measurement of microbial populations, growth rates, substrate assimilation rates, and/or production of SCFAs. A prebiotic index greater than 1.0 indicates that the carbohydrate has a positive effect on probiotic growth. If the prebiotic index is close to 1.0, it indicates that the specific carbohydrate in question is ineffective. The prebiotic index can be computed as shown in the following equation.

1_{preb} = CFU of probiotics in prebiotic carbohydrate/CFU of probiotics in control carbohydrate

16.6 PREBIOTIC ACTIVITY SCORE

$$\text{Apreb} = \frac{(\text{LogP}_{24} - \text{LogP}_0)\text{ prebiotic}}{(\text{LogP}_{24} - \text{LogP}_0)\text{ lactose}} \quad \frac{(\text{Log E}_{24} - \text{Log E}_0)\text{ prebiotic}}{(\text{LogE}_{24} - \text{LogE}_0)\text{ lactose}}$$

where A_{preb} is prebiotic activity score (PAS); LogP is the log of growth (CFU/ml) of probiotic bacteria at 24 h (P_{24}); Log E is the log of growth (CFU/ml) of *E. coli* K12 at 24 h (E_{24}) and 0 h (E_0) of culture on prebiotic and lactose.

According to such definition, substrates having high PAS support good growth of probiotic bacteria, with cell counts (CFU/ml) comparable with that grown-on lactose medium.

The highest PAS were recorded for *L. rhamnosus* (6.68) and *L. rhamnosus* GG (5.46), both on substrate Oligomate 55 (viz. combination of GOS, lactose, and monosaccharides; 55:40:5, w/w) as compared to the use of other sugars and cultures. Conversely, the least PAS (i.e., -0.3) was associated with *L. casei* Shirota on Frutafit (i.e., predominantly inulin). The three strains of *L. casei* exhibited the least PAS in the presence of the tested prebiotics; no appreciable difference was noted between lactulose and Frutafit used as prebiotics. The least PAS was associated when using lactulose in the case of all probiotic cells; the difference was marginal (Figueroa-Gonzalez et al., 2019).

Assessment of the prebiotic index and PAS concepts has validated the authenticity of the newer prebiotic ingredients used in probiotic foods. These are useful tools in defining adequate combinations in colonic food developments.

16.7 IMPORTANT PREBIOTIC SUBSTANCES

Some important prebiotic substances are dealt with individually or, in some cases, a few together.

16.7.1 OLIGOSACCHARIDES

Some oligosaccharides used as prebiotics include FOS, XOS,bo, polydextrose, and GOS (Davani-Davari et al., 2019).

GOS is a milk-based prebiotic that is derived from lactose using the β-galactosidase enzyme. It is a naturally occurring prebiotic in human breast milk that promotes the growth of *Bifidobacteria* and *Lactobacilli* in breast-fed infants. GOS is primarily used commercially in Japan and other parts of Asia (Saxelin et al., 2003). The GOS is characterized by the structure viz., α-D-Glu-(1–4)-[β-D-Gal-1–6-]n, where n ranges from 2 to 5 (Crittenden and Playne, 1996). GOS is synthesized commercially by incubating the pertinent β-galactosidase enzyme in the presence of a high level of lactose, in which the enzymes form complex GOS mixtures via transglycosylation reactions comprising several molecules of differing length, linkage type, and degree of branching. GOS is a non-digestible carbohydrate that exerts similar beneficial health effects as human milk oligosaccharides. Infants fed formulas supplemented with GOS create a gut microbiome that closely simulates the microbes of breast-fed infants as against those fed on formula, deprived of GOS (van Bueren et al., 2017).

16.7.2 FRUCTOOLIGOSACCHARIDE AND INULIN

FOS and inulin are fructose oligomers and polymers with the same generic structure – α-D-Glu-(1–2)-[β-D-Fru-1-2-]n; n ranges from 2 to 4 and up to 60 for FOS and inulin respectively (Crittenden and Playne, 1996). FOS are linear polymers of fructosyl monomers linked through β (2–1) bonds having a degree of polymerization of ≤ 10 and are considered to be a subgroup of inulin. Their structure is similar to corn sweeteners, and they can be used as a fat substitute in food products (Ahmadi et al., 2014).

16.7.3 GUAR GUM

Guar gum, a natural gum sourced from the seeds of *Cyamopsis tetragonolobus*, in a partially hydrolyzed (enzymatically) state, is another prebiotic compound that aids in stimulating the growth of probiotic bacteria and even the native gut microflora (Mudgil et al., 2018).

A soluble dietary fiber with a linear backbone made of 1,4-linked D-mannose residues and 1,6-linked D-galactose residues linked with mannose is called partially hydrolyzed guar gum (PHGG). This galactomannan properly dissolves in water, does not gel, and exhibits prebiotic effects by boosting colonic SCFA levels and the quantity of *Lactobacilli* and *Bifidobacteria* (Ohashi et al., 2015; Yasukawa et al., 2019). In gastrointestinal disorders like irritable bowel syndrome (IBS), PHGG was more effective than a placebo at reducing bloating (Niv et al., 2016), as well as symptoms and quality of life associated with IBS (Parisi et al., 2002; Reider et al., 2020). The microbiota's fermentation of PHGG resulted in an increase in the concentration of SCFAs like acetate, propionate, and butyrate. Colonic epithelial cells use SCFAs as a source of energy, and they also play a variety of regulatory roles in immunity, metabolism, and gut physiology (Koh et al., 2016).

16.7.4 LACTULOSE

Lactulose is a disaccharide having the formula β-D-Gal-D-(1–4) D-Fru (Crittenden and Playne, 1996). In fermented milk (skim milk) cultured with *Lb. acidophilus, Lb. rhamnosus*, and *Bifidobacterium* lactis in co-culture with *Str. thermophilus*, the use of lactulose (4.0% by weight) as a prebiotic led to increased counts of all the probiotic organisms, especially *B. lactis* (bifidogenic effect). Moreover, the acidification rate was hastened to reach the final fermentation pH (i.e., 4.5 pH) in a lesser period compared to control skim milk, devoid of lactulose (Oliveira et al., 2011).

16.7.5 HONEY

Honey is claimed to possess prebiotic activity owing to a variety of oligosaccharides inherent in it. However, the prebiotic effect exerted in the product can vary based on the type of honey used (e.g., plant species visited by the bees in honey production) (Ustunol and Gandhi, 2001). Both Jatai (*Tetragonisca angustula*) and Africanized (*Apis mellifera*) bee honeys, used at a level of 5.0 and 10.0%, favored the viability of probiotic cultures (viz., *Lb. acidophilus* La-5 and *Bifidobacterium* BB12 in bioyogurt) (Caldeira et al., 2018).

16.8 SOME SELECTED SALIENT PREBIOTICS

Some important prebiotic substances are dealt herein individually or few together.

16.8.1 INULIN AS PREBIOTIC

Inulin is a well-known prebiotic composed of fructooligo- and polysaccharides (Manning and Gibson, 2004; Gibson et al., 2000).

Inulin is a soluble and fermentable fructan that is resistant to digestion by α-amylase or other hydrolytic enzymes in the upper segment of the intestinal tract (Villegas and Costell, 2007); it has suitable applications in non-fat fermented milks (Oliveira et al., 2011). Inulin facilitates the digestion of high-protein diets, limits fat absorption, contributes to roughage avoiding constipation, provides satiety without conferring calories, lowers blood cholesterol and triglycerides, controls blood glucose, and decreases the incidence of colon cancer (Kaur et al., 2021). Prebiotics such as inulin markedly enhance the viability of probiotic organisms (Wang, 2009).

A diverse mixture of fructose polymers known as inulin is found widely in nature as a plant's form of storage carbohydrates. The degree of polymerization (DP) of inulin ranges from 2 to 60. Oligofructose, a sub-group of inulin, consists of polymers within a DP $\leq$ 10 (Sangeetha et al., 2005). The structure of these two compounds is similar to that of corn sweeteners. Inulin and oligofructose, often christened as fat replacers, influence the products' texture and mouthfeel. Since they are resistant to hydrolysis in the stomach and small intestine, they are regarded as dietary fiber (Spiegel et al., 1994; Niness, 1999).

Aryana and McGrew (2007) studied the effect of inulin comprising short, medium, and long chain lengths on the sensory acceptance of yogurt with added *Lactobacillus casei.* The yogurt made using *L. casei,* the sample with *L. casei* containing short chain length inulin, and the control (without inulin) product had a superior score for flavor compared to the sample containing longer-chain prebiotics. Regarding texture, the product with long-chain inulin exhibited markedly greater viscosity compared to the control product as well as the yogurt formulation containing oligosaccharides. Inulin served as a fat substitute in yogurt (Niness, 1999; Douglas, 2005).

16.8.2 INULIN AND OLIGOFRUCTOSE

Inulin is a heterogeneous blend of fructose polymers, widely distributed in nature as plant storage carbohydrates. Inulin has a degree of polymerization (DP) ranging between 2 and 60. Oligofructose, a sub-group of inulin, consists of polymers within a DP $\leq$ 10 (Sangeetha et al., 2005). Inulin shares a structure with corn sweeteners, which are frequently used to make ice cream. Inulin and oligofructose, which are categorized as fat replacers, affect the texture and mouthfeel of the products. They are categorized as dietary fiber ingredients because they are resistant to hydrolysis both in the stomach and in the small intestine (Spiegel et al., 1994; Niness, 1999).

Inulin can form inulin microcrystals when sheared in water or milk but is less soluble than oligofructose because of its longer chain length. These inulin microcrystals interact to give the final food product a mouthfeel with a creamy texture (Niness, 1999).

16.8.2 OLIGOSACCHARIDES AS PREBIOTICS

Dietary oligosaccharides are composed of 2–10 monomers linked together through glycosidic linkage, but inulin is often included in spite of a chain length larger than this. The colonic microflora may use nondigestible oligosaccharides as substrates.

Both inulin and FOS (oligofructose) are extracted from plant material (i.e., chicory) or synthesized from sucrose. The role played by inulin and the oligofructose in a food matrix is bi-functional. They tend to provide a richer texture to liquid products and spreads, without increasing the viscosity of milk (Saxelin et al., 2003).

Short-chain polysaccharides purported to be 'prebiotic substances' (e.g., fructo-oligosaccharides, FOS) have favorable effects on probiotic stability. However, the impact of using such carbohydrates on product texture should be taken into consideration when they are to serve as cryoprotectant and/or prebiotics (Mohammadi et al., 2011). Marx et al. (2000) studied metabolization of FOS and showed that inulin polymers of molecular mass > 4500 daltons were not utilized by *Bifidobacteria*, because hydrolysis of long-chain FOS is inhibited by steric hindrance at the active site of the hydrolytic enzymes. During fermentation, *B. longum* utilized short-chain FOS as the carbohydrate source and the growth slowed or ceased once the short-chain oligosaccharides were consumed. The capacity of *Bifidobacteria* to synthesize GOS has been documented in literature (Ambrogi et al., 2021).

16.8.3 SPIRULINA AS PREBIOTIC

Spirulina is a blue-green microalgae (also known as cyanobacteria) that is a photoautotrophic microorganism, rich in protein, essential amino acids, vitamins (E, A, K, B_{12}, B_6, and B_2), fatty acids, especially ω-6 representatives such as γ-linolenic acid, antioxidants, and minerals (including iron) (Akalin et al., 2009). Spirulina exerts various functional effects such as anti-inflammatory, antiviral, decreasing blood lipid levels, anti-tumoral activity, blood sugar regulation, and body weight management (Caire et al., 2000; Gyenis et al., 2005). The use of spirulina in conjunction with probiotics promotes and enhances viable growth as well as rapid acid production by the probiotics (Shirota et al., 1964; Celekli et al., 2018).

Spirulina contains 78.0% proteins, 4.0–7.0% lipids, vitamins, minerals, carbohydrates, and pigments. Due to these phytonutrients, spirulina has protective properties against a number of diseases such as cancer, hypertension, hypercholesterolemia, diabetes, anemia, etc. Research indicates that the extracellular products produced by *Spirulina platensis* significantly promote the growth of Lactic acid bacteria (LAB) such as *Lactococcus lactis, Str. thermophilus, Lb. casei, Lb. acidophilus,* and *Lb. bulgaricus* (Gupta et al., 2017). Especially, *Spirulina platensis* consumes nitrogen from the growth medium and elaborates extracellular carbohydrates and various growth substances that are responsible for stimulating the growth of *Lactobacilli* and other LABs.

L. acidophilus count increased to 2.4 and 3.1 log CFU/g in stored (30 days at 4°C) acidophilus milk made using 0.5 and 1.0 % spirulina powder respectively. The survival of *L. acidophilus* was greater than *S. thermophilus* or even *L. bulgaricus* during the entire storage period (Guldas and Irkin, 2010); a level of 0.5% spirulina powder was found acceptable with regard to the sensory quality of fermented products. The sensory score of yogurts prepared using 0.5% spirulina powder was superior to the one made using 1.0% spirulina powder. The viable counts of LAB in probiotic yogurt containing spirulina powder were ≥ 6.0 CFU/g, much greater than the counts associated with control yogurt even after a 30-day storage period at 4°C (Guldas and Irkin, 2010).

16.8.4 JERUSALEM ARTICHOKE AS PREBIOTIC

The natural raw material known as Jerusalem artichoke contains the sugars inulin, oligofructose, and fructose. Such constituents are claimed to possess both nutritional as well as functional traits, especially beneficial to persons afflicted with type 2 diabetes and obesity (El-Kholy and Mahrous, 2015).

16.8.5 RESISTANT STARCHES AS PREBIOTIC

In vivo studies revealed the ability of resistant starch and non-starch polysaccharides to stimulate *Bifidobacteria. Bifidobacteria* and FOS, *Lactobacillus* GG and inulins, and *Bifidobacteria* and *Lactobacilli* with FOS or inulins are the most common synbiotic combinations currently available (Fuentes-Zaragoza et al., 2011).

16.9 APPLICATION OF PREBIOTICS IN PROBIOTIC AND SYNBIOTIC DAIRY FOODS

16.9.1 FERMENTED MILKS

Use of maltodextrin (16 dextrose equivalent, DE) at a 3.0% rate as a prebiotic ingredient stimulated the probiotic bacteria (*Bifidobacterium longum* ATCC 15707) count and enhanced the sensory score of fermented milk compared to the use of inulin (from Jerusalem artichoke) at the same level (Bisar et al., 2015).

The greatest viability of *Bifidobacterium lactis* BB-12 was obtained in Doogh (a drink made from fermented milk in Iran) when 3.0% of Hi-maize was used as a prebiotic; the least viability of *Bifidobacteria* was noted in control Doogh, devoid of the prebiotic substance (Noubakhti et al., 2009).

Villegas et al. (2009) optimized the acceptability of prebiotic low-fat beverage formulations containing a few types of inulin, namely (a) a high level of short-chain molecules (2 to 10 monomers) and (b) long-chain length (>23 monomers). The beverages were made with varying levels of sucrose and inulin. The differences in the products' sensory acceptance were influenced by the formulation, especially those containing a specific type of inulin. The samples containing short-chain inulin had a somewhat superior acceptability score.

Incorporation of inulin in fermented milk drink at a level of 4.0% stimulated the growth of *L. acidophilus* and *B. longum*, leading to a greater number of probiotics after one month of storage. Such prebiotics improved the β-galactosidase activity during fermentation. Supplementation of this prebiotic led to the product having the highest viable count of *L. acidophilus* (7.92 CFU/ml) at the 10th hour of fermentation. In the case of *B. longum*, a markedly higher viable count was observed in the inulin-supplemented fermented milk than in the sucrose and control (milk without such supplementation). During storage, the presence of inulin in fermented milk maintained the population of *L. acidophilus* up to 7.27 log CFU/ml; the same was observed when using inulin as a prebiotic. Contrary to this, inulin tended to decrease the viability of *B. longum* (decreased to 6.01 log CFU/ml) during storage (Chen et al., 2004).

The firmness of fermented milks made using inulin as a prebiotic along with probiotic cultures such as *B. lactis, Lb. acidophilus,* and *Lb. rhamnosus* was enhanced markedly (increased by 14.0%) as influenced by such prebiotic additive; the period of cold storage also led to an increase in the firmness of fermented milk (i.e., 13.0% increase). In binary co-cultures (i.e., *B. lactis* with *Streptococci*) with *Str. thermophilus*, there was a drastic improvement in the count of probiotic *B. lactis*; the counts increased to 9.1 log CFU/ml from the beginning till the 7th day of refrigerated storage (Oliveira et al., 2011). Incorporation of quinoa flour up to a 3.0% level had a neutral effect on the probiotic activity by *B. animalis* subsp. *lactis* and *L. acidophilus* La-5. However, the nutritive value and the appeal of the fermented milk were improved. The fermented milk made using quinoa flour exhibited markedly lower titratable acidity values when compared to the control beverage at the entire storage period up to 28 days. However, the pH value of the fermented milk control was highest (pH of 4.31) on the 28th day of storage as compared to the pH value of the experimental fermented beverage containing 2.0% and 3.0% quinoa flour (i.e., pH value of 4.25) (Casarotti et al., 2014).

16.9.2 YOGURT

With the co-existence of inulin (at 40.0 mg/g of skim milk base) in probiotic yogurt, the time required by *Str. thermophilus* (ST), *B. animalis* subsp. *lactis* (BL), and ST-BL to complete fermentation (i.e., target pH being 4.5) was shorter than the conditions prevailing in the absence of inulin. There was an enhancement in the biomass growth and levels of lactic and acetic acids and volatile compounds in the presence of inulin (Oliveira et al., 2012).

In yogurt, prebiotics acted as a substrate for the growth and multiplication of probiotic bacteria and thereby improved gastrointestinal functions as well as the immune system. Kumari et al. (2015) noted an increase in the cell numbers of *Bifidobacterium* in yogurt incorporated with rice used as a prebiotic. In a similar note, Amarakoon et al. (2013) reported that cooked rice facilitated the growth and survival of probiotic bacteria, including *Bifidobacteria*.

Incorporation of both FOS and inulin at levels of 1.0–3.0% by weight in yogurt formulation led to an increase in the numbers of live cells of *Str. thermophilus, Lb. acidophilus*, and *Bifidobacterium* spp. as compared to the control product, devoid of such prebiotics. The respective counts were 9.0, 7.8, and 7.7 log CFU/g in yogurt made utilizing 1.0% FOS. Likewise, the *Bifidobacteria* count reached 7.5 log CFU/g in the presence of 1.0% inulin in yogurt. Probiotic yogurt containing resistant starch and FOS as prebiotics had greater apparent viscosity and hardness compared to control yogurt; the hardness tended to increase at a greater pace during refrigerated storage of 21 days as compared to that noted for control yogurt (Gustaw et al., 2011). Utilization of prebiotics such as lactulose (0.25 and 2.5%) and inulin (0.5 and 1.0%) in yogurt formulation did not influence the growth of yogurt starter and *Lb. acidophilus* LA-5. However, such prebiotics stimulated the growth of B. bifidum BB-02 (Ozer et al., 2005). The use of oligofructose was more effective on the growth pattern of *Lb. acidophilus* in yogurt as compared to inulin. However, the count of the prebiotic strain noted during refrigerated storage was lower by 1.0 log unit than in control yogurt; the decrease in the count of probiotic cells was only to the tune of 0.4 and 0.3 log units in the presence of inulin and oligofructose, respectively (Juhkam et al., 2007).

Probiotic yogurt has been prepared utilizing *Lactobacillus* casei Lc 01 culture along with prebiotics such as lactulose, inulin, and oligofructose. Use of inulin at a level of 1.5% led to a beneficial effect on the physicochemical properties (pH, proneness to syneresis) and resulted in superior viability of probiotic cells compared to the use of the other two prebiotic substances at the same level. The probiotic cell count of freshly prepared yogurt was 8.27 log CFU/ml vs. 8.18 and 8.17 log CFU/ml in the case of products containing lactulose and oligosaccharide as prebiotics (Aghajani et al., 2012). The pH of yogurt prepared using lactulose, inulin, and oligofructose was 4.38, 4.17, and 4.38 respectively; inulin tended to yield yogurt with the lowest pH. The control probiotic product (devoid of prebiotic substance) was associated with the lowest probiotic count and had the lowest scores for taste and texture. Yogurt was prepared supplemented with (i.e., 1.0–3.0% by weight) or without inulin to substitute milk fat with the prebiotic substance. Yogurt devoid of inulin had superior sensory attributes (color, appearance, texture, aroma, overall impression), followed by the samples containing an incremental quantity of inulin (Guven et al., 2005).

Mudgil et al. (2016) studied the influence of using partially hydrolyzed guar gum, which served as a soluble fiber in formulating functional yogurt. The guar gum was subjected to enzymatic hydrolysis by *Aspergillus niger*-derived cellulase and freeze-dried to obtain powder for use in yogurt. When compared to control yogurt, guar gum-containing yogurt showed an increase in pH, viscosity, water holding capacity, but had lower titratable acidity; the functional and sensory quality of the product was satisfactory. Hassan et al. (2015) reported that the incorporation of 2.5% guar gum or 0.5% cress seed mucilage improved the quality of set yogurt. However, fermentation time not affected during storage of yogurt at 5°C for up to 15 days. Yogurts possessing acceptable sensory traits (i.e., flavor, appearance, body, and texture) have been prepared utilizing a formulation containing polysaccharides sourced from taro corm (*Arum colocasia*) at 0.3%, mature okra fruit (*Hibiscus esculents*) at 0.1%, and Jew's-mallow (*Corchorus olitorius*) at 0.1% (Hussein et al., 2011); all of these served as prebiotic substances.

A favorable influence of *S. platensis* (spirulina) powder on the survival of the LAB during refrigerated storage of yogurt has been reported. The sensory scores of yogurts containing 0.5% spirulina powder were superior to the product containing 1.0% of such additive. The use of a higher level (i.e., 1.0%) of spirulina powder contributed to a faint green color and algal flavor in the resultant yogurt, decreasing its sensory acceptability (Guldas and Irkin, 2010).

16.9.3 ICE CREAM AND FROZEN YOGURT

Many studies have incorporated prebiotic substances in producing bio-ice cream (i.e., synbiotic product). Inulin and oligofructose as prebiotics, which serve as fat replacers too, possess an array of functional and nutritional traits that may be used to develop innovative healthy foods. Prebiotic ingredients typically have a greater impact on the flavor and texture of ice cream, whereas probiotics mainly affect the flavor of the finished product (Cruz et al., 2010).

The loss of viability of probiotic organisms in ice cream can take place either during the production stage or at the time of frozen storage (Jay et al., 2005). At each stage of ice cream production, the probiotic organisms are prone to be stressed; the factors include detrimental chemical matter in the product formulation (viz., pH/titratable acidity, sugar content), freezing injuries, oxygen toxicity, and even mechanical stresses. At the time of frozen storage of ice cream, heat shock should be avoided. Failure to adopt such practices would lead to periodic partial melting and refreezing of ice, injuring the probiotic cells.

Akin et al. (2007) specifically prepared ice cream with probiotic bacteria (*L. acidophilus* and *B. lactis*) by mixing fortified milk fermented using probiotic strains with ice cream mixes, employing varying sugar levels (i.e., 15.0–21.0%). Cultures were grown at 37°C for 12 hours in UHT. The presence of inulin in the formulation stimulated the growth of *L. acidophilus* and *B. lactis*, which resulted in improved cell viability. Another study reported improvement in the viability of probiotic organisms along with stability to freezing, even under acidic conditions (pH 5.5) in ice cream upon the use of inulin and/or oligofructose (Champagne and Rastall, 2009).

Increasing the amount of inulin in ice cream mixes increased the firmness of the products and their resistance to melting. Inulin served as a stabilizer due to its capacity to bind water (El-Nagar et al., 2002, Akalin and Erişir, 2008). The interactions of the dietary fiber and liquid components of the probiotic ice cream mix were attributed to the high apparent viscosity in probiotic ice cream mixes containing oligofructose or inulin (Akalin and Erisir, 2008).

An acceptable quality vanilla-flavored synbiotic ice cream was prepared utilizing 2.5% inulin as a prebiotic, incorporating *Lb. casei* as the probiotic culture. The LAB count was markedly higher in synic ice cream as compared to control ice cream (without prebiotic); the respective counts were 6.5–6.9 log CFU/g and 3.0 log CFU/g. The freezing of ice cream mix and storage of synbiotic ice cream (at -20°C) was not detrimental to the probiotic count. The probiotic vanilla ice cream was organoleptically rated slightly lower than the score of control vanilla ice cream. The probiotic ice cream containing 5.0 and 10.0% inulin tended to be firmer by 50.0% and 25.0% respectively when compared to control ice cream; experimental ice creams had about 10.0% lower overrun compared to control product (Di Criscio et al., 2010).

Supplementation of yog-ice cream (i.e., frozen yogurt) with FOS at a level of 8 8.0% led to an improvement in the overrun of the product, with a simultaneous decrease in its firmness, especially when probiotic cells of *Lb. acidophilus* (la-5) microencapsulated in alginate beads were used. The decrease in the cells during frozen storage of yog-ice cream was just 1.0 log cycle (i.e., 9.7 log10 CFU/g) at the 60th day of storage owing to the presence of FOS and encapsulation of culture (Ahmadi et al., 2014).

The observed increased apparent viscosity of such ice cream mixes was most likely due to an interaction between inulin and milk proteins. The effect of inulin on improving the viscosity of mixes has been reported by El-Nagar et al. (2002) and Akin (2005) for yogurt-ice cream and probiotic fermented ice cream, respectively.

16.9.4 CHEESE SYSTEM

Synbiotic Petit-Suisse cheese made utilizing 10.0% oligofructose, and a combination of inulin and oligofructose (1:1, at 10.0% by weight) performed best in terms of prebiotic potential, good viable counts of *Lb. acidophilus* and *B. animalis* ssp. *lactis*, and sensory quality as compared to the use of other prebiotic substances (viz., honey, a mixture of honey, and the use of inulin and oligofructose per se) (Cardarelli et al., 2008); the viable *B. animalis* subsp. lactis count (log CFU/g) of cheese at the 28th day of refrigerated storage was 7.37, 7.35, 7.34, and 7.32 respectively for samples containing combined inulin and oligofructose (5.0% each by weight), 10.0% inulin, 10.0% oligofructose, and the control (without prebiotics) product. The sensory overall scores (9-point hedonic scale) of stored (28 days at 4°C) cheeses were 7.43, 7.00, 7.43, and 6.63 respectively for the same products sequentially. A marked increase in the sensory acceptance of Petit-Suisse cheese was noted during storage, especially in samples supplemented with oligofructose and/or inulin (Cardarelli et al., 2008).

The most favored synbiotic goat cheese contained *E. faecium* and oligofructose; other cultures were *Lb. paracasei* and *B. longum*. The cheese formulation involved the use of 1.5% and 3.0% each of inulin and oligofructose. The number of aromatic compounds associated with cheese LP-IO (*Lb. paracasei* with 1.5% each of inulin and oligofructose; LP-IO) was maximum (19 compounds) while the minimum number of aromatic compounds was associated with cheeses LP-O (only oligofructose), BL-C (Bifido without prebiotic), and BL-O (Bifido with oligofructose). The aromatic substances identified were 2-butanone-3-hydroxy, decanoic acid, hexanoic acid, heptanoic acid, and octanoic acid. The number of aromatic compounds tended to increase in all the cheeses during storage. The cheeses made using inulin and oligofructose tended to be softer than those made without such additives. The prebiotics used at 1.5–3.0% levels failed to markedly influence the hardness of cheeses but led to greater adhesiveness and gumminess. At the initial stages of storage, the gumminess of cheeses containing inulin-oligofructose (*E. faecium* EF-IO, LP-IO, BL-IO) was lower than cheeses without prebiotics. The least chewiness (7.72 kg vs. 12.15 kg for control cheese) was associated with cheese containing the inulin-oligofructose blend. The highest color value was noted for cheese BL-O (*B. longum*, 3.0% oligofructose). The prebiotics tended to affect the color scores of cheeses; the control cheese had a superior score. Cheeses made utilizing combined inulin and oligofructose had a lower sensory texture score towards the end of the storage period. Synbiotic cheeses containing *E. faecium* were associated with the highest sensory texture scores. Cheeses with inulin were awarded the least taste scores; a slight metallic taste was perceived by some panelists (Kinik et al., 2017).

Incorporation of the symbiont *Saccharomyces boulardii* (such yeast was isolated from the bark of the Litchi tree) and inulin in the cheeses impacted positively on the quality sensory scores; both the yeast cell and prebiotic substance were encapsulated using sodium alginate and cactus mucilage (Zamora-Vega et al., 2013). Such yeast was resistant to gastric acidity and proteolysis, paving the way for implantation in the gastrointestinal tract in viable numbers (McFarland and Bernasconi, 1996). The cheese stored at 4°C maintained *S. boulardii* count at a level ≥ 10^7 CFU/g for up to 4 weeks of storage; the cheese was categorized as a probiotic. The overall sensory acceptability of synbiotic fresh cheese containing *S. boulardii* and inulin was superior to the control cheese. Araujo et al. (2009) did not notice any marked differences in taste, texture, and overall acceptability between the synbiotic type cottage cheese made with *L. delbrueckii* and inulin and the control cheese when stored for up to 15 days at 4°C.

16.10 *IN VITRO* STUDY WITH PREBIOTICS

The interaction between inulin and milk proteins, namely *Lb. acidophilus* LAFTI L10 (L10), *B. animalis lactis* LAFTI B94 (B94), and *Lb. casei* LAFTI L26 (L26), was most likely the cause of the observed increased apparent viscosity of such ice cream mixtures. All three probiotic microbes had the ability to utilize a range of carbohydrates. Maximal growth of L10 was noted in a basal medium enriched with SOS, followed by FOS and inulin. In contrast, the maximum growth of the B94 strain was noted in a basal medium enriched with SOS, followed by raffinose, FOS, β-glucan hydrolysate, inulin, and Fibregum. The strain L26 exhibited maximum growth in a basal medium enriched with FOS followed by inulin, SOS, β-glucan hydrolysate, and β-glucan concentrate (Su et al., 2007).

16.11 PREBIOTICS AND HUMAN HEALTH

Prebiotic administration has an impact on the host's epithelial, immune, nervous, and endocrine signaling and has a positive impact on the host's health, including bowel function, immune response, glucose and lipid metabolism, bone health, appetite, and feeling full (Cunningham et al., 2021).

Prebiotic compounds (i.e., FOS, GOS) when incorporated into symbiotic products could markedly improve the probiotic culture viability, especially of *Bifidobacteria* in food as well as in the gastrointestinal tract (Gibson et al., 2004; Rycroft et al., 2001).

16.11.1 MINERAL ABSORPTION

Studies involving experimental rats have indicated that non-digestible oligosaccharides such as inulin and oligofructose tend to enhance the absorption and retention of minerals such as Ca, Mg, Fe, and Zn (Andersson et al., 2001; Scholz-Ahrens and Schrezenmeir, 2002; Kaur et al., 2021). Similar results have been reported when using pectin as a prebiotic. Such improved mineral absorption could be attributed to the increased solubility of Ca due to the lower pH of the colonic content induced by fermentation of oligosaccharides. Inulin and oligofructose prebiotics are reported to be indigestible and fermentable fructans that promote the absorption of calcium, leading to improved bone density (Thammarutwasik et al., 2009).

16.11.2 CONTROLLING OBESITY

An increase in satiety and therefore a decrease in appetite led to a decrease in total energy uptake, coupled with a decrease in hepatic *de novo* lipogenesis in human volunteers fed on inulin-type prebiotics (16.0 g per day for a few weeks) as opposed to feeding with maltodextrin as a placebo trial (Cani et al., 2006; Delzenne and Cani, 2010).

16.11.3 LIPID-LOWERING AND BULKING EFFECTS

The effect of inulin as a prebiotic has demonstrated serum lipid-lowering effects in animals; attempts to reproduce similar effects in human subjects have led to conflicting results. According to animal studies, the triacylglyceride-lowering effect of inulin has been attributed to the inhibition of fatty acid synthesis (Andersson et al., 2001). According to reports, the prebiotics inulin and oligofructose are indigestible and fermentable fructans that lowered cholesterol levels and stimulated the development and activity of probiotic cells (Thammarutwasik et al., 2009). Prebiotics also have a favorable influence on plasma lipids (Andersson et al., 2001).

An increase in fecal wet weight of $\geq$ 1 g/g ingested oligofructose and 1.5–2.0 g/g ingested inulin has been reported in the literature (Andersson et al., 2001). Since oligosaccharides are fermented entirely, their bulking effect (i.e., fecal bulk) is expected to take place through an increased microbial mass.

16.11.4 ANTI-OXIDATIVE EFFECT

Consumption of synbiotic food containing 4×10^8 CFU/ml of *Lb. casei* and 400.0 mg of inulin rendered a positive influence on human plasma antioxidant capacity and antioxidant enzyme activities. Additionally, such synbiotic food improved the absorption of minerals, prevented diarrhea, and favored better assimilation of nutrients (Andersson et al., 2001; Buterikis et al., 2008; Kleniewska et al., 2016).

The dairy products that are preferentially enriched with prebiotics include yogurts and yogurt drinks. However, ice cream, frozen yogurt, spreads, fresh cheeses, and milks have also been harnessing such valued substances. Incorporating prebiotics in commonly consumed food will help in ushering the intake of prebiotic ingredients for improved human health. The unpacking of newer prebiotic/probiotic/synbiotic functional foods will allow the food industry to constantly renew itself by introducing a variety of food items with enhanced nutritional value and health benefits to a wide range of consumers. Both in the laboratory as well as in the food industry, value-added dairy foods such as yogurt, cheese, ice cream, frozen yogurt, and other fermented dairy foods have emerged as predominant foods that have served as a vehicle for the probiotic microbes. This, in turn, has paved the way for the production of prebiotic ingredients, probiotic/prebiotic food, and even synbiotic food. The selection of probiotic bacteria that can synthesize prebiotics would be more economical than food utilizing both the probiotics and prebiotics.

KEYWORDS

- **bacteria**
- **carbohydrates**
- **fatty acids**
- **food production**
- **lactulose**
- **nutrition**
- **oligosaccharides**
- **physical strain**
- **prebiotics**
- **probiotics**

REFERENCES

Adebola, O. O., Corcoran, O., & Morgan, W. A. (2014). Synbiotics: The impact of potential prebiotics inulin, lactulose, and lactobionic acid on the survival and growth of lactobacilli probiotics. *Journal of Functional Foods, 10*, 75–84.

Aghajani, A. R., Pourahmad, R., & Adeli, H. R. M. (2012). Evaluation of physicochemical changes and survival of probiotic bacteria in symbiotic yoghurt. *Journal of Food Biosciences & Technology, Islamic Azad University, Science and Research Branch, 2*, 13–22.

Ahmadi, A., Milani, E., Madadlou, A., Mortazavi, S. A., Mokarram, R. R., & Salarbashi, D. (2014). Synbiotic yogurt-ice cream produced via incorporation of microencapsulated *Lactobacillus acidophilus* (LA–5) and fructooligosaccharide. *Journal of Food Science & Technology, 51*(8), 1568–1574.

Akalin, A. S., & Erisir, D. (2008). Effects of inulin and oligofructose on the rheological characteristics and probiotic culture survival in low-fat probiotic ice cream. *Journal of Food Science, 73*(4), 184–188.

Akalin, A. S., Unal, G., & Dalay, M. C. (2009). Influence of *Spirulina platensis* biomass on microbiological viability in traditional and probiotic yogurts during refrigerated storage. *Italian Journal of Food Science, 21*(3), 357–364.

Akin, M. B., Akin, M. S., & Kirmaci, Z. (2007). Effects of inulin and sugar levels on the viability of yogurt and probiotic bacteria and the physical and sensory characteristics in probiotic ice cream. *Food Chemistry, 104*(1), 93–99.

Akin, S. (2005). Effects of inulin and different sugar levels on the viability of probiotic bacteria and the physical and sensory characteristics of probiotic fermented ice cream. *Milchwissenschaft, 60*(3), 297–301.

Akin, S., Ozer, D., & Ozer, B. (2005). Effect of inulin and lactulose on survival of *Lactobacillus acidophilus* LA-5 and *Bifidobacterium bifidum* BB-02 in Acidophilus-Bifidus yoghurt. *Food Science and Technology International, 11*(1), 19–24.

Amarakoon, A. M. S. B. K., Jayawardane, I. N. T., Senaviratne, N. D., Ranadheera, C. S., Silva, K. F. S. T., & Vidanarachchi, J. K. (2013). Development of a rice incorporated symbiotic yogurt. In J. K. Vidanarachchi & S. M. C. Himali (Eds.), *Proceedings of the 23rd Annual Student Research Sessions* (p. 5). Department of Animal Science, University of Peradeniya, Sri Lanka.

Ambrogi, V., Bottacini, F., O'Callaghan, J., Casey, E., van Breen, J., Schoemaker, B., Cao, L., Kuipers, B., O'Connell, M. M., Schoterman, M., & van Sinderen, D. (2021). Infant-associated bifidobacterial β-galactosidases and their ability to synthesize galacto-oligosaccharides. *Frontiers in Microbiology, 12*, 949.

Andersson, H., Asp, N. G., Bruce, A., Roos, S., Wadstrom, T., & Wold, A. E. (2001). Health effects of probiotics and prebiotics: A literature review on human studies. *Scandinavian Journal of Nutrition/Näringsforskning, 45*(1), 58–75.

Araujo, E. A., Carvalho, A. F., Eleana, S. L., Furtado, M. M., & Moraes, C. A. (2009). Development of a symbiotic cottage cheese added with *Lactobacillus delbrueckii* UFV H2b20 and inulin. *Journal of Functional Foods, 2*(1), 85–89.

Aryana, K. J., & McGrew, P. (2007). Quality attributes of yogurt with *Lactobacillus casei* and various prebiotics. *LWT–Food Science and Technology, 40*(10), 1808–1814.

Bisar, G. H., El-Saadany, K. H., Khattab, A., & El-Kholy, W. M. (2015). Implementing maltodextrin, polydextrose, and inulin in making a symbiotic fermented dairy product. *British Microbiology Research Journal, 8*(5), 585–603.

Buterikis, G., Matusevikcius, P., Januskevicius, A., Jankowski, J., Mikulski, D., Blok, J., & Reid, G. (2008). Evaluation of reuterin production in urogenital probiotic *Lactobacillus reuteri* RC-14. *Journal of Applied and Environmental Microbiology, 74*, 4645–4649.

Caire, G. Z., Parada, J. L., Zaccaro, M. C., & Cano, M. M. S. (2000). Effect of *Spirulina platensis* biomass on the growth of lactic acid bacteria in milk. *World Journal of Microbiology and Biotechnology, 16*(6), 563–565.

Caldeira, L. A., Alves, E. E., Ribeiro, A. de M. F., Junior, V. R. R., Antunes, A. B., dos Reis, A. F., Gomes, J. da C., de Carvalho, M. H. R., & Martinez, R. I. E. (2018). Viability of probiotic bacteria in bioyogurt with the addition of honey from Jatai and Africanized bees. *Pesquisa Agropecuária Brasileira, 53*(2), 206–211.

Cani, P. D., Joly, E., Horsmans, Y., & Delzenne, N. M. (2006). Oligofructose promotes satiety in healthy humans: A pilot study. *European Journal of Clinical Nutrition, 60*(5), 567–572.

Cardarelli, H. R., Buriti, F. C. A., Castro, I. A., & Saad, S. M. I. (2008). Inulin and oligofructose improve sensory quality and increase the probiotic viable count in potentially synbiotic petit-suisse cheese. *LWT–Food Science and Technology, 41*(6), 1037–1046.

Casarotti, S. N., & Penna, A. L. B. (2015). Acidification profile, probiotic in vitro gastrointestinal tolerance, and viability in fermented milk with fruit flours. *International Dairy Journal, 41*(1), 1–6.

Casarotti, S. N., Carneiro, B. M., & Penna, A. L. B. (2014). Evaluation of the effect of supplementing fermented milk with quinoa flour on probiotic activity. *Journal of Dairy Science, 97*(10), 6027–6035.

Celekli, A., Alslibi, Z. A., & Bozkurt, H. (2018). Use of spirulina in probiotic fermented milk products. *International Journal of Advanced Science, Engineering and Technology, 6*(3), 42–46.

Champagne, C. P., & Rastall, R. A. (2009). Some technological challenges in the addition of probiotic bacteria to foods. In D. Charalampopoulos & R. A. Rastall (Eds.), *Prebiotics and Probiotics: Science and Technology* (pp. 763–806). Berlin: Springer.

Chen, M. J., Chen, K. N., & Lin, C. W. (2004). Optimization of the viability of probiotics in a fermented milk drink by the response surface method. *Asian-Australasian Journal of Animal Sciences, 17*(5), 705–711.

Crittenden, R., & Playne, M. (1996). Production, properties, and applications of food-grade oligosaccharides. *Trends in Food Science & Technology, 7*(11), 353–361.

Cruz, A. G., Cadena, R. S., Walter, E. H. M., Mortazavian, A. M., Granato, D., Faria, J. A. F., & Bolini, H. M. A. (2010a). Sensory analysis: Relevance for prebiotic, probiotic, and synbiotic product development. *Comprehensive Reviews in Food Science and Food Safety, 9*(4), 358–373.

Cruz, A. G., Faria, J. A. F., Andrade, R. R., Moura, M. R. L., Carvalho, L. M. J., & Granato, D. (2010b). Prebiotic in dairy products: Benefits to health and technology features. In J. Krause & O. Fleischer (Eds.), *Industrial Fermentation: Food Processes, Nutrient Sources and Production Strategies* (pp. 1–14). New York: Nova Publisher.

Cunningham, M., Azcarate-Peril, M. A., Barnard, A., Benoit, V., Grimaldi, R., Guyonnet, D., Holscher, H. D., Hunter, K., Manurung, S., Obis, D., Petrova, M. I., Steinert, R. E., Swanson, K. S., van Sinderen, D., Vulevic, J., & Gibson, G. R. (2021). Shaping the future of probiotics and prebiotics. *Trends in Microbiology* (in press).

Davani-Davari, D., Negahdaripour, M., Karimzadeh, I., Seifan, M., Mohkam, M., Masoumi, S. J., Berenjian, A., & Ghasemi, Y. (2019). Prebiotics: Definition, types, sources, mechanisms, and clinical applications–A review. *Foods, 8*(3), 92–118.

Delzenne, N. M., & Cani, P. D. (2010). Nutritional modulation of gut microbiota in the context of obesity and insulin resistance: Potential interest of prebiotics. *International Dairy Journal, 20*(4), 277–280.

Di Criscio, T., Frantianni, A., Minogna, R., Cinquanta, L., Coppola, R., Sorrentino, E., & Panfili, G. (2010). Production of functional probiotic, prebiotic, and symbiotic ice creams. *Journal of Dairy Science, 93*(10), 4555–4563.

do Espirito-Santo, A. P., Cartolano, N. S., Silva, T. F., Soares, F. A. S. M., Gioielli, L. A., Perego, P., Converti, A., & Oliveira, M. N. (2012). Fibers from fruit by-products enhance probiotic viability and fatty acid profile and increase CLA content in yogurts. *International Journal of Food Microbiology, 154*(3), 135–144.

Douglas, L. (2005). Prebiotics overview. GTC nutrition handout. Available from: http://www.npicenter.com/anm/templates/newsNWTemp.aspx?articleid=10723&zoneid=49S (accessed on 19 July 2024).

Dwivedi, S., Sahrawat, K., Puppala, N., & Ortiz, R. (2014). Plant prebiotics and human health: Biotechnology to breed prebiotic-rich nutritious food crops. *Electronic Journal of Biotechnology, 17*(5), 238–245.

El-Kholy, W. M., & Mahrous, H. (2015). Biological studies on bio-yoghurt fortified with prebiotics obtained from Jerusalem artichoke. *Food and Nutrition Sciences, 6*(16), 1552–1564.

El-Nagar, G., Clowes, G., Tudorica, C. M., & Kuri, V. (2002). Rheological quality and stability of yogurt-ice cream with added inulin. *International Journal of Dairy Technology, 55*(2), 89–93.

FAO/WHO. (2007). *FAO Technical Meeting on Prebiotics: Food Quality and Standards Service (AGNS)*. Food and Agriculture Organization of the United Nations (FAO), Rome, Italy: September 15–16, 2007. FAO Technical Meeting Report.

Fazilah, N. F., Ariffa, A. B., Khayat, M. E., Rios-Solis, L., & Halim, M. (2018). Influence of probiotics, prebiotics, symbiotic, and bioactive phytochemicals on the formulation of functional yogurt. *Journal of Functional Foods, 48*, 387–399.

Figueroa-Gonzalez, I., Hernandez-Sanchez, H., Rodriguez-Serrano, G., Gomez-Ruiz, L., García-Garibay, M., & Cruz-Guerrero, A. (2010). Antimicrobial effect of *Lactobacillus casei* strain Shirota co-cultivated with *Escherichia coli* UAM0403. *Revista Mexicana de Ingeniería Química, 9*(1), 11–16.

Figueroa-Gonzalez, I., Quijano, G., Ramirez, G., & Cruz-Guerrero, A. (2011). Probiotics and prebiotics - perspectives and challenges. *Journal of the Science of Food and Agriculture, 91*(8), 1341-1348.

Figueroa-Gonzalez, I., Rodríguez-Serrano, G., Gomez-Ruiz, L., García-Garibay, M., & Cruz-Guerrero, A. (2019). Prebiotic effect of commercial saccharides on probiotic bacteria isolated from commercial products. *Food Science and Technology, Campinas, 39*(3), 747–753.

Fuentes-Zaragoza, E., Sanchez-Zapata, E., Sendra, E., Sayas, E., Navarro, C., Fernandez-Lopez, J., & Perez-Alvarez, J. A. (2011). Resistant starch as prebiotic: A review. *Starch–Starke, 63*(7), 406–415.

Fuller, R., & Gibson, G. R. (2008). Probiotics and prebiotics: Microflora management for improved gut health. *Clinical Microbiology and Infection, 4*(9), 477–480.

Gibson, G. R., & Roberfroid, M. B. (1995). Dietary modulation of the human colonic microbiota: Introducing the concept of prebiotics. *Journal of Nutrition, 125*(6), 1401–1412.

Gibson, G. R., Rabiu, B., Rycroft, C. E., & Rastall, R. A. (2004). Trans-galactooligosaccharides as prebiotics. In C. Scott & J. O'Brien (Eds.), *Handbook of Functional Dairy Products* (pp. 99–109). Boca Raton: CRC Press.

Guldas, M., & Irkin, R. (2010). Microflora of yoghurt and acidophilus milk. *Mljekarstvo, 60*(4), 237–243.

Gulewicz, P., Ciesieolka, D., Frias, J., Vidal-Valverde, C., Frejnagel, S., Trojanowska, K., & Gulewicz, K. (2003). Simple method of isolation and purification of T-galactosides from legumes. *Journal of Agricultural and Food Chemistry, 48*(8), 3120–3123.

Gullon, B., Pintado, M. E., Fernandez-Lopez, J., Perez-Alvarez, J. A., & Viuda-Martos, M. (2015). In vitro gastrointestinal digestion of pomegranate peel (Punica granatum) flour obtained from co-products: Changes in the antioxidant potential and bioactive compounds stability. *Journal of Functional Foods, 19*(Part A), 617–628.

Gupta, S., Gupta, C., Garg, A. P., & Prakash, D. (2017). Prebiotic efficiency of blue-green algae on probiotics microorganisms. *Journal of Microbiology & Experimentation, 4*(4), 11–14.

Gustaw, W., Kordowska-Wiater, M., & Koziol, J. (2011). The influence of selected prebiotics on the growth of lactic acid bacteria for bio-yoghurt production. *Acta Scientiarum Polonorum, Technologia Alimentaria, 10*(4), 455–466.

Guven, M., Yasar, K., Karaca, O. B., & Hayalaglu, A. A. (2005). The effect of inulin as a fat replacer on the quality of set-type low-fat yogurt manufacture. *International Journal of Dairy Technology, 58*(3), 180–184.

Gyenis, B., Szigeti, J., Molnar, N., & Varga, L. (2005). Use of dried microalgal biomasses to stimulate acid production and growth of Lactobacillus plantarum and Enterococcus faecium in milk. *Acta Agraria Kaposvariensis, 9*(2), 53–59.

Hassan, L. K., Haggag, H. F., El-Kalyoubi, M. H., El-Aziz, M. A., & El-Sayed, M. M. (2015). Physico-chemical properties of yoghurt containing cress seed mucilage or guar gum. *Annals of Agricultural Sciences, 54*(1), 1–8. https://doi.org/10.1016/j.aoas.2014.11.021.

Hoppert, K., Zahn, S., Janecke, L., Mai, R., Hoffmann, S., & Rohm, H. (2013). Consumer acceptance of regular and reduced-sugar yogurt enriched with different types of dietary fiber. *International Dairy Journal, 28*(1), 1–7.

Hussein, M. M., Hassan, F. A. M., Abdel Dayam, H. H., Salama, A., Enab, A. K., & Abd El-Galil, A. A. (2011). Utilization of some plant polysaccharides for improving yoghurt consistency. *Annals of Agricultural Sciences, 56*(2), 97–103.

Ishwarya, S. P., & Prabhasankar, P. (2014). Prebiotics: Application in bakery and pasta products. *Critical Reviews in Food Science and Nutrition, 54*(4), 511–522.

Jay, J. M., Loessner, M. J., & Golden, D. A. (2008). *Modern Food Microbiology*. Springer Science & Business Media, pp. 156–161.

Jimenez-Vera, R., Monroy, O., Corona-Cruz, A., & Garcia-Garibay, M. (2008). Construction of a model of the human proximal colon. *World Journal of Microbiology and Biotechnology, 24*(12), 2767–2774.

Juhkam, K., Elias, P., Roasto, M., & Tamme, T. (2007). Viability of Lactobacillus acidophilus in yoghurt containing inulin or oligofructose during refrigerated storage. *Milchwissenschaft, 62*(1), 52–54.

Kaur, A. P., Bhardwaj, S., Dhanjal, D. S., Nepovimova, E., Cruz-Martins, N., Kuca, K., Chopra, C., Singh, R., Kumar, H., Sen, F., Kumar, V., Verma, R., & Kumar, D. (2021). Plant prebiotics and their role in the amelioration of diseases. *Biomolecules, 11*(3), 440–465.

Kinik, O., Kesenkaş, H., Ergonul, P. G., & Akan, E. (2017). The effect of using pro and prebiotics on the aromatic compounds, textural and sensorial properties of symbiotic goat cheese. *Mljekarstvo, 67*(1), 71–85.

Kleniewska, P., Hoffmann, A., Pniewska, E., & Pawliczak, R. (2016). The influence of probiotic Lactobacillus casei in combination with prebiotic inulin on the antioxidant capacity of human plasma. *Journal of Oxidative Medicine and Cellular Longevity, 2016*. https://doi.org/10.1155/2016/1340903.

Koh, A., De Vadder, F., Kovatcheva-Datchary, P., & Bäckhed, F. (2016). From dietary fiber to host physiology: Short-chain fatty acids as key bacterial metabolites. *Cell, 165*(6), 1332–1345. https://doi.org/10.1016/j.cell.2016.05.041.

Kumar, S. G. V., Singh, S. K., Goyal, P., Dilbaghi, N., & Mishra, D. N. (2005). Beneficial effects of probiotics and prebiotics on human health. *Die Pharmazie–An International Journal of Pharmaceutical Sciences, 60*(3), 163–171.

Kumari, A. G. I. P., Ranadheera, C. S., Prasanna, P. H. P., Senevirathne, N. D., & Vidanarachchi, J. K. (2015). Development of a rice incorporated symbiotic yogurt with low retrogradation properties. *International Food Research Journal, 22*(5), 2032–2040.

Liu, Z., Wang, W., Huang, G., Zhang, W., & Ni, L. (2016). In vitro and in vivo evaluation of the prebiotic effect of raw and roasted almonds (Prunus amygdalus). *Journal of the Science of Food and Agriculture, 96*(5), 1836–1843.

Maftei, N. M. (2019). Probiotic, prebiotic and symbiotic products in human health. In *Frontiers and New Trends in Science of Fermented Food and Beverages*. London: IntechOpen.

Manning, T. S., & Gibson, G. R. (2004). Prebiotics. *Best Practice & Research Clinical Gastroenterology, 18*(2), 287–298.

Martin, F., Cachon, R., Pernin, K., De Coninck, J., Gervais, P., Guichard, E., & Cayot, N. (2011). Effect of oxidoreduction potential on aroma biosynthesis by lactic bacteria in nonfat yogurt. *Journal of Dairy Science, 94*(2), 614–622.

Martinez-Villaluenga, C., Frias, J., Gomez, R., & Vidal-Valverde, C. (2006). Influence of addition of raffinose family oligosaccharides on probiotic survival in fermented milk during refrigerated storage. *International Dairy Journal, 16*(7), 768–774.

Marx, S. P., Winkler, S., & Hartmeier, W. (2000). Metabolization of I2(2,6)-linked fructose-oligosaccharides by different bifidobacterial. *FEMS Microbiology Letters, 182*(1), 163–169.

McFarland, L., & Bernasconi, P. V. (1993). Saccharomyces boulardii: A review of an innovative biotherapeutic agent. *Microbial Ecology in Health and Disease, 6*(4), 157–171.

McFarlane, G. T., Steed, S., & Macfarlane, S. (2008). Bacterial metabolism and health-related effects of galacto-oligosaccharides and other prebiotics. *Journal of Applied Microbiology, 104*(2), 305–344.

Mohamadi, R., & Mortazavian, A. M. (2011). Technological aspects of prebiotics in probiotic fermented milks. *Food Reviews International, 27*(2), 192–212.

Mohammadi, R., Mortazavian, A. M., Khosrokhavar, R., & da Cruz, A. G. (2011). Probiotic ice cream: Viability of probiotic bacteria and sensory properties. *Annals of Microbiology, 61*(3), 411–424.

Mortazavian, A. M., Mohammadi, R., & Sohrabvandi, S. (2012). Delivery of probiotic microorganisms into gastrointestinal tract by food products. In *New Advances in the Basic and Clinical Gastroenterology* (Chapter 6). London: IntechOpen, 121–146.

Mudgil, D., Barak, S., & Khatkar, B. S. (2016). Development of functional yoghurt via soluble fiber fortification utilizing enzymatically hydrolyzed guar gum. *Food Bioscience, 14*(1), 28–33.

Mudgil, D., Barak, S., Patel, A., & Shah, N. (2018). Partially hydrolysed guar gum as a potential prebiotic source. *International Journal of Biological Macromolecules, 112*, 207–210.

Mussatto, S. I., & Mancilha, I. M. (2007). Non-digestible oligosaccharides. *Carbohydrate Polymers, 68*(3), 587–597.

Nagpal, R., Yadav, H., Puniya, A. K., Singh, K., Jain, S., & Marotta, F. (2007). Potential of probiotic and prebiotics for symbiotic functional dairy foods: An overview. *International Journal of Probiotics and Prebiotics, 2*(2/3), 75–84.

Niness, K. R. (2009). Nutritional and health benefits of inulin and oligofructose: Inulin and oligofructose: What are they? *American Society for Nutritional Sciences*, 1402S–1406S.

Niv, E., Halak, A., Tiommny, E., Yanai, H., Strul, H., Naftali, T., & Vaisman, N. (2016). Randomized clinical study: Partially hydrolyzed guar gum (PHGG) versus placebo in the treatment of patients with irritable bowel syndrome. *Nutrition & Metabolism, 13*(1), 1–7.

Noubakhti, A. R., Ehsani, M. R., Mousavi, S. M., & Mortazavian, A. M. (2009). Influence of lactulose and Hi-maize addition on viability of probiotic microorganisms in freshly made symbiotic fermented milk drink. *Milchwissenschaft, 64*(2), 191–193.

Ohashi, Y., Sumitani, K., Tokunaga, M., Ishihara, N., Okubo, T., & Fujisawa, T. (2015). Consumption of partially hydrolyzed guar gum stimulates Bifidobacteria and butyrate-producing bacteria in the human large intestine. *Beneficial Microbes, 6*(4), 451–455.

Oliveira de Souza, R. P., Florence-Franco, A. C., Perego, P., Oliveira, M. N., & Converti, A. (2011b). Use of lactulose as prebiotic and its influence on the growth, acidification profile and viable counts of different probiotics in fermented skim milk. *International Journal of Food Microbiology, 145*(1), 22–27.

Oliveira de Souza, R. P., Perego, P., de Oliveira, M. N., & Converti, A. (2011a). Effect of inulin as prebiotic and symbiotic interactions between probiotics to improve fermented milk firmness. *Journal of Food Engineering, 107*(1), 36–40.

Oliveira de Souza, R. P., Perego, P., de Oliveira, M. N., & Converti, A. (2012). Prebiotic effect of inulin on the growth and organic acid profile of *Bifidobacterium lactis* in co-culture with *Streptococcus thermophilus*. *Chemical Engineering Transactions, 27*, 277–282.

Ong, L., & Shah, N. P. (2009). Probiotic Cheddar cheese: Influence of ripening temperatures on survival of probiotic microorganisms, cheese composition, and organic acid profiles. *LWT–Food Science and Technology, 42*(7), 1260–1268.

Ozcan, O., Ozcan, T., Ersan, Y. L., Akpinar-Bayizit, A., & Delikanli, B. (2016). The use of prebiotics of plant origin in functional milk products. *Food Science and Technology, 4*(2), 15–22.

Pandey, K. R., Naik, S. R., & Vakil, B. V. (2015). Probiotics, prebiotics and symbiotic–A review. *Journal of Food Science and Technology, 52*(12), 7577–7587.

Parisi, G. C., Zilli, M., Miani, M. P., Carrara, M., Bottona, E., Verdianelli, G., Battaglia, G., Desideri, S., Faedo, A., Marzolino, C., & Tonon, A. (2020). High-fiber diet supplementation in patients with irritable bowel syndrome (IBS): A multicenter, randomized, open trial comparison between wheat bran diet and partially hydrolyzed guar gum (PHGG). *Digestive Diseases and Sciences, 47*(8), 1697–1704.

Parvez, S., Malik, K. A., Kang, S. A., & Kim, H. Y. (2015). Probiotics and their fermented food products are beneficial for health. *Journal of Applied Microbiology, 100*(6), 1171–1185.

Reider, S. J., Moosmang, S., Tragust, J., Trgovec-Greif, L., Tragust, S., Perschy, L., Przysiecki, N., Sturm, S., Tilg, H., Stuppner, H., & Rattei, T. (2020). Prebiotic effects of partially hydrolyzed guar gum on the composition and function of the human microbiota–Results from the PAGODA trial. *Nutrients, 12*(5), 1257.

resent and future in food science and technology. *Food Research International, 42*(1), 8–12.

Roberfroid, M. (2002). Functional foods: Concepts and application to inulin and oligofructose. *British Journal of Nutrition, 87*(Suppl. 2), S139–S143.

Rycroft, C. E., Jones, M. R., Gibson, G. R., & Rastall, R. A. (2001). A comparative in vitro evaluation of the fermentation properties of prebiotic oligosaccharides. *Journal of Applied Microbiology, 91*(5), 878–887.

Sangeetha, P. T., Ramesh, M. N., & Prapulla, S. G. (2005). Recent trends in the microbial production, analysis, and application of fructooligosaccharides. *Trends in Food Science and Technology, 16*(10), 442–457.

Sanz, M. L., Polemis, N., Morales, V., Corzo, N., Drakoularakou, A., Gibson, G. R., & Rastall, R. A. (2005). In vitro investigation into the potential prebiotic activity of honey oligosaccharides. *Journal of Agricultural and Food Chemistry, 53*(8), 2914–2921.

Saxelin, M., Korpela, R., & Mayra-Makinen, A. (2003). Introduction: Classifying functional dairy products. In T. Mattila-Sandholm & M. Saarela (Eds.), *Functional Dairy Products* (Chapter 11). Woodhead Publishing Ltd., Cambridge, England, 1–16.

Schaller-Povolony, L. A., & Smith, D. E. (2001). Viscosity and freezing point of a reduced-fat ice cream mix as related to inulin content. *Milchwissenschaft, 56*(1), 25–29.

Scholz-Ahrens, K. E., & Schrezenmeir, J. (2002). Inulin, oligofructose and mineral metabolism–Experimental data and mechanism. *British Journal of Nutrition, 87*(Suppl. 2), S179–186.

Schrezenmeir, J., & de Vrese, M. (2001). Probiotics, prebiotics, and synbiotics–approaching a definition. *American Journal of Clinical Nutrition, 73*(Suppl. 2), 361S–364S.

Sendra, E., Fayos, P., Lario, Y., Fernandez-Lopez, J., Sayas-Barbera, E., & Perez-Alvarez, J. A. (2008). Incorporation of citrus fibers in fermented milk containing probiotic bacteria. *Food Microbiology, 25*(5), 13–21.

Sharma, P., Singh, A., & Mathur, N. (2017). Review synbiotics–Combination of probiotics and prebiotics. *IOSR Journal of Biotechnology and Biochemistry, 3*(4), 19–24.

Shirota, M., Nagamatsu, N., & Takechi, Y. (1964). Method for cultivating lactobacilli. *U.S. Patent No. 3,123,538*. Cited by Ehsani, J., Mortazavian, A. M., Khomeiri, M., & Nejad, A. G. (2015). Effects of artichoke (*Cynara scolymus L.*) extract addition on microbiological and physico-chemical properties of probiotic yogurt. *Journal of Microbiology, Biotechnology and Food Sciences, 4*(6), 536–541.

Spiegel, J. E., Rose, R., Karabell, P., Frankos, V. H., & Schmitt, D. F. (1994). Safety and benefits of fructooligosaccharides as food ingredients. *Food Technology, 48*(1), 85–89.

Su, P., Henriksson, A., & Mitchell, H. (2007). Selected prebiotics support the growth of probiotic mono-cultures in vitro. *Anaerobe, 13*(3–4), 134–139.

Sveje, N. (2007). Prebiotics and probiotics–Improving consumer health through food consumption. *Nutracoss*, 28–31.

Thammarutwasik, P., Hongpattarakera, T., Chantachum, S., Kijroongrojana, K., Itharat, A., Reanmongkol, W., Tewtrakul, S., & Buncha, O. (2009). Prebiotic–A review. *Songklanakarin Journal of Science and Technology, 31*(4), 1–8.

Tuohy, K. M., Probert, H. M., Smejkal, C. W., & Gibson, G. R. (2003). Using probiotics and prebiotics to improve gut health. *Drug Discovery Today, 8*(15), 692–700.

Ustunol, Z., & Gandhi, H. (2001). Growth and viability of commercial *Bifidobacterium* spp. on honey-sweetened skim milk. *Journal of Food Protection, 64*(11), 1775–1779.

van Bueren, A. L., Mulder, M., van Leeuwen, S., & Dijkhuizen, L. (2017). Prebiotic galactooligosaccharides activate mucin and pectic galactan utilization pathways in the human gut symbiont *Bacteroides thetaiotaomicron*. Cited from: www.nature.com/articles/srep40478.pdf.

Villegas, B., & Costell, E. (2007). Flow behavior of inulin-milk beverages. Influence of inulin average chain length and of milk fat content. *International Dairy Journal, 17*(7), 776–781.

Villegas, B., Carbonell, I., & Costell, E. (2007). Inulin milk beverages: Sensory differences in thickness and creaminess using r-index analysis of the ranking data. *Journal of Sensory Studies, 22*(4), 377–393.

Wang, Y. (2009). Prebiotics: Pnovember

WGO Global Guidelines. (2011). Probiotics and prebiotics. *World Gastroenterology Organization Global Guidelines*, 3–28. Cited from: www.worldgastroenterology.org/guidelines/global-guidelines/probiotics-and-prebiotics (accessed on 19 July 2024).

Wichienchot, S., Jatupornpipat, M., & Rastall, R. A. (2010). Oligosaccharides of pitaya (dragon fruit) flesh and their prebiotic properties. *Food Chemistry, 120*(3), 850–857.

Yasukawa, Z., Inoue, R., Ozeki, M., Okubo, T., Takagi, T., Honda, A., & Naito, Y. (2019). Effect of repeated consumption of partially hydrolyzed guar gum on fecal characteristics and gut microbiota: A randomized, double-blind, placebo-controlled, and parallel-group clinical trial. *Nutrients, 11*(9), 2170.

Zamora-Vega, R., Montanez-Soto, J. L., Venegas-Gonzalez, J., Bernardino-Nicanor, A., Cruz, L. G., & Martínez-Flores, H. E. (2013). Development and characterization of a symbiotic cheese added with *Saccharomyces boulardii* and inulin. *African Journal of Microbiology Research, 7*(23), 2828–2834. https://doi.org/10.5897/AJMR2013.5566.

Zare, F., Boye, J. I., Orsat, V., Champagne, C. P., & Simpson, B. K. (2011). Microbial, physical, and sensory properties of yogurt supplemented with lentil flour. *Food Research International, 44*(8), 2482–2488.

Zare, F., Champagne, C. P., Simpson, B. K., Orsat, V., & Boye, J. I. (2012). Effect of the addition of pulse ingredients to milk on acid production by probiotic and yogurt starter cultures. *LWT–Food Science and Technology, 45*(2), 155–160.

CHAPTER 17

Prebiotic Food Product Development

HARSHA HIRDYANI

Unique Palate Sdn Bhd, Kuala Lumpur, Malaysia

The regular consumption of prebiotics has been shown to have several benefits. In many nations, the market for functional foods is growing, particularly for goods with added prebiotics.

17.1 INTRODUCTION

Viscosity, prebiotic resistance to thermal treatments, solubility, sweetness intensity provided by the prebiotic, and synergy with sweeteners are all important prebiotic properties to consider in product development. Consumption of prebiotics can sometimes cause a mild laxative effect, so there is a daily limit. Fermented milk and yogurts are the most common carriers of cultures of prebiotics and probiotics, but with the latest food trends, new products such as powdered milk, butter, mayonnaise, baby foods, cereal bars, jams, fermented vegetables, fermented meat products, bakery products, and many others are being introduced in the international market.

17.2 PREBIOTIC AS A FUNCTIONAL FOOD

Development of functional foods with prebiotic ingredients has become a trend in the food industry. It has become a very promising market due to its proven scientific benefits. This market is very promising for both economic and health-related reasons.

Prebiotic Foods for a Healthy Gut: Scientific Studies to Manage Metabolic Disorders.
Mini K. Sheth and Shruti Dwivedi (Eds.)

Customers are becoming more aware of appropriate nutrition and health, and they are increasingly seeking foods that offer multiple health benefits (Burgain et al., 2011). To meet the increasing needs of this market, a few prebiotics like inulin and fructooligosaccharides are being fortified into a wide variety of natural diet foods and beverages, including cereals, dairy products, breads, and dietary supplements (Coman et al., 2012). Prebiotics have a broad spectrum of health-promoting effects, which is why the food and pharmaceutical industries are leaning towards them. Prebiotics are used in the development of starter cultures, maintenance of intestinal health, cancer prevention, obesity reduction, treatment of constipation, and promotion of a healthy immune system (Delgado et al., 2011). Prebiotic supplementation and biotechnological synthesis of prebiotics from food waste are some recent trends. Researchers from all over the world are looking for substitutes to make it easier to obtain these ingredients so they can be more widely known and consumed.

Prebiotics play a beneficial role in the entire physiology of the intestines as they have the potential to promote bacterial activity and growth during fermentation in the colon.

Prebiotics are advantageous because of the bioprocess in which they produce microbiome biomass, which helps in reducing constipation (Mann et al., 2007). They function as fat and sugar substitutes and help in the absorption of calcium, lipids, prevent diarrhea, and other diseases like colon cancer, in addition to providing the sensory qualities in food.

17.3 RESISTANT STARCH

Resistant Starch (RS) is a type of natural prebiotic that is made up of starch and starch degradation products that are not absorbed in the small intestine of normal healthy individuals but instead get fermented by the colon's natural microflora to produce short-chain fatty acids. RS acts as a nutraceutical and if consumed provides multiple health benefits. The RS content of cereals and cereal products was determined by examining the effects of processing and storage on the RS content. The RS content of raw cereals varied from 0.53 g% (pearl millet) to 2.09 g% (rice). Shallow frying was followed by roasting, baking, and boiling, all of which increased the RS content. Frying and steaming both lower the RS content (Vaidya and Sheth, 2011).

17.4 REGULATORY ISSUES AND LABELING

Many countries do not require an approval of prebiotics usage as there is no established or implemented system for health claims, though authorities should have access to scientific support upon request. The following countries have specific policies regarding the use of prebiotics and the claims that can be made about them.

17.4.1 *UNITED STATES*

The 2010 Dietary Recommendations Advisory Committee reviewed systematic reviews of prebiotics and probiotics and health published since 2004 and concluded that although the gut microbiota's function is a significant and fairly new area of research, there is not sufficient data to confirm and recommend prebiotics or probiotics as dietary supplements (USDA Carbohydrates, 2010).

In the "Food Labeling: Revision of the Nutrition and Supplement Facts Labels; Proposed Rule," the FDA proposed changes to the voluntary declaration of starch and oligosaccharides and the declaration of indigestible carbohydrates as dietary fibers on the nutrition facts panel (Policy and Regulatory Development Related to Prebiotics, 2014).

17.4.2 *EUROPE*

In France, a dosage of inulin and FOS 5 g/d has been approved by the French Agency for Food, Environmental, and Occupational Health and Safety (AFSSA 2004; AFSSA, 2005). The following are the allowed claims: (1) "a bifidogenic effect at a daily dose of 5 g oligofructose;" (2) "A daily dose of 5 g/d of native inulin from chicory is bifidogenic (stimulates the growth of intestinal Bifidobacteria);" (3) "A daily dose of 5 g/d of native inulin from chicory is prebiotic." A daily dose of 5 g/d of native inulin from chicory is said to help the colon's healthy intestinal flora. United Kingdom. Inulin was approved as a prebiotic component of Vitaalbrood in the Netherlands at a ratio of 5 g inulin/100 g. (ANNCI 2008).

17.4.3 *JAPAN*

Prebiotics have been commercialized in Japan over the last 15 years after their classification as foods for specific health uses (FOSHU) by the Japanese Ministry of Health, Labor, and Welfare. "Foods containing an ingredient claiming its physiological effects on the human body are officially approved" and are eligible for FOSHU status (FOSHU, 2010). On-label health claims like "suitable for enhancing the control and regulation of the digestive system" or "improves bowel movement" or equivalent are permitted for all products containing dietary fiber and/or oligosaccharides.

17.4.4 SOUTH EAST ASIA

In South East Asia, the Agri-Food and Veterinary Authority of Singapore has approved Inulin and FOS as prebiotics (Agri-Food and Veterinary Authority of Singapore, 2011). In Malaysia as well, they have been approved and are allowed to be added to all foods except for infant formula (1.25 g/portion) (Tee and Sulong, 2010). Nutrient function claims in Malaysia describe the physiological role of these prebiotics in growth, development, and regular bodily functions (Nik Daud, 2011; Ministry of Health Malaysia, 2010). The following Malaysian nutrient function claims for prebiotics are permitted: "Inulin/FOS is prebiotic," "Inulin/FOS helps increase intestinal *Bifidobacteria* and helps maintain a good intestinal environment," and "Inulin/FOS is bifidogenic." The Thai FDA's Ministry of Health in Thailand has developed criteria and guidelines for the evaluation of safety, quality, and efficacy of probiotics and prebiotics in food products (FDA Thailand, 2011).

17.4.5 KOREA

For inulin and chicory extracts, the following health claims are permitted in Korea: A level of 9–10 g inulin/day helps to maintain a healthy blood cholesterol level and a healthy postprandial glucose level; a dosage of 8–20 g inulin/day helps to maintain a healthy bowel function (Amy et al., 2012).

17.4.6 SOUTH AMERICA

In Brazil, inulin and FOS have received approval as prebiotics at a dosage of 2.5 g/portion with a suggested daily intake of 5 g. In Chile, inulin and FOS have received approval as prebiotics at a dosage of 1.5 g/portion and a minimum of 3 g/day (Amy et al., 2012).

17.4.7 SOUTH AFRICA

In South Africa, prebiotics have been referred to as novel fibers and have been registered by the South African Health Products Regulatory Authority. They have been shown to have a physiologically beneficial effect on health. According to the guidelines, the following requirements should be satisfied for a substance to be considered a prebiotic: resistance to gastric acidity, hydrolysis by mammalian enzymes, GI absorption; fermentation by intestinal microbiota; stimulation of the growth of the entire native population of Bifidobacteria; and the selective stimulation of other native GI microbiota that support health and well-being.

17.5 EXTRACTION OF OLIGOSACCHARIDES

17.5.1 PREBIOTIC OLIGOSACCHARIDES FROM PLANTS

Fruits and vegetables that naturally contain prebiotics include onion, garlic, chicory, asparagus, artichoke, and banana. Animal colostrum and human milk also contain minimal amounts of prebiotics in the form of free sugars or glycoconjugates (Bucke and Rastall, 1990). A few years ago, it was discovered that dragon fruit contains oligosaccharides, which are a potential prebiotic (Wichienchot et al., 2010). Other plant materials, such as resistant starch, non-starch polysaccharides like pectin, cellulose, and hemicellulose, gums, and xylan, can also be used to produce prebiotics. Prebiotics can also be developed chemically from polysaccharides, including starch or sugars, by using the right enzymes. For example, GOS can be produced from lactose using β-galactosidase, FOS can be made from fructose monomers polymerized by fructosyltransferase, and isomaltooligosaccharide from transglucozylation of liquefied starch. Lactulose can be made by chemical processes (Vernazza et al., 2006). However, the ongoing search for reliable natural sources of prebiotics has been prompted by consumer preferences for natural products or ingredients.

Table 17.1 provides a list of several types of prebiotics, including oligomers with their sources that are best recognized (e.g., FOS, inulin, and GOS from plants) and the emerging prebiotics (IMO, XOS, and lactitol) (Al-Sheraji et al., 2013).

TABLE 17.1 Prebiotics From Plant Sources

Prebiotic	Sources	References
Fructooligosaccharide	Asparagus, sugar beet, garlic, chicory, onion, Jerusalem artichoke, wheat, honey, banana, barley, tomato and rye	(Sridevi et al., 2014)
Xylo oligosaccharides	Bamboo shoots, fruits, vegetables, milk, honey and wheat bran, coconut husk	(Panesar et al., 2013; Jnawali et al., 2017)
Galactooligosaccharides	Human's milk and cow's milk	(Torres et al., 2010)
Cyclodextrins	Water-soluble glucans	(Al-Sheraji et al., 2013)
Raffinose oligosaccharides	Seeds of legumes, lentils, peas, beans, chickpeas, mallow composite, and mustard	(ElSayed et al., 2014)
Lactulose	Lactose (Milk)	(Awad et al., 2014)
Palatinose	Sucrose	(Al-Sheraji et al., 2013)
Isomalt oligosaccharides	Honey, sugarcane juice	(Barea-Alvarez et al., 2014)
Lactosucrose	Lactose	(Zhou et al., 2014)
Malt oligosaccharides	Starch	(Beeren and Hindsgaul, 2014)
Soybean oligosaccharide	Soybean	(Fei et al., 2014)
Enzyme-resistant	Potato starch	(Lim et al., 2014)
Arabinoxylooligosaccharides	Wheat bran	(François et al., 2014)

17.5.2 PREBIOTIC OLIGOSACCHARIDES FROM ALGAE

According to the taxonomic group, different types of polysaccharides can be found in macroalgae. Polysaccharides with relatively high molecular weight are hydrophilic because they are rich in hydroxyl (OH) groups. They form intra-chain H-bond networks, which makes them rigid and stiff, making them suitable to be used as thickeners. Their regularity promotes interaction with external ions as well as inter-chain H-bonding (O'Sullivan et al., 2010). Different carbohydrates such as ulvans, carrageenan, and alginates are extracted from macroalgae and are widely used as functional ingredients in the food and pharmaceutical industries. Key polysaccharides in macroalgae such as Chlorophyta, Phaeophyta, and Rhodophyta and their prebiotic effect on the intestinal microflora are summarized in Table 17.2 (Warrand, 2006; O'Sullivan et al., 2010).

TABLE 17.2 Prebiotics from Marine Macroalgae

Prebiotic	Sources	Prebiotic effect
Alginate-(1–4) linked a-Lguluronic acid and b-Dmannuronic acid and pyranose residues	Brown seaweed species (Phaeophyta) but also bacterial sources (Azotobacter)	Alginate can alter the colonic microflora and the quantities of short chain fatty acids produced. Faecal *Bifidobacteria* levels increased while the number of some potentially pathogenic bacteria decreased during alginate consumption
Carrageenan- 1,3-linked b-Dgalactose and 1,4-linked a-D-galactose	Red seaweed species (Rhodophyta)	Degraded carrageenans may cause ulcerations in the GIT and gastrointestinal cancer
Ulvan- uronic acids, rhamnose (sulfated or not), xylose and the disaccharide composed of b -D-glucuronic acid-(1–4)-a-Lrhamnose	Green seaweed species (Ulva) (Chlorophyta)	Fermentated by *Bifidobacteria* and serve as growth promotors that are involved in the intestinal epithelial growth and wound repair

Probiotics have both interesting nutritional qualities and significant technological properties (Huebner et al., 2007, 2008). Some of them can be processed at an industrial level using renewable resources. They have a significant impact on foods and have shown to improve the organoleptic characteristics, enhancing both taste and mouthfeel, and improving nutritional quality. Studies show that commercialized products, particularly milk derivatives, contain probiotic bacteria within the minimal recommended limits, ensuring their viability for the duration of their shelf lives (Barreto et al., 2003). Prebiotics, as functional food ingredients, must be chemically stable to food processing treatments such as high temperature, low pH, and Maillard reaction conditions (Bohm et al., 2005).

17.5.3 DAIRY PRODUCTS

Dairy products have been used for a long time now to produce synbiotic foods and are a potential vehicle for prebiotics and probiotics. Prebiotics such as inulin, oligofructose, honey, and probiotics like *Lactobacillus acidophilus* and *Bifidobacterium lactis* were combined and tested in vitro to assess the potential of prebiotics in novel petit-suisse cheese formulations. It was observed that prebiotics make cheese ferment more quickly and produce more lactic acid, which encourages the growth of *bifidobacteria* and *lactobacilli*.

The addition of probiotics as a starter culture to control cheese resulted in a high level of lactic acid production (Rodrigues, 2012).

The physicochemical, physico-mechanical, and microbiological properties of unflavored yogurt were investigated, and it was observed that oligofructose had no effect on the pH, proteolysis, or viability of *Streptococcus thermophilus* or *Lactobacillus bulgaricus* after 28 days of refrigerated storage (Cruz, 2013). According to the survival analysis, 25% of consumers rejected the yogurt with various levels of oligofructose, and only 2.58% wt. V-1 oligofructose was acceptable.

In a study based on fermented probiotic milk, on day 1 of storage, the physicochemical composition and on days 1, 7, 14, and 21 microfiltered cow's milk samples fermented both with and without inulin were tested for rheological properties. The addition of inulin increased the total solid content and acidity of the prebiotic fermented milk (Debon et al., 2010). A dairy fruit beverage was developed with 9% glucan and green banana flour and had shown high acceptability by consumers (Guergoletto, 2010). Mumtaz et al. (2008) described the enrichment of yogurt with XOS at various levels. Chemical and enzymatic hydrolysis methods were used to extract the XOS used in the experiment from almond shells. Yogurt containing XOS was compared to yogurt containing a stabilizer (gelatin, 0.4% w/w) in addition to XOS for various characteristics. Moisture content, pH, acidity, and total solids were studied in the products. The use of stabilizer and rate of XOS incorporation had a significant effect on these attributes. The results also revealed that adding up to 3.5% XOS had no effect on taste or overall acceptability, but higher levels contributed to the aftertaste.

17.5.4 CEREALS AND CEREAL PRODUCTS

Gluten-free prebiotic breads were developed using arabinoxylan oligosaccharides from cereals (AXOS) which had shown high acceptability in terms of the breads' softness, traditional bread aroma, sweetness, and crumb color (Tuohy et al., 2001). A similar study replacing sugar with 20 g, 5 g, and 6 g of inulin respectively was also conducted on breads, cookies, and porridge that demonstrated very high acceptability (Parnami and Sheth, 2010). Another study in which three commonly consumed Indian products—dal, kadi, and khichdi—were examined by substituting with FOS at 10%, 10%, and 5% respectively, had also shown great acceptability (Gupta and Sheth, 2011).

Yacon potato (*Smallanthus sonchifolius*), a rich source of FOS and inulin, is also being used for the development of new products. Processing yacon yields pulps, juices, and flours that can be used as food additives. By incorporating yacon potatoes into breads, Rolim et al. (2011) confirmed a high-fiber food with a low glycemic index and prebiotic effect. The bread also showed the count of *Bifidobacteria* and *Lactobacilli* between 10^7 and 10^{10} ($UFCmL^{-1}$), which may be capable of colonizing the intestines given that the probiotic effects are only produced when 10^6 $UFCmL^{-1}$ or more are consumed. Researchers also examined the effects of adding yacon potato flour to chocolate cakes as sugar; it showed favorable effects on the cakes' chemical composition, glycemic index, and prebiotic capacity (Padilha et al., 2010).

In 2013, Mantzouridou studied fructooligosaccharides (FOS) and guar gum (PHGG) which were partially hydrolyzed, added to biscuits, and tested on volunteers. In a related study, Padma Ishwarya et al. (2013) investigated the effects of adding FOS to biscuits in syrup and powder form to offer both nutritional and technological benefits. FOS in powder form was more stable and, in comparison, had fewer negative effects on the characteristics of the biscuit. The impact of adding fructooligosaccharide on the sensory characteristics of two Indian fried foods, lilva kachori and vegetable paratha, was examined. When compared to the traditional recipes, fructooligosaccharide added to one serving of the stuffing for lilva kachori and vegetable paratha at 10 g, 15 g, and 5 g, 10 g, and 15 g, respectively showed that up to 15 g of fructooligosaccharide can be added to vegetable paratha and up to 5 g to lilva kachori. Fructooligosaccharides may be added to stuffed fried snacks to satisfy consumer demand for healthy foods (Jain et al., 2013).

Four food items were prepared: *chapati*, *thepla*, *dhokla*, and *patra*. In chapati and thepla, wheat flour was replaced with FOS at four levels: 3 g, 5 g, 8 g, and 10 g. FOS was substituted for Bengal gram in dhokla and patra at three levels, namely 5, 8, and 10 g. All organoleptic characteristics of chapati and thepla significantly decreased with an increase in substitution level of up to 10 g (Mahindra and Sheth 2013). The research has paved the way for the development of commonly consumed FOS-infused cereal and pulse-based food products.

17.5.5 FRUIT AND VEGETABLE-BASED PRODUCTS

Papaya and banana bars were developed using inulin (IN) and fructooligosaccharide (FOS) both in liquid as well as in powder in various proportions. The control bars contained 70% fruit pulp, 30% sugar, and 0.3% citric acid, and the treated bars contained 70% fruit pulp, 15% sugar, 15% IN and FOS, and 0.3% citric acid. Papaya bars with 90% FOS in powder form + 10% IN and banana bars with 90% FOS in liquid form + 10% IN were most acceptable (Megala and Hymavathi, 2011).

Fruit juice concentrates of apple and orange were analyzed with different proportions of sucrose and FOS (oligofructose + inulin 86 g/100 g, Glucose + fructose + sucrose 14 g/100 g) to achieve uniform sweetness. Fruit juice beverages were prepared (Ghavidel et al., 2014). One sample contained only sugar, samples number two, three, and four contained 2/3, 1/2, and 1/3 sucrose, respectively, and 1/3, 1/2, and 2/3 FOS. Sample number five contained only FOS. Upon sensory analysis, results concluded that there were no appreciable differences between the FOS-fortified fruit juice beverages and the control after two months of storage at 4°C, although improvements in flavor and odor were observed when the sugar and FOS ratios were increased. In another research, the physicochemical and sensory properties of fruit juices fortified with FOS were investigated over a 6-month period at ambient and refrigerated temperatures. FOS-fortified fruit beverages were acceptable even after 6 months of storage at ambient or refrigerated temperatures (Renuka et al., 2009).

Cashew juice and carbonated water in equal proportions were mixed to develop a functional beverage. Shelf life studies were conducted upon addition of FOS (2% w/v) to the mixture by applying heat treatments and preservatives. Sensory studies showed good positive results, and the product was shelf-stable for 30 days (Curi et al., 2013). FOS has also demonstrated good acceptability when added at 5% to lemon water and orange juice (Gupta and Sheth, 2011).

17.5.6 SOUPS

Different levels of FOS—2.5%, 4.5%, 5.5%, 6.5%, and 7.5%—were used to develop tomato soup with FOS incorporation. Results showed that tomato soup's color and appearance significantly improved at a 7.5% level without affecting its organoleptic qualities (Sheth and Gupta, 2013).

17.6 PREBIOTICS IN EDIBLE COATING

Inulin, polydextrose, glucose oligosaccharides, and wheat dextrin were selected as the four soluble fibers, as prebiotic components for the immobilization of *L. rhamnosus* GG in a combination of gelatin-based matrices and plasticizer glycerol to develop prebiotic edible films by encapsulating probiotic living cells. The viability of *L. rhamnosus* GG was significantly increased when glucose oligosaccharides and polydextrose were subjected to air drying (by 300% and 75%, respectively), whereas inulin and wheat dextrin decreased viable counts by 33% and 80%, respectively. Inulin, on the other hand, was found to be the most effective at controlling the sub-lethal effects on *L. rhamnosus* GG during storage. Prebiotic supplementation of edible films increased the storage stability of *L. rhamnosus* GG in all cases (Soukoulis et al., 2014).

17.7 EFFECT OF PREBIOTICS ON THE SENSORY ACCEPTANCE OF FOODS

Sensory analysis studies are important in the development of any food products. A chocolate dairy dessert was developed using prebiotics and different high-intensity sweeteners in place of sucrose (Morais et al., 2014). Relative sweetness analysis depicted that sweeteners had more sweetening power in comparison to the prebiotic chocolate dairy dessert with 8% sucrose. Cruz et al. (2013) sought to determine the effect of oligofructose concentrations on the physicochemical, rheological, and microbiological properties of unflavored yogurt. Refrigerated storage for 28 days had no effect on the pH, proteolysis, or viability of *Streptococcus thermophilus* or *Lactobacillus bulgaricus* ($p > 0.05$).

Peach flavored drinkable yogurt was developed to see the prebiotic (fructooligosaccharide) effects on the sensory properties and consumer acceptability. The prebiotic yogurts did not differ significantly from their comparable controls, indicating that a prebiotic can be added without affecting acceptance (Gonzalez et al., 2011).

FOS has a slightly sweet taste and thus it is likely to blend well with a huge range of products. In comparison to sucrose, which has a sweetness of about 30–35% in pure form. Its sweetening profile is the same as that of sugar. The flavor of fructooligosaccharide is very clear, with no aftertaste. It blends beautifully with delicate aromas and even enhances fruit flavors (Franck, 2002, 2008). FOS was also used to partially replace sucrose in fruit juices without affecting the overall quality (Prapulla et al., 2009).

17.8 COMMERCIALIZATION AND SCALING UP

Prebiotics have demonstrated advantages for human health in terms of gastrointestinal, cardiovascular, and other systems as well as growth, development, cell differentiation, physiological function, and antioxidant activity. According to Euromonitor International, the food and beverage markets associated with health and well-being reached US$750 billion in 2013. In response to global market trends, many industries have invested in scientific research and technological innovation in the development of products referenced by their functional properties. This segment includes various types of functional ingredients for bakery foods, cereals, drinks, and dairy products, as well as the food supplement segment. In 2013, there were 91 new records in the category of functional foods, compared to 178 between 2011 and 2012. Surprisingly, this figure is increasing year after year.

17.9 FUTURE PERSPECTIVES

Population-based research on prebiotic consumption is costly. Establishing partnerships between business and government could be a way to finance these initiatives. Funding requirements might also be met by raising the field of prebiotics' visibility among federal organizations. Finding a strategy that will work best with regulatory bodies like the European Food Safety Authority is currently one of the most crucial regulatory issues (EFSA). Prebiotics research is undoubtedly gaining traction due to the wide range of preventive and therapeutic possibilities it offers.

KEYWORDS

- **food preparation**
- **food production**
- **Kuala Lumpur**
- **laxatives**
- **meat production**
- **pharmaceutical industry**
- **prebiotics**
- **probiotics**
- **product development**
- **sweeteners**

REFERENCES

AFSSA. (2004). French approval for inulin as prebiotic. Available from: http://www.afssa.fr/Documents/NUT2004-sa0365.pdf (accessed on 19 July 2024).

Agri-Food and Veterinary Authority of Singapore. (2011). Food labelling and advertisements. Agri-food and veterinary authority of Singapore. Available from: https://www.sfa.gov.sg/docs/default-source/default-document-library/food-regulations-(1-feb- 2019).pdf?sfvrsn=cabf094_0.

Al-Sheraji, S. H., Ismail, A., Manap, M. Y., Mustafa, S., Yusof, R. M., & Hassan, F. A. (2013). Prebiotics as functional foods: A review. *Journal of Functional Foods, 5*, 1542–1553.

Brownawell, A. M., Caers, W., Gibson, G. R., Kendall, C. W. C., Lewis, K. D., Ringel, Y., & Slavin, J. L. (2012). Prebiotics and the health benefits of fiber: Current regulatory status, future research, and goals. *The Journal of Nutrition, 142*(5), 962–974.

Awad, R., Hagrass, A., Salama, W. M., Elmalek, F. A., & Eldardiry, A. I. (2014). Lactulose production from milk permeate and its performance in healthy functional frozen yoghurt. *World Journal of Dairy and Food Sciences, 9*, 1–9.

Barea-Alvarez, M., Benito, M. T., Olano, A., Jimeno, M. L., & Moreno, F. J. (2014). Synthesis and characterization of isomaltulose-derived oligosaccharides produced by transglucosylation reaction of *Leuconostoc mesenteroides* dextransucrase. *Journal of Agricultural and Food Chemistry, 62*, 9137–9144.

Barreto, G. P. M., Silva, N., Silva, E. N., Botelho, L., Yim, D. K., Almeida, C. G., & Saba, G. L. (2003). Quantificação de *Lactobacillus acidophilus*, bifidobactérias e bactérias totais em produtos probióticos comercializados no Brasil. *Brazilian Journal of Food Technology, 6*(1), 119–126.

Beeren, S. R., & Hindsgaul, O. (2014). A fluorescence assay that detects long branches in the starch polysaccharide amylopectin. *Chemical Communications, 50*, 1530–1532.

Böhm, A., Kaiser, I., Trebstein, A., & Henle, T. (2005). Heat-induced degradation of inulin. *European Food Research and Technology, 220*, 466–471.

Bucke, C., & Rastall, R. A. (1990). Synthesizing sugars by enzymes in reverse. *Chemistry in Britain, 26*, 675–678.

Burgain, J., Gaiani, C., Linder, M., & Scher, J. (2011). Encapsulation of probiotic living cells: From laboratory scale to industrial applications. *Journal of Food Engineering, 104*(4), 467–483.

Coman, M. M., Cecchini, C., Verdenelli, M. C., Silvi, S., Orpianesi, C., & Cresci, A. (2012). Functional foods as carriers for SYNBIO®, a probiotic bacteria combination. *International Journal of Food Microbiology, 157*(3), 346–352.

Cruz, A. G., Cavalcanti, R. N., Guerreiro, L. M. R., Sant'Ana, A. S., Nogueira, L. C., et al. (2013). Developing a prebiotic yogurt: Rheological, physico-chemical, and microbiological aspects and adequacy of survival analysis methodology. *Journal of Food Engineering, 114*, 323–330.

Curi, P. N., Almeida, A. B., et al. (2017). Optimization of tropical fruit juice based on sensory and nutritional characteristics. *Food Science and Technology, Campinas, 37*(2), 308–314.

Davim, S., Andrade, S., Oliveira, S., Pina, A., Barroca, M. J., & Guiné, R. P. F. (2015). Development of fruit jams and juices enriched with fructooligosaccharides. *International Journal of Fruit Science, 15*(1), 100–116.

Debon, J., Prudencio, E. S., & Petrus, J. C. C. (2010). Rheological and physico-chemical characterization of prebiotic microfiltered fermented milk. *Journal of Food Engineering, 99*, 128–135.

Delgado, G. T. C., Tamashiro, W. M. S. C., Maróstica Junior, M. R., Moreno, Y. M. F., & Pastore, G. M. (2011). The putative effects of prebiotics as immunomodulatory agents. *Food Research International, 44*(10), 3167–3173.

Elsayed, A. I., Rafudeen, M. S., & Golldack, D. (2014). Physiological aspects of raffinose family oligosaccharides in plants: Protection against abiotic stress. *Plant Biology, 16*, 1–8.

Fei, B. B., Ling, L., Hua, C., & Ren, S. Y. (2014). Effects of soybean oligosaccharides on antioxidant enzyme activities and insulin resistance in pregnant women with gestational diabetes mellitus. *Food Chemistry, 158*, 429–432.

Food and Drug Administration Thailand. (2011). *Prebiotics [in Thai]*. Medical and Health Product Bulletin, *5*(1), 1–31. Available from: https://food.fda.moph.go.th/media.php?id=509371120177258496&name=No.%20339%20Us e%20of%20Probiotic%20Microorganisms%20in%20Foods.pdf.

Food Safety and Quality Division. (2010). *Guide to Nutrition Labelling and Claims*. Ministry of Health Malaysia.

Franck, A. (2002). Technological functionality of inulin and oligofructose. *British Journal of Nutrition, 87*, 287–291.

Franck, A. (2008). Food applications of prebiotics. In G. Gibson & M. B. Roberfroid (Eds.), *Handbook of Prebiotics* (pp. 437–448). CRC Press/Taylor & Francis Group.

François, I. E., Lescroart, O., Veraverbeke, W. S., Marzorati, M., Possemiers, S., Hamer, H., Windey, K., Welling, G. W., Delcour, J. A., & Courtin, C. M. (2014). Effects of wheat bran extract containing arabinoxylan oligosaccharides on gastrointestinal parameters in healthy preadolescent children. *Journal of Pediatric Gastroenterology and Nutrition, 58*, 647–653.

Ghavidel, R. A., Karimi, M., & others. (2014). Effect of fructooligosaccharide fortification on quality characteristic of some fruit juice beverages (apple & orange juice). *International Journal of Farm & Allied Sciences, 3*(2), 141–146.

Gonzalez, N. J., Adhikari, K., & Sancho-Madriz, M. F. (2011). Sensory characteristics of peach-flavoured yoghurt drinks containing prebiotics and symbiotic. *LWT—Food Science and Technology, 44*, 158–163.

Guergoletto, K. B., Magnani, M., Martin, J. S., Andrade, C. G. T. J., & Garcia, S. (2010). Survival of *Lactobacillus casei* (LC-1) adhered to prebiotic vegetal fibers. *Innovative Food Science and Emerging Technologies, 11*, 415–421.

Gupta, N., & Sheth, M. (2011). Acceptability trials of fructooligosaccharides (FOS) added soup and beverages. *Asian Journal of Home Science, 6*(2), 131–136.

Health Canada. (2009). *Guidance Document: The Use of Probiotic Microorganisms in Food*. Ottawa: Health Canada.

Huebner, J., Wehling, R. L., & Hutkins, R. W. (2007). Functional activity of commercial prebiotics. *International Dairy Journal, 17*, 770–775.

Huebner, J., Wehling, R. L., Parkhurst, A., & Hutkins, R. W. (2008). Effect of processing conditions on the prebiotics. *International Dairy Journal, 18*(3), 287–293.

Jain, N., Sheth, M., Asudani, A., & Parnami, S. (2013). Sensory attributes of Indian fried foods incorporated with different levels of fructooligosaccharides. *International Journal of Applied Biology and Pharmaceutical Technology, 4*(3), 255–259.

Japanese Ministry of Health Labour and Welfare. (2010). *Food for Specified Health Uses (FOSHU)*. Available from: http://www.mhlw.go.jp/english/topics/foodsafety/fhc/02.html (accessed on 19 July 2024).

Lim, Y. M., Barnes, M. B., Gras, S. L., McSweeney, C., Lockett, T., Augustin, M. A., & Gooley, P. R. (2014). Esterification of high amylose starch with short chain fatty acids modulates degradation by *Bifidobacterium* spp. *Journal of Functional Foods, 6*, 137–146.

Mahendra, & Sheth, M. (2013). Feasibility and acceptability of fructooligosaccharide substituted popular Indian foods. *Nutrition & Food Science, 43*(3), 204–212.

Mann, J., Cummings, J. H., Englyst, H. N., Key, T., Liu, S., Riccardi, G., Summerbell, C., Uauy, R., van Dam, R. M., Venn, B., Vorster, H. H., & Wiseman, M. (2007). FAO/WHO scientific update on carbohydrates in human nutrition: Conclusions. *European Journal of Clinical Nutrition, 61*(1), S132–S137.

Mantzouridou, F., Karousioti, A., & Kiosseoglou, V. (2013). Formulation optimization of a potentially prebiotic low-in-oil oat-based salad dressing to improve *Lactobacillus paracasei* subsp. *paracasei* survival and physicochemical characteristics. *Food Science and Technology, 53*, 560–568.

Megala, P., & Hymavathi, T. V. (2011). *International Scholarly and Scientific Research & Innovation, 5*(11), 658–663.

Morais, E. C., Morais, A. R., Cruz, A. G., & Bolini, H. M. A. (2014). Development of chocolate dairy dessert with addition of prebiotics and replacement of sucrose with different high-intensity sweeteners. *Journal of Dairy Science, 97*(5), 2600–2609.

Moshfegh, A. J., Friday, J. E., Goldman, J. P., & Ahuja, J. K. C. (1999). Presence of inulin and oligofructose in the diets of Americans. *Journal of Nutrition, 129*, 1407–1411.

Mumtaz, S., Rehman, S., Huma, N., & others. (2008). Xylooligosaccharide enriched yoghurt: Physicochemical and sensory evaluation. *Pakistan Journal of Nutrition, 7*(4), 566–569.

Nelson, A. L. (2002). *High-Fiber Ingredients*. Eagan Press.

Netherlands Nutrition Centre. (2008). *Assessment Report of Netherlands Nutrition Centre on Inulin in Vitaalbrood*. Available from: http://www.voedingscentrum.nl/resources2008/beoordelingsrapport_vitaalbrood_florapdf.pdf.

Nik Daud, N. I. (2011). Malaysian legal requirements for innovative foods. *World of Food Science*. Available from: http://www.worldfoodscience.org/cms/?pid=1004358.

O'Sullivan, L., Murphy, B., McLoughlin, P., Duggan, P., Lawlor, P. G., Hughes, H., & Gardiner, G. E. (2010). Prebiotics from marine macroalgae for human and animal health applications. *Marine Drugs, 8*, 2038–2064.

Padilha, V. M., Rolim, P. M., Salgado, S. M., Livera, A. S., Andrade, S. A. C., & Guerra, N. B. (2010). Perfil sensorial de bolos de chocolate formulados com farinha de yacon (*Smallanthus sonchifolius*). *Food Science and Technology (Campinas), 30*(3), 735–740.

Padma Ishwarya, S., & Prabhasankar, P. (2013). Fructooligosaccharide–Retention during baking and its influence on biscuit quality. *Food Bioscience, 4*, 68–80.

Panesar, P. S., Kumari, S., & Panesar, R. (2013). Biotechnological approaches for the production of prebiotics and their potential applications. *Critical Reviews in Biotechnology, 33*, 345–364.

Parnami, S., & Sheth, M. (2011). Indian fermented milk (dahi) fortified with probiotic bacteria and inulin improves serum lipid, blood glucose levels, and gut microflora. *Journal of The Indian Academy of Geriatrics, 7*(1), 12–21.

Jnawali, P., Kumar, V., Tanwar, B., Hirdyani, H., & Gupta, P. (2017). Enzymatic production of xylooligosaccharides from brown coconut husk treated with sodium hydroxide. *Journal of Food Science and Technology, 9, 1757–1766.* https://doi.org/10.1007/s12649-017-9963-4.

Prapulla, S. G., & Aachary, A. A. (2009). Value addition to spent osmotic sugar solution (SOS) by enzymatic conversion to fructooligosaccharides (FOS), a low-calorie prebiotic. *Innovative Food Science & Emerging Technologies, 10*(2), 284–288.

Prapulla, S., Renuka, G. B., Kulkarni, S. G., & Vijayananad, P. (2009). Fructooligosaccharide fortification of selected fruit juice beverages: Effect on the quality characteristics. *Food Science and Technology, 42*, 1031–1033.

Moura, P. (2015). Development of prebiotic food products and health benefits. *Food Science and Technology (Campinas), 35*(1), 3–10.

Renuka, B., Kulkarni, S. G., & Vijayanand, P. (2009). Fructooligosaccharide fortification of selected fruit juice beverages: Effect on the quality characteristics. *LWT–Food Science and Technology, 42*(5), 1031–1033.

Rodrigues, D., Rocha-Santos, T. A. P., Gomes, A. M., Goodfellow, B. J., & Freitas, A. C. (2012). Lipolysis in probiotic and synbiotic cheese: The influence of probiotic bacteria, prebiotic compounds, and ripening time on free fatty acid profiles. *Food Chemistry, 131*, 1414–1421.

Rolim, P. M. (2015). Development of prebiotic food products and health benefits. *Food Science and Technology (Campinas), 35*(1), 3–10.

Soukoulis, C., Jobbehda, S. B., Yonekura, L., Parmenter, C., & Fisk, I. D. (2014). Stability of *Lactobacillus rhamnosus* GG in prebiotic edible films. *Food Chemistry, 159*(100), 302–308.

Sridevi, V., Sumathi, V., Guru Prasad, M., & Kumar, M. S. (2014). Fructooligosaccharides–type prebiotic: A review. *Journal of Pharmacy Research, 8*, 321–330.

Tarrega, A., Rocafull, M., & Costell, E. (2010). Effect of blends of short and long-chain inulin on the rheological and sensory properties of prebiotic low-fat custards. *Food Science and Technology, 43*, 556–562.

Tee, E. S., & Sulong, F. (2010). Make effective use of nutrition information on food labels. Ministry of Health Malaysia. Available from: https://hq.moh.gov.my/nutrition/wp-content/uploads/2024/02/01.Buku-MDG-2020_26Feb2024.pdf.

Torres, D. P., Gonçalves, M. D. P. F., Teixeira, J. A., & Rodrigues, L. R. (2010). Galacto-oligosaccharides: Production, properties, applications, and significance as prebiotics. *Comprehensive Reviews in Food Science and Food Safety, 9*, 438–454.

Tuohy, K. M., Kolida, S., Lustenberger, A. M., & Gibson, G. R. (2001). The prebiotic effects of biscuits containing partially hydrolyzed guar gum and fructo-oligosaccharides–a human volunteer study. *British Journal of Nutrition, 86*, 341–348.

Vaidya, S., & Sheth, M. (2011). Processing and storage of Indian cereal and cereal products alters its resistant starch content. *Journal of Food Science and Technology, 48*(5), 622–627.

Van Den Ende, W. (2013). Multifunctional fructans and raffinose family oligosaccharides. *Frontiers in Plant Science, 4*, 1–11.

Vernazza, C. L., Rabiu, B. A., & Gibson, G. R. (2006). Human colonic microbiology and the role of dietary intervention: Introduction to prebiotics. In G. R. Gibson & R. A. Rastall (Eds.), *Prebiotics: Development & Application* (pp. 1–28). John Wiley & Sons, Ltd.

Warrand, J. (2006). Healthy polysaccharides. *Food Technology and Biotechnology, 44*, 355–370.

Wichienchot, S., Jatupornpipat, M., & Rastall, R. A. (2010). Oligosaccharides of pitaya (dragon fruit) flesh and their prebiotic properties. *Food Chemistry, 120*(3), 850–857.

Zhou, X., Ruan, Z., Huang, X., Zhou, Y., Liu, S., & Yin, Y. (2014). The prebiotic lactosucrose modulates gut metabolites and microbiota in intestinal inflammatory rats. *Food Science and Biotechnology, 23*, 157–163.

CHAPTER 18

Standardized Prebiotic Recipes

MINI K. SHETH and SHRUTI DWIVEDI

Department of Foods and Nutrition, The Maharaja Sayajirao University of Baroda, Gujarat, India

Since FOS only contains about 1.5 kcal/g of energy and is only about one third as sweet as sucrose, it has a low sweetness intensity. The digestive enzymes barely hydrolyze it, and the gut flora, particularly probiotics, use it as an energy source. Because of its qualities, it makes a good replacement for fat and sugar, giving food therapeutic value.

18.1 INTRODUCTION

FOS and XOS can replace sugar and fat within Food Safety and Standards Authority of India's permissible limits (FSSAI). This chapter describes the savory and sweet foods that have undergone standardization through the addition and replacement of FOS and XOS at concentrations ranging from 5% to 20%.

Prebiotic Foods for a Healthy Gut: Scientific Studies to Manage Metabolic Disorders.
Mini K. Sheth and Shruti Dwivedi (Eds.)

18.2 PANEER BUTTER MASALA

18.2.1 INGREDIENTS

Paneer – 100 gms
Onions, finely chopped – 2 medium
Ginger, chopped – 1 tsp
Garlic – 3–4 Cloves
Tomatoes – 2 medium
Cashew Nuts – 6–8
Bay Leaf – 1 small piece
Green chilies – 2 nos.
Full Fat Milk – 1/4th cup
Garam Masala Powder – 1/2 tsp
Coriander Powder – 1 tsp
Red Chili Powder – 1/2 tsp
Sugar – 20 g
Fresh Cream – 1 tbsp
Butter – 1 tbsp
Oil – 1 tbsp
Salt to taste
XOS – 8 g

18.2.2 PREPARATION

- Ground onion, ginger and garlic in a grinder to make paste. Grind soaked cashew nuts into a smooth paste with 2 tbsps. of water.
- Blanch and blend tomatoes into a tomato puree.
- Heat cooking oil and butter in a non-stick frying pan over medium flame.
- Sauté onion pastes and bay leaf until it turns light brown for approx. 4–5 minutes.
- Add green chilies and red chili powder and sauté for 30–40 seconds. Add and cooked cashew nut paste for 2 minutes.
- Add tomato puree, sugar and XOS. Cook until oil started to separate from the puree.
- Add and mixed well coriander powder and garam masala.
- Add 1/2 cup milk, 1/2 cup water and salt cook until oil comes on the surface.

- Add paneer cubes and cook for approx. 3–4 minutes or until the desired consistency of gravy.
- Add fresh cream and mix properly. Turn off the flame.
- Paneer butter masala is ready garnish with milk cream or cube of butter and serve.

18.3 POTATO CURRY

18.3.1 INGREDIENTS

Potato – 80 g
Tomato – 70 g
Ginger – 2 g
Green chili – 1/2
Water – 100
Oil – 5 g
Cumin seeds – ½ pinch
Asafetida – ½ pinch
Salt – to taste
FOS – 7.5 g

18.3.2 PREPARATION

- Cut and fry tomatoes in oil and spices till they form thick gravy now add peeled sliced potatoes and FOS.
- Pressure cooks all the ingredients for 15 min at 95°C.

18.4 DAAL

18.4.1 INGREDIENTS

Toor Dal – 40 g
Water – 160 g
Oil – 5 g
Turmeric powder – 1/2tsp
Salt – To taste
Cumin seeds – ½ pinch
Asafetida – ½ pinch
FOS – 7.5 g

18.4.2 PREPARATION

- Pressure cooking 250 g of bengal gram *Dal* with water, FOS, turmeric powder and salt for 15 minutes.
- Season by cumin seed, red chili powder and asafetida sautéed in ghee.

18.5 KADI

18.5.1 INGREDIENTS

Buttermilk – 250 g
Besan – 7 g
Oil – 5 g
Salt – to taste
Fenugreek seeds – 0.5 g
Mustard seeds – 0.5 g
Cumin seeds – 0.5 g
Asafetida – 1 pinch
Chili powder – 1/2tsp
FOS – 7.5 g

18.5.2 PREPARATION

- Add besan to buttermilk.
- Prepare seasoning from oil, asafetida, turmeric and chili powder.
- Add the besan and buttermilk mix to the seasoning, stir continuously till the boiled on slow flame add FOS.

18.6 PRAWN PATIA

18.6.1 INGREDIENTS

Vegetable oil – 2 tbsps
Finely chopped medium sized onion – 1 no.
Chopped green chilies – 4 nos.
Tomato puree – 100 g
Ginger-garlic paste – 1 tbsp
Red chili powder – 1 ½ tsp
Turmeric powder – 1 tsp

Cumin powder – 1 tsp
Granulated sugar – 30 g
Vinegar – 1 ½ tbsp
Salt – as preferred
Sugar – 20 gm
Medium sized deveined frozen prawns – 1 cup
XOS – 8 g

18.6.2 PREPARATION

- Thaw the frozen prawns.
- Wash, peel and chop onion and chilies.
- Heat mustard oil in a skillet over medium-high heat until simmering. Add chopped onions and stir until it turns light brown. Add green chilies and stir for a minute.
- Add tomato puree and continue stirring until oil is separated from the onion tomato mix. Add Ginger-garlic paste, red chili, turmeric and cumin and stir ensuring that the mixture doesn't stick to the bottom.
- Add sugar, XOS, vinegar and water cook till the rawness of the tomato is gone. season with salt to taste. Add prawns, stir and cooked for 10 more minutes until the prawns get cooked.

18.7 *KHICHDI*

18.7.1 INGREDIENTS

Toor Dal – 25 g
Rice – 25 g
Water – 150 g
Ghee – 5 g
Salt – to taste
Turmeric – two pinches
Red chili powder – two pinches
Cumin seeds- two pinches
FOS – 7.5 g

18.7.2 *PREPARATION*

- Cut and fry tomatoes in ghee and spices till they become soft, to this add washed rice, Dal and water.
- Pressure cook all the ingredients together for 15 min at 95° C add FOS.

18.8 *CHAPATI*

18.8.1 *INGREDIENTS*

Wheat flour – 100 g
Water – 100 ml
FOS – 6 g

18.8.2 *PREPARATION*

- Knead wheat flour with water and 6 gm FOS to get a desired consistency of dough.
- Flatten and roast the *chapati*.

18.9 *THEPLA*

18.9.1 *INGREDIENTS*

Wheat flour – 100 g
Bengal gram flour – 30 gm
Fenugreek leaves – 50 g
Water – 80 ml
Oil – 10 ml
Garlic-ginger and chili paste – 10 g
Spices – 2 g
FOS – 10 g

18.9.2 *PREPARATION*

- Knead all the ingredients together to make the desirable dough.
- Make balls, flatten and shallow fry at medium flame.

18.10 VEGETABLE PARATHA

18.10.1 INGREDIENTS

Wheat flour – 120 g
Potatoes – 2 small
Salt – 6 g
Oil – 15 ml
Water –75 ml
Corn flour powder: 5 g
FOS – 10 g

18.10.2 PREPARATION

- Knead together wheat flour, salt, oil, and water.
- Divide into equal parts, make balls and roll.
- Wash all the vegetables and chop in small pieces.
- Cook in oil for 15 min add corn flour powder, FOS, salt and spices.
- Fill 30 g stuffing in one paratha and roll again.
- Shallow fry on medium flame for 3 min with ½ tsp. of oil.

18.11 *DHOKLA*

18.11.1 INGREDIENTS

Bengal gram flour – 100 g
Semolina mix-
Curd – 70 g
Spices – 2 g
Oil – 10 ml
Water – 20 ml
Baking soda – ½ teaspoon
FOS – 16 gm

18.11.2 PREPARATION

- Mix the batter and stirred for 10 min.
- Add baking soda and leaven for 20 mins.

18.12 *PATRA*

18.12.1 INGREDIENTS

Bengal gram flour: 100 gm
Water: 50 ml
Spices: 2 g
Tamarind: 5 gm
Colocasia leaves: 60 g
Oil: 10 ml
Gingelly seed: 2 g
Coriander leaves: 5 gm
FOS: 10 g

18.12.2 PREPARATION

- Prepare the batter using Bengal gram flour, water, spices, tamarind and FOS.
- Layer the batter on Colocasia leaves roll it and keep for steaming 20 mins.
- Further cut into small pieces and sauté using oil, gingelly seeds and coriander leaves.

18.13 TOMATO SOUP

18.13.1 INGREDIENTS

Tomato – 120 g
Water – 110
Garlic – 1 clove
Ginger – 2 g
Green chili – 1/2
Potato – 16 g
Onion – 20 g
Sugar – 2 g
Salt – To taste
FOS – 7.5 g

18.13.2 PREPARATION

- Tomatoes, onion, potato, garlic cloves, ginger, green chili, sugar, FOS and salt were together pressure cooked for10 min at 95°C.
- The puree was cooled, blended and sieved. The soup thus obtained was then boiled up to 3 min

18.14 *LILVA KACHORI*

18.14.1 INGREDIENTS

Maida – 80 g
Whole wheat flour – 20 g
Oil – 20 ml
Salt – ¼ tsp.
Water – 60 ml
FOS – 5 gm
Fresh green peas (without shell)
Til seeds – 2 g
Sugar: 5 g

18.14.2 PREPARATION

- Knead well whole wheat flour, oil, FOS and salt
- Divide into equal parts for making balls.
- Balls were rolled and put into the mold.
- Crush fresh green peas.
- Add till seeds, green chilies, garam masala, 5 g sugar and salt sauté for 2 min.
- Fill the stuffing in the flattened dough kept in the mold with cover.
- Close the mold to give the desirable shape.
- Deep fry the *kachori* in fresh hot oil (1600–1700 C) for one minute.

18.15 *DUDHI MUTHIYA*

18.15.1 INGREDIENTS

Flour mix (whole wheat, maize and split chickpea in ratio of 3:1:1) – 50 gm
Ginger-garlic-chili paste – 1.25 g
Fresh fenugreek leaves – 5 g
Fresh coriander leaves – 5 g
Sugar – 5 g
Salt – 2.5 g
Curd – 50 g
Turmeric – 1 pinch
Red chili powder – 1 pinch
Oil – 10 ml
Sesame seeds – 2 pinches
Asafetida – 1 pinch
FOS – 7.5 g

18.15.2 PREPARATION

- Mix properly all the ingredients along with FOS and knead in form of dough by adding 30 ml of water.
- Shape the dough into cylindrical rolls and steam for 15 minutes.
- As they cool down, slice the rolls into 1icnh x 1inch dimension.
- Heat 10 ml of oil in a small pan and 2 pinches of sesame seeds and 1 pinch of asafetida as they start cracking, pour the entire content along with oil on steamed *Dudhi Muthiya* as a garnish.

18.16 VEGETABLE *CHEELA*

18.16.1 INGREDIENTS

Yellow split moong dal – 20 gm
White split urad dal – 20 gm
Chopped spinach – 10 gm
Chopped onion – 5 g
Fresh coriander leaves – 2.5 g
Cumin seeds – 0.5 g
Turmeric – 0.5 g
Salt – 2 g
FOS – 7.5 g

18.16.2 *PREPARATION*

- Soak daal mixes overnight and grind.
- Add chopped spinach, onion, fresh coriander leaves, cumin seeds, turmeric and salt.
- Make a spreadable batter by adding FOS and water slowly avoiding lumps.
- Pour this batter in three parts on hot plate and shallow fry using 5 ml of oil on medium flame to get a golden crispy layer.

18.17 *HANDWA*

18.17.1 *INGREDIENTS*

Parboiled rice, split red gram dal and split chickpea dal (flour) – 100 gm
Chopped spinach – 10 g
Chopped onion – 5 g
Fresh coriander leaves – 2.5 g
Curd – 15 g
Sugar – 5 g
Salt – 5 g
Turmeric – 1.5 gm
Oil – 30 ml
FOS – 7.5 g

18.17.2 *PREPARATION*

- To the mix flour add chopped spinach, onion, fresh coriander leaves, curd, sugar, salt, turmeric and oil. Add FOS.
- Allow the batter to sit for 2 hours.
- Pour into baking tray and bake at 180 °C for 20 min.

18.18 VEGETABLE MINI *SAMOSA*

18.18.1 *INGREDIENTS*

Refined wheat flour – 75 g
Potato – 80 g
Green chili – 8 g
French beans – 33 g
Carrots – 33 g
Spring onion – 8 g

Oil – 15 ml
Salt – 2 g
Fresh coriander leaves – 8 g
Garam masala – 1 pinch
Oil – 2.5 ml
Salt – 2 g
Water – 15 ml
FOS – 7.5 g

18.18.2 PREPARATION

- Kneed together refined wheat flour, oil, salt and water.
- Form dough balls from flour.
- Roll out 4 strips from one dough ball.
- Chop the vegetables finely and cook in hot oil, Add a pinch of garam masala, red chili powder, turmeric powder. to get desired flavor and texture.
- Roll the dough thin and flat as much as possible into shape of large circle. Cut into strips of 2-inch width across length.
- Fold each strip initially from one corner to form a pocket of triangle shape.
- Add cooked stuffing to this pocket. Cover the corners from all ends.
- Pour 300 ml of oil into deep pan for frying. Heat the oil on high flame till it reaches its smoking point.
- Deep fry the vegetable mini *samosas* on medium flame for 2–3 minutes until golden crispy.

18.19 BLACK RICE *KHEER*

18.19.1 INGREDIENTS

Black rice – 30 g
Milk – 250 ml
Sugar – 20 g
Almonds chopped – ½ tbsp
Cashew nuts chopped– ½ tbsp
Raisins – 1 tsp
XOS – 10 gm

18.19.2 PREPARATION

- Wash black rice under running water till water runs clear. soak rice in enough water overnight. Chop nuts and kept aside.
- Take milk in a heavy bottom pan and turn the heat on medium. Let the milk come to a boil.
- Add the rice once the milk started boiling. Allow to simmer on low-medium heat 25–30 minutes with continuous stirring.
- Add sugar, XOS, chopped cashews, almonds and raisins mix well. Allow to simmer for 8–10 minutes. Turn off the stove when the *Kheer* is ready.

18.20 *GAJAR KA HALWA*

18.20.1 INGREDIENTS

Ghee, melted – ½ cup
Carrot, grated – 1 kg
Evaporated milk – 1 cup
Sugar – 20 g
Khoya – 50 g
Almonds, soaked in hot water, peeled – 15
Cardamom powder – ½ tsp
XOS – 10 g

18.20.2 PREPARATION

- Boiled grated carrots and milk in a large heavy based pan. Cook carrots and stir constantly, until all the milk dries up.
- Add ghee to the carrots. Add khoya to it and sauté for a few minutes.
- Add sugar and XOS and continue to cook and stir till it dries.
- Remove from heat when still moist and stir in the cardamom powder and almonds reserving a few to garnish.

18.21 *DUDHI HALWA*

18.21.1 INGREDIENTS

Grated bottle gourd – 60 g (1/2 cup)
Ghee – 3.5 g (1 tsp)
Milk – 160 ml (1 ½ cup)
Sugar – 6 g (1 ½ tsp)
Almond – 3 pc
Cardamom – 1–2 pc
Liquid FOS – 12 g (2 ½ tsp)

18.21.2 PREPARATION

- Heat ghee in a pan, add grated bottle gourd and sauté till it turns translucent.
- Add milk and bring to a boil.
- Cook gently for 20 minutes, while stirring occasionally.
- When the milk has almost evaporated, add the sugar and FOS and mix well.
- When the mixture has become dry, add cardamom powder, and mix well.
- Garnish with almonds. Serve hot.

18.22 *GULAB JAMUN*

18.22.1 INGREDIENTS

Cow's *mava* – 25 g
Refined flour – 2.5 g
Water – 2tsp
Sugar – 20 g
Water – 40 ml
Oil for frying – 200 ml
Liquid FOS – 20 g

18.22.2 PREPARATION

- Take water and sugar in a pan. Add FOS after sugar has dissolved, cook on a medium flame till the syrup is of one string consistency.

- Combine mava and refined flour in a bowl and knead well using water into a smooth dough.
- Divide the mixture into equal portions and roll each into round balls. Ensure that there are no cracks on the surface. Keep aside.
- Heat ghee in a deep non-stick pan. Fry the balls on a slow flame till they turn brown in color from all the sides.
- Drain well and immediately put in warm sugar syrup.
- Soak the *gulab jamuns* in the sugar syrup for at least 30 minutes.

18.23 *JALEBI*

18.23.1 INGREDIENTS

Refined flour – 20 g (1/8 cup)
Curd – 5 g (1 tsp)
Sugar – 17.5 g (1/8 cup)
Water – 25 ml (1/8 cup)
Liquid FOS – 25 g (1/8 cup)
Saffron – 3 strands
Ghee – 200 ml for frying

18.23.2 PREPARATION

- Combine sugar and water in a pan, mix well, add FOS when the sugar has dissolved. Heat till the sugar syrup is of one thread consistency.
- Combine saffron strands and 2 tsp of sugar syrup in a bowl and add to the sugar syrup. Mix well. Keep aside.
- Combine curd with the refined flour in a bowl and mix well
- Add water if required to make a thick batter and keep aside for 30 minutes.
- Pour the batter in a squeeze tomato ketchup bottle with a small hole and fill with the batter.
- Heat the ghee in a pan and press the bottle to form coil shaped *jalebi.*
- Deep-fry till the jalebis turn golden brown in color from both the sides.
- Place the fried jalebis into the warm syrup. Soak in sugar syrup and then remove from it.

18.24 *KAJU KATLI*

18.24.1 INGREDIENTS

Cashew nut – 30 g (1/3 cup)
Sugar – 12.5 g (1 tbsp)
Water – 7.5 ml (1 ½ tsp)
Liquid FOS – 5.5 g (1 tsp)
Silver warq – 1

18.24.2 PREPARATION

- Grind cashew nut to powder using mixer grinder.
- Dissolve the sugar in water. Add FOS when the sugar is dissolved. Cook till the syrup is of one thread consistency.
- Now add the cashew nut powder to the syrup and cook till the mixture leaves the sides of the pan.
- Quickly turn onto a greased marble and roll thinly with a greased roller. Spread the silver warq over it and cut into diamond shaped katlis.

18.25 *KAJU PISTA* ROLL

18.25.1 INGREDIENTS

Ingredients for cashew roll:
Cashew nut – 25 g (1/3 cup)
Sugar – 8.5 g (1 tbsp)
Water – 4 ml (1 tsp)
Liquid FOS – 4 g (1 tsp)
Ingredients for Pista roll:
Pista – 15 g (1/4 cup)
Sugar – 7 g (1 ½ tsp)
Water – 4 ml (1 tsp)
Liquid FOS – 3 g (1/2 tsp)
Silver warq – 1

18.25.2 PREPARATION

- Grind cashew nut and pistachio separately to make powder using mixer grinder.
- Dissolve the sugar in water and add FOS when the sugar is dissolved. Simmer to make syrup of one string consistency.
- Now add the cashew nut powder to the syrup and cook till the mixture leaves the sides of the pan.
- Repeat the step 2 and 3 for pistachio.
- Quickly turn onto a greased marble separately and roll thinly. Let it cool for 2 minutes.
- Keep the pistachio layer on cashew nut layer. Place it at one end of the cashew nut layer and roll up the cashew sheet enclosing the pistachio roll.
- Gently press silver warq on the rolls. Cut into equal sized cylindrical pieces and serve.

18.26 *KHOPRA PAK*

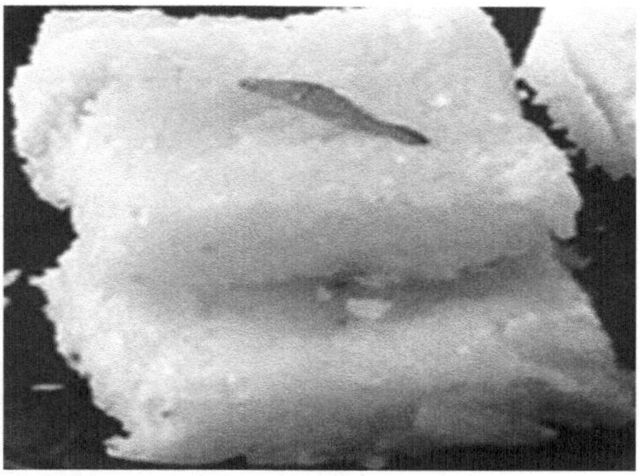

18.26.1 INGREDIENTS

Grated coconut – 16 g (1/3 cup)
Sugar – 12 g (1 tbsp)
Water – 10 ml (1 tsp)
Liquid FOS – 12 g (2 ½ tsp)

18.26.2 PREPARATION

- Take water and sugar in a non-stick pan. Add FOS when the sugar is dissolved. Heat till one string consistency is obtained.
- Add the grated coconut and sauté on a medium flame.
- Cook till the mixture leaves the sides of the pan.
- Transfer the mixture into a greased plate and spread it evenly using a flat spoon.
- Keep aside to let it cool completely. Cut into equal square pieces and serve.

18.27 *MALAI PEDA*

18.27.1 INGREDIENTS

Milk – 280 ml (1 1/3 cup)
Sugar – 7.5 g (2 tsp)
Citric acid – 5 pc
Corn flour – 5 g (2 tsp)
Liquid FOS – 7.5 g (1 ½ tsp)

18.27.2 PREPARATION

- Take milk in a non-stick pan and bring to a boil, while stirring in between.
- Reduce the flame to medium and cook till the milk reduces to half its quantity, while stirring occasionally and scraping the sides of the pan.
- Add the sugar and FOS to the milk. Mix well and continue to cook.
- Combine the corn flour and water in a bowl and mix well. Keep aside. Also, combine the citric acid and 1 tbsp of water in a bowl and mix well.
- Now add both the mixtures to the milk, mix well and cook till the mixture resembles mava.
- Transfer it to a plate, spread with the help of a at ladle and keep aside to cool for 15 minutes.
- Divide the mixture into equal portions and shape each portion into a at round in between your palms.

18.28 *MOHANTHAL*

18.28.1 INGREDIENTS

Gram flour – 25 g (1/3 cup)
Ghee – 7.5 g (1 ½ tsp)
Powdered sugar – 12 g (1 tbsp)
Liquid FOS – 8 g (1 ½ tsp)

18.28.2 PREPARATION

- Take ghee in a non-stick pan.
- Add gram flour to it and roast till it turns brown on a low flame.
- Turn off the flame. Let the mixture cool for 2 minutes.

- Add powdered sugar and FOS to the mixture. Mix well.
- Transfer the mixture to a greased plate and spread evenly using a spatula. Let it set.
- Cut into equal sized square pieces.

18.29 *RASGULLA*

18.29.1 INGREDIENTS

Cow's Milk – 250 ml (1 cup)
Lemon juice – 1 – 2 tsp
Sugar – 25 g (3 tbsp)
Water – 150 ml (1 ¼ cup)
Liquid FOS – 25 g (1 ½ tsp)

18.29.2 PREPARATION

- Take milk in a deep non-stick pan and bring to a boil.
- Switch of the flame and wait for 1 minute, while stirring occasionally.
- Add lemon juice gradually and keep stirring gently. Allow it to curdle. Strain using a muslin cloth and wash chena in running water. Drain out extra water.
- Take water in a vessel, add sugar and FOS, bring to a boil, stirring occasionally.
- Knead the chena well using palms till it is smooth and free of lumps.
- Divide the chena into equal portions and roll each portion into smooth balls.
- Put the balls into the sugar water and steam for 15 minutes. Cover the vessel.
- Switch off the flame and transfer it into a bowl, refrigerate, and serve chilled.

18.30 *RAVA SHEERA*

18.30.1 INGREDIENTS

Semolina – 25 g (1/4 cup)
Ghee – 6 g (1 ½ tsp)
Water – 100 ml (1/2 cup)
Sugar – 18 g (1/8 cup)

Cardamom – for garnish
Almond – 2–3 pc (for garnish)
Liquid FOS – 12 g (1 tbsp.)

18.30.2 PREPARATION

- Heat the ghee in a broad non-stick pan, add the semolina and sauté on a slow flame for 5 minutes, while stirring occasionally.
- Heat water in another vessel.
- Add hot water, mix well, and cook while stirring continuously.
- Add the sugar, FOS and mix well. Cover and cook on a slow flame for 5 minutes, while stirring occasionally.
- Add the cardamom powder, mix well, and cook on a medium flame.
- Serve hot, garnished with almonds.

18.31 *SHRIKHAND*

18.31.1 INGREDIENTS

Curd – 150 g (1 cup)
Powder sugar – 5.6 g (1 ½ tsp) Liquid
Cardamom – 1 pc
Almonds – 1–2 pc
FOS – 13 g (1 tbsp)

18.31.2 PREPARATION

- Hang the curd in a muslin cloth in a cool and dry place for approximately 45 minutes until all the liquid (whey) has drained off.
- Remove the hung curd into a bowl. Add sugar, FOS and cardamom powder to it and churn using a hand blender.
- Refrigerate for 30 minutes.
- Serve cold, garnished with almonds.

18.32 *SUKHADI*

18.32.1 INGREDIENTS

Wheat flour – 20 g (1/4 cup)
Ghee – 11 g (1 tbsp.)
Jaggery – 15 g (1/8 cup)
Liquid FOS – 6 g (1 tsp)

18.32.2 PREPARATION

- Heat the ghee in a non-stick pan. Add the wheat flour and sauté on a medium flame till it turns golden brown in color.
- Switch of the flame. After 2 minutes add jaggery, mix well.
- Once the jaggery starts melting, add FOS. Mix well.
- Grease a plate with ghee. Transfer the prepared mixture in the plate and allow it to cool.
- Once set cut into equal sized squares pieces and serve.

18.33 WHEAT FLOUR *SHEERA*

18.33.1 INGREDIENTS

Wheat flour – 20 g (1/3 cup)
Ghee – 11 g (2 tsp)
Water – 80 ml (1/3 cup)
Sugar – 19 g (1/8 cup)
Almonds – 2–3 pc (for garnish)
Liquid FOS – 6 g (1 tsp)

18.33.2 PREPARATION

- Heat ghee in a non-stick pan, add the wheat flour and sauté on a medium flame for 10 minutes or till it turns brown in color and the ghee separates.
- Heat water in another vessel.
- Add the hot water to the mixture of wheat flour, mix well, and cook while stirring continuously.

- Once the water has almost dried, add the sugar and FOS, mix well, and cook on a slow flame for another 5 minutes or till the sugar dissolves, while stirring continuously.
- Serve hot garnished with almond.

18.34 *KHEER*

18.34.1 INGREDIENTS

Milk – 340 g
Rice – 16 g
Water – 40 g
Sugar – 25 g
Cardamom powder – 1–2 pinches
Chopped almonds – 10 g
Raisins – 6 g
FOS – 7.5 g

18.34.2 PREPARATION

- To pasteurized full-cream milk, add rice and water, and boil for 30 minutes on a low flame.
- Add sugar, FOS, cardamom powder, of almonds and raisins.

18.35 *AMBIL*

18.35.1 INGREDIENTS

Rice flour – 80 g
Water – 520 ml
Buttermilk – 350 ml
Color and flavor – as per choice

18.35.2 PREPARATION

- Cook the cereal flour in water for 10 minutes.
- Let it cool at room temperature.
- Add buttermilk, blend in electric mixer.
- Sieve the liquid to remove all lumps.
- Keep it overnight for fermentation.

- Add salt and jeera or flavor as per your choice.
- Add FOS and refrigerate.
- Serve cold.

Ambil can be modified by replacing rice flour with pearl millet flour, barley flour or finger millet flour.

KEYWORDS

- **cashew nut**
- **coriander**
- **fats**
- **flours**
- **ginger**
- **oil**
- **sugar**
- **sugars**
- **temperature**
- **tomato**

Index

C

F

I

N

O

Z